Environmental Science in a Changing World

Environmental Science in a Changing World

Editor: Paige Tucker

Callisto Reference
www.callistoreference.com

Callisto Reference,
118-35 Queens Blvd., Suite 400,
Forest Hills, NY 11375, USA

Visit us on the World Wide Web at:
www.callistoreference.com

ISBN: 978-1-64116-275-3 (Hardback)

The publisher's policy is to use permanent paper from mills that operate a sustainable forestry policy. Furthermore, the publisher ensures that the text paper and cover boards used have met acceptable environmental accreditation standards.

Printed in China.

Cataloging-in-Publication Data

Environmental science in a changing world / edited by Paige Tucker.
p. cm.
Includes bibliographical references and index.
ISBN 978-1-64116-275-3
1. Environmental sciences. 2. Ecology. I. Tucker, Paige.
GE105 .E58 2020
363.7--dc23

Table of Contents

Preface

This book aims to highlight the current researches and provides a platform to further the scope of innovations in this area. This book is a product of the combined efforts of many researchers and scientists, after going through thorough studies and analysis from different parts of the world. The objective of this book is to provide the readers with the latest information of the field.

Environmental science is an interdisciplinary subject which encompasses biological, physical and information sciences such as, ecology, chemistry, zoology, oceanography, soil science and many others. It also includes atmospheric science. It uses all these scientific fields to study the environment and find solutions to environment related problems. Environmental science provides a quantitative and integrated approach to the study of environmental systems. Environmental science aims to understand the earth processes, mitigation of pollution, alternative energy systems and also the effects of global climate change. It examines the environmental problems and finds the appropriate solutions. The four primary components of this discipline include ecology, geoscience, atmospheric sciences and environmental chemistry. All these components can be further classified into focused branches to facilitate the understanding of our environment. This book traces the progress of this field and highlights some of its key concepts and applications. It strives to provide a fair idea about this discipline and to help develop a better understanding of the latest advances within this field. Those in search of information to further their knowledge will be greatly assisted by this book.

I would like to express my sincere thanks to the authors for their dedicated efforts in the completion of this book. I acknowledge the efforts of the publisher for providing constant support. Lastly, I would like to thank my family for their support in all academic endeavors.

Editor

1

Households solid waste generation and management behavior in case of Bahir Dar City, Amhara National Regional State, Ethiopia

Kassahun Tassie Wegedie[1]*

*Corresponding author: Kassahun Tassie Wegedie, Department of Agricultural Economics, Bahir Dar University, College of Agriculture and Environmental Science, Bahir Dar, Ethiopia
E-mail: tasiekassahun@gmail.com

Reviewing editor: Xiaolei Zhang, Queen's University Belfast, UK

Abstract: Solid waste management remains a serious problem in most developing nations, even although it consumes a larger portion of the municipal budget. This article examines factors that might influence solid waste generation and solid waste management behavior of households in Bahir Dar city. Stratified and systematic random sampling techniques were employed to select 196 households for primary data collection. The data were subjected to both descriptive and econometric technique of analysis. Multiple linear regression analysis was used to identify households' solid waste generation dominant factors. The result shows that mean weight waste generation by sample households was 0.22kg per capita per day (0.28kg in the core zone, 0.17kg in the outer zone and 0.20kg in the middle zone). About 112(57%) of the households are served by door-to-door solid waste collection service. Only 53(27%) households received solid waste management services by municipality waste collectors weekly but 18(9%), 29(15%), and 12(6%) households received solid waste management services within 15 days, 21 days and once in a month, respectively. About 84(43%) households never received solid waste collection, transportation and final disposal in to landfill services until this interview was

ABOUT THE AUTHOR

Kassahun Tassie Wegedie holds bachelor's degree in natural resource Economics and management/ NREM/from Mekelle University, Ethiopia in July/ 2010. After this he was done in different agriculture and trade development and regulation offices in Amhara National Region State. After two years experience he got in country scholarship in Bahir Dar University and now he holds masters degree in developmental economics in October/2016. The author joined Department of Agricultural Economics in Agriculture and Environmental Sciences College of Bahir Dar University since 2016. Economic valuation, food security, livelihood analysis, impact analysis, choice experiment, poverty analysis and determinates of adoption are the author's interest areas of research.

PUBLIC INTEREST STATEMENT

Currently in Ethiopia, solid waste management is growing beyond the capacity of the municipal governors. The increasing volume of waste pushes a high demand on the waste retrieval services. These challenges are reflected in most cities of the country. Due to population growth and urbanization difficulties are created to respond to the required demand in waste management. Hence, the country (Ethiopia) has started to organize and participate in micro and small enterprises in solid waste management activities. Bahir Dar is a fast growing and center of tourism regional city in Ethiopia that also embarked on solid waste collection service through micro and small enterprises. They work in a public private partnership that aims solutions for solid waste collection and disposal. Since 2010, micro and small enterprises have delivered a waste collection, transportation and dispose in to landfill site services to the residential and commercial sites to minimize the problem. But still most residents dispose their waste on open spaces which may cause social, economic, political and environmental problems for living organisms.

made. Among the sample households, only 139(69%) of households having separate bins for storage of different types of waste and the remaining households do not have any separate bins. The multiple linear regression results showed that the quantity of solid waste generation of households in Bahir Dar city was significantly affected by household size and household aggregate income.

Subjects: Environment & Society; Environment & Economics; Environmental Change & Pollution

Keywords: solid waste management; multiple linear regression models; households; Ethiopia

1. Introduction

Economic growth in developing countries has led to the massive movements of people from rural areas to urban areas for searching better life (Lichter & Brown, 2011). As a result, the number of persons added to the urban population in developing countries increase challenges faced by governments, urban planners, non-governmental service providers and urban residents (Baqui, 2009). Due to this, there is increasing solid waste generation in urban centers (Alam & Ahmade, 2013). Solid waste is not only rising in quantity but also changing in composition (from less organic matter to more paper, packing materials, plastics, glass, metal and other substances), which is exacerbated by low collection rates (Solomon, 2011). A survey by the United States Agency for International Development (USAID) in 2009 purged the rate of solid waste generated in Africa at approximately 0.5 kg per person per day. From these only a small amount of these solid wastes are properly disposed of in designated landfills. The rest are either dumped openly without recourse to treatment systems or left in public dumpsters with no one to properly dispose them (Pradhan, 2009). To overcome these problems implementation of effective waste management practices has been identified as essential for economic development in low income countries in particular (Zerbock, 2003).

In developing countries, improper handling and disposal of solid wastes contribute to high level of mortality and morbidity even if produce less per capita solid waste (Medina, 2002), because of cities and municipalities cannot cope up with the accelerated rate of waste production and composition in terms of technology, institutional arrangement and cost effectiveness of solid waste management (Modak et al., 2010; Zurbrügg et al., 2004). Due to these, improper municipal solid waste disposal and management causes all types of pollution, i.e. air, soil and water pollution (Alam & Ahmade, 2013). This indicated that human health and ecosystem service is threatened by improper solid waste management system (Pradhan, 2009). Even though the overall goal of urban solid waste management is to collect, treat and dispose of solid waste generated by all urban dwellers, 30%–60% of all the urban solid wastes are uncollected, and less than 50% of the population is served (Monyoncho, 2013). As a result, inefficient municipal solid waste management system increase disease transmission, contaminate ground and surface water, create greenhouse gas emissions, damage ecosystem services, discourages tourism and other business activities (Alam & Ahmade, 2013; Chinasho, 2015; Ejaz et al., 2010; Fenta, 2017).

In line with this, solid waste management is becoming a major public health and environmental concern in urban areas of Ethiopia that only 2% of the population received solid waste collection, transportation and landfill disposal services (Kassa, 2010). Accordingly, Bahir Dar city is one of the highly expanding and rapidly growing cities in Ethiopia (Chinasho, 2015; Unep, 2010). With the current annual population growth rate of 6.6% continues, the city population will double within 11 years and increase pressure on municipal solid waste management in order to outbreak disease (Alam & Ahmade, 2013; Lohri et al., 2014). Thus the need for adequate solid waste management is unquestionable (Abate & Goshu, 2017; Lohri et al., 2014). A total of 98.8 tons per day solid waste was generated from Bahir Dar city. But the city municipality collects and disposes only 58% of the total municipal solid waste and the 42% disposed to the environment (Fikreyesus et al., 2011; Unep, 2010). This implies that small proportions of the urban dwellers are served and large quantity of solid waste left uncollected.

Only few studies in the area of solid waste management and generation are done in Bahir Dar city. Some of these studies include: "assessment of liquid waste management" (Fesseha, 2012); "Financial sustainability in municipal solid waste management cooperatives"(Lohri et al., 2014); "plastic bag waste generation rate"(Ababa, 2007); "sustainability of solid waste collection and transport services by micro and small enterprises"(Edmealem, 2003); "solid waste characterization and quantification for the development of an integrated solid waste management" (FFE, 2010; UNEP, 2010b), and "household demand for improved solid waste management services"(Endalew & Tassie, 2018). But such studies do not guarantee to discus households' solid waste generation and management behavior in different zones in the study area.

2. Objectives

The general objective of this study was to estimate households' solid waste generation and management behavior in the case of Bahir Dar city, Amhara National Regional state, Ethiopia and the following specific objectives:

- To assess households solid waste generation and management behavior in the study area.
- To identify the determinate factors of households' quantity of solid waste generation in the case of Bahir Dar city.

3. Methodology

3.1. Study area description

Bahir Dar city is located in the north western part of Ethiopia as shown on the Figure 1, and is faster growing city in the country (UNEP, 2010a). It has a total population of 96,140 in 1994 and 230,344 in 2007 (CSA, 2007), and estimated 297,749 in 2014 (FFE, 2010). Currently the city is

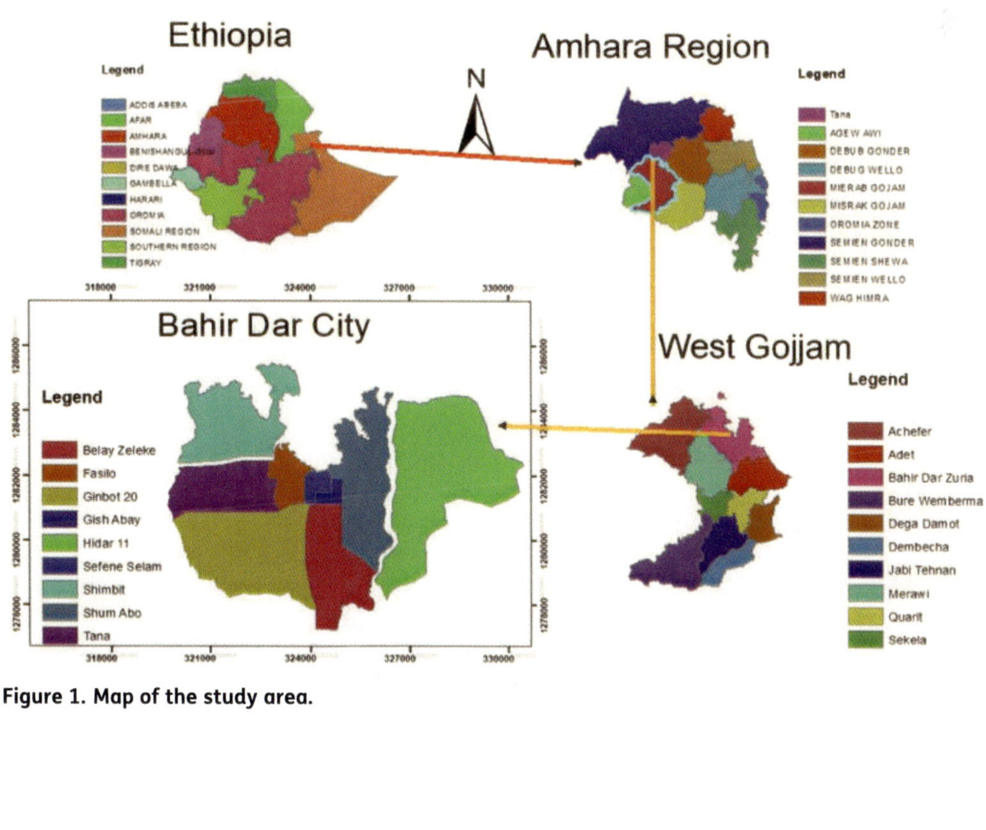

Figure 1. Map of the study area.

serving as a regional capital of Amhara National Regional State (ANRS) in the Federal Democratic Republic of Ethiopia. It has become one of the major tourist destinations of the country with a variety of attractions in the nearby Lake Tana (Ethiopia's largest lake and famous for churches and monasteries on the lake's 37 Islands) and Blue Nile river (FFE, 2010). A global position of the city is between 15.620 N latitude and 37.420 E longitude, and enjoys tropical type of climate with an average annual temperature of 19.60C and the average elevation of the city is estimated 1801 m above sea level.

The city has expanded rapidly throughout the 20th century. Today, waste discharges into Lake Tana have become a serious and highly visible problem (FFE, 2010).

In Bahir Dar city, urbanization is increasing at an alarming rate (287,756) in 2014 and it is 301,425 in 2015) (CSA, 2007). It is putting immense pressure on municipal services. One of the problems of the urbanization and increasing population is the increased generation of waste (Table 1). Solid waste generation per day continuously increase and estimated to be 212.2 tones solid waste will be generated in 2022, and which is greater than when it compared with Mekele and Awassa cities. **Therefore, this calls for urgent need for an appropriate need of determinant factors of quantity solid waste generation by assessing households' solid waste management behavior in Bahir Dar city.**

3.2. Data sources

All required data were collected from both primary and secondary data sources. The study was conducted primary data sources from sample households using a structured questionnaire by applying face to face interview to reduce non response rate and incompleteness of data. Secondary data were collected from published and unpublished materials.

3.3. Sampling techniques and sample size

A researcher selected the study area purposely having the information from officials of city municipality office of ANRS (Appendix 1). The sample size would be determined by considering resource constraints & representativeness of sample. A list of households was generated from Kebelle administrations to form the sampling frame. The choice of sampling technique (probability or non-probability) depends on purpose of the study. So, the objective of this study is to estimate households' solid waste generation and management behavior in the case of Bahir Dar city. For such a quantitative research, probability sampling technique is appropriate as compared to non-probability sampling technique because of it gives for every sample household equal chance of being interviewed.

3.3.1. First stage: selection of sample rural kebele administrations

The study area divided in to 17 administrative Kebelles[1] and after 2007 reforms these Kebelles arranged with a total of 52 zones (smallest legal administrative units). So, for the purpose of this study all zones categorized into three main groups, i.e. Core, Middle and Outer depending on population density, settlement, geographical location and land use pattern to make homogenous for primary data collection from sample units. Thus it employed a multi-stage sampling technique to make sample representative and reduce heterogeneity characteristics of study population, and to made conclusion and recommendation for study area population. To make the sample representative, sample zones were randomly from each stratum (i.e. core, middle and outer zones).

3.3.2. Second stage: selection of households

Sample households are the main primary data sources of this study. From the total of 5353 households of from six samples zones, a total of 196 households, which was proportionally allocated for the respective sample zones, were included in the sampling frame population (Table 2). Then to maintain the representativeness of the study, respondents were picked systematically at a distance of every ith household/.$i = \frac{N}{n}$ Where, "N" is the total number HHs in the sampling frame list adopted from kebele administrative offices and "n" is the number of sample households determined by using scientific formula. A critical component of sample size formulas is the estimation of variance in the

Table 1. Municipal solid waste generation projection per day of Bahir Dar city

Year	2010	2011	2012	2013	2014	2015	2016	2017	2018	2019	2020	2021	2022
Population	218,975	233,427	248,833	265,256	287,756	301,425	321,319	342,526	365,333	389,232	414,921	442,306	471,498
Waste generation (tons)	98.5	105	112	119.4	129.5	135.6	144.6	154	164.4	175.2	186.7	199	212.2

Source: (FFE, 2010)

Table 2. Summary of household sample size from each stratum

Group of Kebelles	Kebelles in each stratum	Sample Kebele	Sample zones from each Sample kebele	Number of households in sample zone		Total	Number of sample households in each sample zone		Total
				A, E,C2	C, B, D1		A, E,C2	C, B, D1	
Kebelles close to the center/inner	01,02,03,04,05,06,12	Kebele 04	A & C	897	1007	1904	33	37	70
Kebelles located in the middle	07,08,09,10,15,17	Kebele 17	E & B	908	1002	1910	33	37	70
Periphery/Outer	11,13,14,16	Kebele 11	C_2 &D_1	628	911	1539	23	33	56
Total	17 Kebelles	3 Kebeles	6 zones	2433	2920	5353	89	107	196

Source: Bahir Dar city administration office, 2016

primary variables of interest in the study (Cochran, 2007). Two ways of estimating population variances for sample size determinations, (1) use pilot study results and (2) use data from previous studies of the same or a similar population. For categorical dependent variable, 5% margin of error is acceptable, and for continuous dependent variable 3% margin of error is acceptable (Krejcie & Morgan, 1970). The formula that we used for determining sample size is the following:

$$n = \frac{NZ^2PQ}{d^2(N-1) + Z^2PQ} \tag{1}$$

Where: n = sample size of housing units; Q = 1-P; N = Total numbers of housing units; P = residential housing unit variable; Q = housing units used for commercial activities, offices public centers, etc. Z = Standardized normal variable and valued that corresponds to 95% confidence interval equal to 1.96; d = Allowable error (0.05).

According to data obtained from city housing development office, there are total of 52,386 legal housing units (N): Out of these, more than 85% (P = 0.85) are residential and the rest 15% (Q = 0.15) are used for commercial activities, offices and for others (Report, 2017). **Therefore, based on the above specified formula the required sample size becomes**:

$$n = \frac{52386 * 1.96^2 * 0.85 * 0.15}{(0.05^2)(52386-1) + (1.96^2)(0.85)(0.15)} = \frac{25658.87}{131.4523} \approx 196$$

Therefore, $\mathbf{n}$ **= 196** was the minimum sample size of households for reliable results. Finally, by using proportional allocation method the researcher was decided to take sample households from selected zones. These sample households was drawn for data collection using simple random sampling method.

3.4. Data collection method

The name and number of the household head was collected from the sample frame list of 2016 for 6 selected zones that government administrative offices used for distribution of basic consumption goods (sugar, palm oil, wheat flour, etc). After preparing the list of the household head sample households were selected randomly using the random Table. Two survey instruments were developed to gather the necessary information (i.e. checklist and questionnaire). The major objective of the checklist was to get a general overview of the existing solid waste management system in operation and also helped to identify problems related to solid waste management in the city. The questionnaire was finalized after a pre-test. To minimize strategic bias, an introductory statement was included for respondents. The questions were on demographic characteristics of the households, information on waste generation by types, waste disposal practices, and door-to-door collection systems, Family total income, sources of income, education level and possession of domestic amenities were also among the questions asked. The questions were filled by visiting the selected households door-to-door from September—October 2017. The households were visited twice to complete the questionnaire. On the first day socioeconomic information were collected and households requested to deposit the wastes in different plastic bags and the next day the wastes were weighted.

3.5. Data analysis methods

The data were entered in the computer and analyzed using different statistical tools. Data were grouped for the Core, Middle and Outer zones. Some of the information were analyzed using simple statistical tools such as mean, percentage and also analyzed econometrically by using STATA version 13. The multiple linear regression econometric models were used for the analysis of household solid waste generation determinant factors.

3.6. Determinant factors of households' solid waste generation

This section, mainly discusses the empirical analysis of the relationships of quantity and composition of household solid waste to selected social and economic variables. The analysis is based on data for components of household wastes, by type of material judged having recycling potential.

3.6.1. Specification of econometric model

Households' quantity of solid waste generation and composition will depend upon a number of important determinants factors. In this study HH's quantity of waste generated is treated as the dependent variable, and it is taken as continuous variable. Normally, waste is a function of consumption. The relationship between waste and consumption activities may be expressed as (Richardson et al., 1978):

$$W = \beta C \quad (2)$$

Where,

- W = vector of components of solid waste
- β = Vector of technical waste transformation coefficients relating the types and quantities of solid waste to each consumption activity
- C = is a vector of consumption activities selected by the household.

Any particular waste may be generated by the consumption of more than one commodity. Here no attempt is made to identify the technical waste transformation coefficients associated with the individual products. It mainly tries to compare the relationship between different types of waste generation and socioeconomic variables affecting the quantity of waste generated.

The major determinants of household consumption activities are assumed to be household monthly total income (TOTI), size of the household (TOTPOP), educational status of the household (GRAD); and extra land area in the house compound (EXTLA), solid waste management services availability in the area (SER).

The choice of econometric model depends on the nature of the dependent variable, i.e. nominal, ordinal, interval and/ratio scale. Since the dependent variable quantity of waste generated per household per day (**TOTW**) is continuous, the appropriate econometric model is multiple linear regression models. The model specified as follows:

$$\text{TOTW} = \beta_0 + \beta_1\text{TOTI} + \beta_2\text{TOTPOP} + \beta_3\text{GRAD} + \beta_4\text{EXTLA} + \beta_5\text{SER} + \beta_6\text{AGEHH} + e \quad (3)$$

Where:

- TOTW = quantity of waste per household per day (Kg)
- TOTI = Monthly income of the household in Ethiopian Birr (EBR.)
- TOTPOP = Household size (numbers of persons within family)
- GRAD = Educational status (number of college graduates)
- EXTLA = Extra land area within the compound of the selected household (ha.)
- SER = solid waste management services availability in the area
- AGE = age of the households(year)

Here household is assumed as a production unit producing solid wastes.

The hypothesis is as follows:

(i) **Increase in income (TOTI)** is expected to increase the demand for convenience factors and services embodied in commodities. The sign of the coefficient is expected to be positive for total quantity of waste generated. Consumption increases as income increase and so does waste generated. ($C = \alpha + \phi Yi$ Where C = household consumption, α = autonomous consumption, ϕ = slope of consumption function and Y = income).

(ii) **A larger household size (TOTPOP)** is expected to generate higher quantity of waste since more households are included in the unit; thus, the sign is also expected to be positive.

(iii) **Educated household members (GRAD)** work in the office and stay outside of the house for a long time. So the waste generation will be low. However, the generation of packing waste may be higher in case of a fully employed family as they have less time to prepare food. As such, they consume more packed food. So, number of educated households has positive relationship with total quantity of waste generated.

(iv) It is assumed that higher the **extra land area within the compound (EXTLA)** less the waste generated by the household. It is also assumed that the household with extra land area may dispose some of the waste in their land, which may not be counted in the total volume of waste generated. Thus extra land area and the total quantity of waste are inversely related and the sign of the coefficient of extra land area will be negative.

(v) **Solid waste management services availability (SER)** in the area is expected to generate higher quantity of waste since the quantity of waste disposed properly by collectors in case they increase demand that results waste.

(vi) **Age of households (AGE)** is expected to have a positive relation with quantity of waste generated because when age increase consumption becomes so does due to demand increases.

4. Result and discussion

4.1. Households solid waste generation

The Table 3 shows that average waste generation by households is 0.22kg/capita/day (0.28kg in the core zone, 0.17kg in the outer zone and 0.20kg in the middle zone). Waste generation is higher in the core zone and lower in the outer zone. This may be due to more economic activities such as trade and public services like transport are common in the core zone. In the core zone, people have no space even though they generate more wastes per household per day than other zones.

From the above table, average waste generated/household/day in kilogram column (2) gains when total waste generated column (6) divided by the total number of sample (column (8)). The same way for each zones gaining average waste generated/day/household.

Table 4 shows about the types and proportion of solid waste generated in the city. Kitchen waste is the dominate type of waste generated in the area and it accounts for nearly 82.2% of the total waste. Packing waste 9% is next to kitchen waste. Plastic waste is account 3.2%; whereas paper waste is 2.4% only. The remaining 3.2% are accounts other types of wastes, which include battery, dust, can, electronics wastes, etc. The type of waste generated across different zones are different because of factors might increase quantity of solid waste generated. In general 0.32 kg of solid

Table 3. Per capita waste generation by the households

Zone stratum (1)	Waste generation (Kg/HH/day) (2)	Std. Dev. (3)	Min (4)	Max (5)	Total waste (Kg) (6)	SW generation/ day/capita (Kg) (7)	Sample HHs (8)
All zones	0.91	1.27	0.35	5.15	178.36	0.22	196
Core zone	0.98	1.26	0.43	5.15	68.6	0.28	70
Middle zone	0.91	1.28	0.17	5.99	63.7	0.20	70
Outer zone	0.85	1.28	0.11	3.63	47.6	0.17	56

Source: Field survey, 2017

Table 4. Types of waste and their proportion in all zones

Waste types	Average waste generated per HH per day (Kg.)				Std. Dev	Min	Max	Total waste (Kg)	Per Capita Waste per day (Kg.)
	Core	Middle	outer	Total					
Kitchen waste	0.77	0.62	0.50	0.63	0.75	0.30	4.5	146.6(82.2)	0.16
Packing waste	0.2	0.11	0.11	0.14	0.23	0	3.02	16.1(9)	0.02
Plastic waste	0.21	0.11	0.10	0.14	0.22	0	2.05	5.7(3.2)	0.01
Paper waste	0.05	0.06	0.01	0.04	0.13	0	2.00	4.3(2.4)	0.01
Other waste	0.06	0.05	0.02	0.05	0.22	0	3.02	5.7(3.2)	0.01
Total waste	1.29	0.95	0.49	0.91	1.33	0.35	5.15	178.36(100)	0.22

Figure within brackets indicate percentage; (Source: field survey, 2017)

waste and 4.5 kg waste was generated in minimum and maximum respectively. But there is a difference for each type of wastes generated in the city.

4.2. Existing waste management practices

4.2.1. Solid waste separation practice

Table 5 shows solid waste separation practices of sample households. Some of solid wastes that are separated by households are agricultural waste and food wastes for the purpose of using it as home garden; grass, leaf, waste of sugarcane, waste of cattle's (after dried by sun), wood scrap to use them as fuel and used papers are reused as toilet paper by family. As the researcher tried to observe households solid waste separation activities, only solid wastes that are sellable to quraleos,[2] exchangeable to liwach[3] and to some extent organic wastes. Among the sample households, only 61(31%) of the households reported having separate bins for storage of different types of waste and the remaining households do not have any separate bins. Among the zones the households having separate bins is the highest 37(61%) in the middle zone, lowest 11(18%) in the outer zone and moderate 13(21%) in core zone.

From the survey result presented in the Table 5, 45(23%) of households have been done solid waste separation practice. Out of the total sample, 151(77%) households did not practice solid waste separation because of different reasons in the study area. Some of these are, lack of spaces, cannot see the importance, cannot afford separated bins, separation consuming time, no ready

Table 5. Separation practices of the households

Zones	Separate bin		Separation practice	
	Yes	No	Yes	No
All zones	61(31)	135(69)	45(23)	151(77)
Core	13(21)	57(68)	15(33)	55(36)
Middle	37 (61)	33(24)	22(49)	48(32)
`Outer	11(18)	45(33)	8(18)	48(32)

Source: field survey, 2017; Figures within brackets indicate percentage and absolute number indicate the number of households market for recyclables' materials, etc. The separation practice is highest in the middle zone 22 (49%), moderate in the core zone 15(33%) and lowest in the outer zone 8(18%).

4.2.2. Door-to-Door collection

About 112(57%) of the households are served by the door-to-door solid collection, transportation and final landfill disposal services. The majority of them receive this service by paying the fee and few are receiving the service not because of the fee but because of the location of their house along the road. Municipal solid waste collectors provide the service without receiving any bonus since they are the employees of the municipality. Door to door solid waste collection system of the city is insignificant both in spatial coverage and efficiency.

The perception of the households towards the present collection system was also captured through the questionnaire. The cleanliness of environment of an area depends on the regularity in the collection of waste by the municipality.

As expected the majority of respondents are not satisfied with the existing solid waste management in Bahir Dar city. Table 6 shows that most households 158(80.6%) were not satisfied with the present door-to-door collection system. Only 38(19.4%) households were satisfied and also that households who gates solid waste management services are not satisfied by the current solid waste management services. Only 18(9%) households feel that the disposal of such collected waste is environmentally safe, 51(26%) households feel that the disposal of such collected waste is not environmentally safe, and the remaining 127(64.8%) households could not explain whether it is environmentally safe or not. Table 6 also shows the zone-wise details regarding the knowledge on the disposal of the collected waste. The percentage of unsatisfied households is higher in the middle zone relative to the core and outer zones. This may related with spatial coverage of service provision of the municipality who gives more focus for core and in some extent middle zones of the city because of its economic activities related to transport, trade and other public services.

4.2.3. Waste collection and collectors

Table 7 shows the detail regarding the waste collection and collectors in the study area. Among the sample zones, municipal collectors collect solid waste from resident households from door-to-door collection by using hand carts and the collect one site which is called collection site for transportation purpose to land fill site. A city municipality gives responsibility for small scale enterprises to collect, transport and disposes into landfill. But these enterprises hire labor because of the volume and composition of waste increase over time due to population, urbanization, economic activities and other factors.

Ethiopia has Environmental policy and solid waste management proclamation, SWM proclamation (No. 513/2007). The bases for the Environmental Policy of Ethiopia (EPE, 1998) are articles 92.1 and

Table 6. Door-to-door collection and disposal of waste

Zones	HHs gain door to door collection		Satisfied with the present collection system		Environmentally safe disposal of the collected waste		
	Gain	Not gain	Satisfied	Not Satisfied	Yes	No	Don't know
All zones	112(57)	84(43)	38(19.4)	158(80.6)	18(9)	51(26)	127(64.8)
Core	56(50)	14(16.7)	22(57.9)	48(30.4)	13(72.2)	17(33.3)	40(31.5)
Middle	42(37.5)	28(33.3)	13(34.2)	57(36.1)	4(22.2)	21(41.2)	45(35.4)
Outer	14(12.5)	42(50)	3(7.9)	53(33.5)	1(5.6)	13(25.5)	42(33.1)

Source: field survey, 2017: Absolute number indicate the number of households and Figure within bracket indicate percentages

Table 7. Waste collectors and collection frequency

Zones	Collectors			Collection frequency for HH gain service					Never gain
	Wage workers	Street sweeping	Private firms	Weekly	Every 15 days	Every 21 days	Per month	Total HHs	
All zone	74(100)	131(100)	216 (100)	53(27)	18(9)	29(15)	12(6)	112(57)	84(43)
Core	34(46)	60(45.8)	97(45)	35(18)	6(3)	8(4)	0(0)	49(25)	21(11)
Middle	22 (29.7)	45(34.4)	75(35)	11(5)	10(5)	17(9)	4(2)	42(23)	28(14)
Outer	18(24.3)	26(19.8)	44(20)	7(4)	2(1)	4(2)	8(4)	21(11)	35(18)

Source: field survey, 2017; Where, Pvt = Private small scale enterprises & Absolute number indicate the number of households and Figure within brackets indicate percentages.

92.2 of the constitution of the Federal Democratic Republic of Ethiopia (FDRE). These articles enshrine the following rights: Article 92.1"government shall endeavor to ensure that all Ethiopians live in a clean and healthy environment" and Article 92.2 "government and citizens shall have the duty to protect the environment. Some of the articles of the EPE directly or indirectly address solid waste management issues. However the regional state has not yet proclaimed any law or act on environmental issues. It is rather using the federal laws but facing difficulty on enforcement and monitoring at regional level (FFE, 2010). Even though there are laws and regulations at federal level and also applied by the regional state, their implementations and enforcements are very weak.

As we explained above the municipality planed to provide services for resident once per week in regular by small scale enterprises and dream light Plc. But the result shows only 53(27%) households received solid waste management services weekly. While 18(9%) households, 29(15%) households, 12(6%) households received solid waste management services within 15 days, 21 days and once in a month, respectively, and 84(43%) residents never received solid waste collection, transportation and final disposal in to landfill services until this interview was made. These indicate a huge faillarity of environment protection from pollution. In general solid waste management services provided by the municipality waste collectors is not fair even in waste management frequency.

4.2.4. Solid waste management methods when no service supply

Table 8 shows the detail regarding management practices of sample households. Households that are not received door-to-door collection are managing their waste in different ways. The only solid waste disposal option of household is restricted to two choices. The first one is simply burning, burying, or dumping of solid waste in their compounds. While the second option is throwing of

Table 8. Waste management in households, which have no Door-to-Door collection system

Management practices	Total N° of HH	Core	Middle	Outer
Throwing in the container	8 (4)	6(8)	4(5)	2(3)
Throwing in the road	31(16)	35(50)	10(14)	6(10)
Throwing in open field	26(13)	11(16)	10(15)	6(11)
Burying in own land	43(22)	6(8)	17(24)	13(24)
Prepare the compost from waste	22(11)	4(6)	10(14)	5(9)
Cattle feeding	4(2)	0(0)	2(3)	1(1)
Burn	48(25)	6(8)	16(23)	18(32)
Throw in the river	14(7)	2(4)	1(2)	6(10)
Total N° of HH with different practices	196 (100)	70(100)	70(100)	56(100)

Source: field survey, 2017; Absolute number indicate the number of households and Figure within brackets indicate percentage solid waste at roadsides, open fields, nearby rivers, bridges and gullies. Even households who received solid waste management services practice illegal wastes disposal because of they did not get the services up on the plan of the municipality (once a week for residents; per day for star hotels; and per 3 days for cafe and restaurants). Hence in order to assess the routine method of solid waste disposal practices of households and to know the destination of uncollected solid wastes, the researchers asked sample respondents about their common disposal system.

From the total sample, 130(66.3%) households practice illegal solid waste disposal, and only the rest 63(32.1%) households wait until solid waste collectors come. This indicates even households who receive service from the municipality disposed wastes illegally. It has been observed and concluded that still open dumping of waste is the most prevailing activity practiced by residents of Bahir Dar city and these makes high probability of environmental pollution, breeding grounds of insects, pests and infectious diseases and also produce toxic gases, which spread odor around the dumping places and block drainage channels.

From Table 8, it is clear that 48(25%) of households practiced burning of the waste to get exonerate the collected waste and 22(11%) of households practice burry inside the compound and simply dispose the waste in the compound. Such practice poses high risk on the local environment. These result shows even households who have been get such solid waste management services from municipality practice illegal solid waste disposal because of service provision period with in long period interval and irregular.

Table 8 also shows the zone-wise details regarding residents' solid waste management methods. The percentage of waste throwing into the road is higher in the core zone relative to them middle and outer zones. This may related with absence of open spaces and rivers for waste disposal options. When we seen throwing waste in to the river outer zones is higher than other zones.

It has been observed and concluded that still open dumping of waste is the most prevailing activity practiced by residents of Bahir Dar city and these makes high probability of environmental pollution, breeding grounds of insects, pests and infectious diseases and also produce toxic gases, which spread odor around the dumping places and block drainage channels. Therefore, the participation of the whole community in the municipal solid waste management program of the city is very essential and highly expected to bring changes on the existing low status of municipal solid waste management.

4.2.5. Econometric results

The estimation result of waste generation and socioeconomic model has been shown in Table 9. It mainly tries to compare the relationship between households' quantity of solid waste generation and socioeconomic variables affecting the quantity of waste. Econometric analysis may provide

Table 9. Factors affecting households' solid waste generation

Waste component	Intercept	TOTI	AGEH	TOTPO	EXTLA	GRAD	SER	R^{-2}	F
TOTW (all zones)	−2.72	0.25	0.51	0.50	0.12	0.14	0.25	0.25	22
	(7.6)	(6.24)**	(1.36)	(8.5)**	(1.58)	(2.7)	(6.24)		
TOTW (Core)	−2.25	0.19	0.54	0.55	−0.84	−0.13	0.19	0.38	16
	(-3.8)	(2.79)*	(2.55)	(5.8)**	(-1.9)	(-0.37)**	(2.79)		
TOTW (Middle)	−2.9	0.26	−0.08	0.59	0.04	−0.32	0.26	0.27	11
	(-3.6)	(3.04)	(2.5)	(5.5)**	(0.47)	(-3.72)	(3.04)		
TOTW (Outer)	−3.26	0.45	0.75	0.38	0.09	0.15	0.45	0.27	14
	(-6.2)	(6.08)*	(-2.12)	(2.79)	(1.04)	(0.58)	(6.08)		

Figure within bracket indicate't value; Source: field survey, 2017
***Significance at 1%; **Significance at 5%; and *Significance at 10%

better information and clearer focus on the factors that affect households quantity of solid waste generation responses such that policy recommendations can be made based on result conclusions. Before estimation was done, data exploration is an important step.

To start with, to check whether multicollinearity is present or not a simple correlation coefficient matrix was conveyed. Multicollinearity is a serious problem when correlation coefficient is 0.8 and above (VIF exceeds 10) which is based on rule of thumb (Gujarati & Porter, 1999). There is no problem of multicollinearity (VIF = 2.7). The goodness of fit for the model has been tested in this study with some diagnostic tests which fulfill the following criteria of good results. First, the a R^2 value (which is a measure of goodness of fit of the estimated regression model) of 0.51 depicts a good fitting of the model, which defines that 51% of the variation in change of the waste generation of the households could be explained by the independent variables in the model. The F-test shows that the estimated regression is quite meaningful in the sense that the dependent variable is related to each of the specified explanatory variables. The linear relation of the model is highly significant (the p value for the F-test is less than 0.0001). Second, the signs for the estimated coefficients are consistent.

This study result shows that income and household size were positive and highly significant. The positive coefficient on income variable (TOTI) at 5 percent level of significance, indicates that holding all other variables constant, higher income people are generating more waste than the lower income people. The positive relationship between these two variables is generally supported by the previous literature (ining and Ebreo, 1992; Mohai, 1985; Owens et al., 2000). This result seems reasonable since increase in income is expected to increase the demand for convenience factors and services embodied in commodities. The positive coefficient on households size, at 5 percent level of significance, indicates that holding all other variables constant, large family are generating more waste than the small family. A larger household size is expected to generate higher quantity of waste since more households are included in the unit; thus, the sign is also expected to be positive. The coefficient of education variable is positive but it is not significant. This related with educated households have an opportunity to leave more time out of home for field work. This result supported by, a number of other studies found that the effect is positive but statistically not significant (ining and Ebreo, 1992., Rafia et al., 2008; Owens et al., 2000.). Extra land area has positive but insignificant effect is consistent with (Owens et al., 2000) . In the study area, there are very few households with extra land area; because of urbanization growth, high population density and have high economic value of houses in the city.

The regression coefficient of the household size is higher relative to the coefficient of the total income in all the zones except middle zone. However, the coefficient of income is higher than the coefficient of the household size in the middle zone. Extra land area has positive but insignificant effect in all the zones. A variable family size is significant in both core and middle zones but not in the outer zone. Level of education is significant and has a negative effect on quantity of waste generated only in the core zone.

5. Conclusion and recommendation

Households in the study area do not receive adequate solid waste management services and they are adversely affected by improper solid waste collection. The per capita waste generation was 0.22 kg/person/day in all the zones of the city. The per capita waste generation is found to be the lowest in the outer zone and highest in the core zone. Thus, the low per capita waste generation in the outer zone may be due to the household sorting of waste more intensively in the outer zone than in other zones. Most households feel that the lack of stiff penalty and non-execution of law is the basic problem for the effective management of waste. Thus, provision of strong penalties and effective execution of the law will be the major tool to reduce the problem of solid waste management in Bahir Dar city. It is found that environmental awareness is very low among the residents of the city. Thus, stringent regulations with environmental awareness programs for household sorting and composting can reduce the volume and quantity of waste for land filling. The waste component relationship shows that size of the household and income are the major factor determining the total quantity of the waste in all the zones. It was also found that education has a negative effect on waste generation. Based on our research findings, the following points are recommended:

- There is no any type Biogas plants used by many factories in and around Bahir Dar city to generate electricity from solid wastes generated from resident households. Government could make it mandatory for factories/poultries/farms generating a certain amount of solid waste to have biogas plants. Certain tax incentives could also be given to those complying with these new regulations.
- The study also revealed a strong positive relationship between household family size and quantity of waste generated. It is recommended that measures be embarked upon to encourage people especially the younger generation who are yet to have children of their own to have smaller family sizes rather than larger ones.
- As pointed out several times in the discussion above the present solid waste management system in Bahir Dar city is less than adequate. Organizational strengthening, better management of the conservancy section would certainly help to change this gloomy condition of solid waste management and bring it to acceptable standards. Manpower, number of vehicles, tricycles and rickshaw vans could be increased.

Funding
The author received no direct funding for this research.

Competing interests
The author declares no competing interests.

Author details
Kassahun Tassie Wegedie[1]
E-mail: tasiekassahun@gmail.com
[1] Department of Agricultural Economics, Bahir Dar University, College of Agriculture and Environmental Science, Bahir Dar, Ethiopia.

Notes

1. **"Kebele"** refers a legal administrative area includes different smaller administrative areas such as zones which is the lowest legal administrative area.
2. "Quraleos"- persons who practice exchange of recycled wastes in other goods .
3. "Liwach"- exchange recycled solid waste in terms of goods used for different purposes.

References

Ababa, A. (2007). *Ayana yehuala, Addis Ababa University*.

Abate, B., & Goshu, G. (2017). Waste management in lake tana basin—Case of rapidly urbanizing Bahir Dar City. In *Social and ecological system dynamics* (Vol. 6, pp. 581–593). Siena: Springer.

Alam, P., & Ahmade, K. (2013). Impact of solid waste on health and the environment. *International Journal of Sustainable Development and Green Economics (IJSDGE)*, *2*(1), 165–168.

Bahir Dar city office report (2017). *Bahir Dar city house construction and development office annual report of 2008/2009 fiscal year.*

Baqui, A. (2009). *Global urbanization: Trends, patterns, determinants, and impacts*. Maryland: Bloomberg Publishing.

Chinasho, A. (2015). Review on community based municipal solid waste management and its implication for climate change mitigation. *American Journal of Scientific and Industrial Research*, *6*(3), 41–46.

Cochran, W. G. (2007). *Sampling techniques*. John Wiley & Sons.

CSA. (2007). *(Central Statistical Authority of Ethiopia) Summery and statistical report of the population and housing censes result of Ethiopia.*

Edmealem, B. (2003). *Assessment of the sustainability of solid waste collection and transport services by MSEs, in the case of Bahir Dar city, Ethiopia.*

Ejaz, N., Akhtar, N., Hashmi, H., & Naeem, U. A. (2010). Environmental impacts of improper solid waste management in developing countries: A case study of Rawalpindi city. *The Sustainable World*, 379–388.

Endalew, B., & Tassie, K. (2018). Urban households' demand for improved solid waste management service in Bahir Dar city: A contingent valuation study. *Cogent Environmental Science*, *4*, 1426160.

EPE. (1998). *Environmental policy of Ethiopia.*

Fenta, B. A. (2017). Waste management in the case of Bahir Dar City near Lake Tana shore in Northwestern Ethiopia: A review. *African Journal of Environmental Science and Technology*, *11*(8), 393–412.

Fesseha, H. M. (2012). Liquid waste management: The case of Bahir Dar, Ethiopia. *Ethiopian Journal of Health Development*, *26*(1), 49–53.

FFE. (2010). *"Forum for environment assessment of solid waste management system of Bahir Dar city, for the development of an ISWM Plan.*

Fikreyesus, D., Mika, T., Getane, G., Bayu, N., & Mahlet, E. (2011). *Ethiopia solid waste and landfill: Country profile and action plan.* Community Development Research Sponsored by Global Methane Initiative. Retrieved from https://www.globalmethane.org/documents/landfills

Gujarati, D. N., & Porter, D. C. (1999). Essentials of econometrics application 5th edition.

Ining and Ebreo. (1992). Predicting recycling behavior from global and specific environmental attitudes and changes in recycling opportunities. *Journal of Applied Social. Psychology*, World Resources Institute, United Nations, *22*, 1580–1607.

Kassa, Z. (2010). *The challenges of solid waste management in urban areas, the case of Debre markos town.* Ethiopia: Addis Ababa University.

Krejcie, R. V, & Morgan, D. W. (1970). Determining sample size for research activities. *Educational and Psychological Measurement*, *30*(3), 607-610.

Lichter, D. T., & Brown, D. L. (2011). Rural America in an urban society: Changing spatial and social boundaries. *Annual Review of Sociology, 37*, 565–592.

Lohri, C. R., Camenzind, E. J., Zurbrügg, C. (2014). Financial sustainability in municipal solid waste management–Costs and revenues in Bahir Dar, Ethiopia. *Waste Management, 34*(2), 542–552.

Medina, M. (2002). Globalization, development, and municipal solid waste management in third world cities. *Institute of Advance Studies, Mexico*, 1–23.

Modak, P., Jiemian, Y., Hongyuan, Y., Mohanty, C.R. (2010). *Municipal solid waste management: Turning waste into resources*. Shanghai Manual–A Guide for Sustainable Urban Development in the 21st Century.

Mohai, P. (1985). Public concern and elite involvement in environmental conservation issues. *Social Science, 66*, 820–838.

Monyoncho, G. (2013). Solid waste management in urban areas Kenya: A case study of Lamu town. Department of Real Estate and Construction Management, University of Nairobi, Kenya.

Owens, J., Dickerson, S., Macintosh, D. L. (2000). Demographic covariates of residential recycling efficiency. *Environment and Behavior, 32*, 637–650.

Pradhan, U. M. (2009). *Sustainable solid waste management in a mountain ecosystem*. Darjeeling, West Bengal, India.

Rafia A., Hanaki, K., Kurisu, K. H. (2008). *Factors affecting waste generation and willingness to recycle: A study in a waste management program in Dhaka city, Bangladesh*.

Richardson, A., & Havlicek Jr, R. J. (1978). Economic analysis of the composition of household solid wastes. *Journal of Environmental Economics and Mangement, 5*, 103–111.

Solomon, C. (2011). *Assessment of municipal solid waste management service in Dessie town*.

SWM proclamation (513/2007). Ethiopia has Environmental policy and solid waste management proclamation.

Tegegne, G. (1999). Willingness-to-pay for environmental protection: An application of contingent valuation method (CVM) in Sekota district, northern Ethiopia. *Ethiopian Journal of Agricultural Economics, 2*, 1.

UNEP. (2010a). *Assessment of the solid waste management system in Bahir Dar Town and the Gaps identified for the Development of an ISWM Plan*.

UNEP. (2010b). *Solid waste characterization and quantification of Bahir Dar City for the development of an integrated solid waste management Plan*.

Unep, J. (2010). *Assessment of the solid waste management system in Bahir Dar town and the gaps identified for the development of an ISWM plan*. Forum for Environment.

Zerbock, O. (2003). *Urban solid waste management, waste reduction in developing countries*. USA: School of Forest Resource and Environmental Science, Michigan Technological University.

Zurbrügg, C., Drescher, S., Patel, A., Sharatchandra, H. C. (2004). Decentralised composting of urban waste – An overview of community and private initiatives in Indian cities. *Waste Management, 24* (7), 655–662.

2

Public policy and technology choices for municipal solid waste management a recent case in Lebanon

Michel Soto Chalhoub[1]*

*Corresponding author: Michel Soto Chalhoub, Civil and Environmental, Notre Dame University, Lebanon
E-mail: mchalhoub@live.com

Reviewing editor: Keng Yuen Foo, River Engineering and Urban Drainage Research Centre, Universiti Sains Malaysia - Kampus Kejuruteraan Seri Ampangan, Malaysia

Abstract: Municipal solid waste (MSW) is a natural accrual in various communities resulting from daily activities, which offers challenges and opportunities. Developed countries lean towards "upstream avoidance" type of solutions, while less developed countries use "downstream remediation" solutions. In between these two options, many hybrid solutions involve triage and final disposal of post-treated reduced MSW mass.

The aim of the present article is to briefly review the evolution of policies related to MSW, to address an illustrative case of a recent crisis in Lebanon, and to present the development of a multivariate analysis model on public attitudes towards MSW management. It explores incineration with air pollution control and energy recovery. Non-incineration solutions, including anaerobic digestion, may be more appealing provided the full process, its byproducts and risks, are better understood. Furthermore, the latter solution requires a higher level of investment, expertise,

ABOUT THE AUTHOR

Dr. Michel Soto Chalhoub is an engineer and management consultant. His research addresses societal and environmental concerns. Michel is an industry practitioner in the U.S., Europe and the Middle East who served in senior management positions in major U.S. corporations such as Parsons Corp., Cap Gemini Ernst & Young, and InterPublic. Winner of several awards for innovative design, his projects include river protection, docking cells, theme parks, petrochemical facilities, and artificial reefs. He works at the nexus of engineering and public policy to provide strategic recommendations to decision-makers on issues that impact communities including municipal solid waste and water resource management. Michel holds a B.Sc. in Civil Engineering from l'Ecole Superieure des Ingenieurs, a M.Sc. and Ph. D. from the University of California, Berkeley, and a Master in Public Policy from Harvard University. He is a licensed Professional Engineer (P.E.) and Structural Engineer (S.E.) in the State of California, published extensively in academic journals, and served on emergency response teams to natural hazards and Code development committees.

PUBLIC INTEREST STATEMENT

Municipal solid waste (MSW) is critical to public health, yet several countries still face crises related to its management. The article proposes a framework for MSW analysis in the context of sustainable development, discusses a recent case in Lebanon, and presents an empirical regression analysis on public attitudes. Results show that simplified solutions must be considered including clean technology incineration with energy recovery (IER), along with other approaches such as anaerobic digestion. Lessons could be learned from European and Far Eastern countries that lack vast areas of land. Focus interviews and a quantitative regression analysis show end-user concerns about the ability of the public sector to manage MSW, and believe that sorting in households has had little effect on the final outcome. End-users have little trust in privatization given past monopolies on the MSW cycle, requiring capacity building in the public sector. It is also concluded that the upstream involvement of local communities in the design of a country-level environmental strategy would be key to success.

continuous quality improvement and supervision, especially in light of public management weaknesses in less developed countries.

A multivariate linear regression analysis model is developed and presented to describe public attitudes regarding MSW management. Linear regression was used to model a direct relationship between a response variable and several explanatory variables. A simplifying assumption is made to test the attractiveness of incineration with energy recovery (IER) including a process for air pollution control. In the multivariate analysis, the dependent variable is a composite index that describes the extent to which respondents agree with a potential IER solution. The independent variables include (1) the extent of public awareness-building about upstream triage, (2) privatization, (3) quality management in daily operations, (4) challenges in implementation and (5) reliability of public management authorities. The multivariate analysis showed that there is a statistically significant and positive correlation with a requisite quality in daily operations, and in technological upfront investment. It showed a statistically significant and negative correlation with the reliability of public management and privatization. As for public awareness about upstream household level triage, there was no statistically significant correlation. This last result seems counter-intuitive, but it may be due to the fact that respondents have low expectations about the impact of their actions on government decisions. The general public, faced with socio-economic problems, may not see household triage as a determining factor. Public perception is that triage ends up with a MSW collection and transport with a mixed dumping at final destination.

The analysis shows that prevention is preferred to remediation and that it requires supporting public policies to make it practical, especially in less developed countries. Bio-processes are well-accepted and may become part of the local practice once their long-term effects become better understood to mitigate potential risks. The IER option is perceived as technologically and financially viable and that clear public policies should be designed and implemented.

Subjects: Bioscience; Environmental Studies & Management; Mathematics & Statistics; Engineering & Technology; Urban Studies

Keywords: municipal solid waste; technology options; public policy; incineration with energy recovery; environmental impact

1. Introduction and brief policy background

Municipal solid waste (MSW) is one of the burning issues that are highly debated among scientists and policymakers, as it affects public health and the use of resources, especially in developing countries (Guerrero, Maas, & Hogland, 2013). Simply put, MSW is a byproduct material closely intertwined with our daily lives and is a result of basic living conditions. Therefore, the evaluation of MSW management requires taking into consideration socio-economic factors in tandem with technological options (Aye & Widjaya, 2006; Cheng, Chan, & Huang, 2002). Given the uncertainty related to each phase of the MSW life cycle, techniques were developed to assign weights to each parameter related to the involvement of stakeholders—public managers, households and communities (De Feo & De Gisi, 2010a).

The objective of the article is to develop and present a mathematical model that illustrates public attitudes towards MSW management. Prior to that, the article provides an overview of the evolution of MSW management policies, and discusses a recent waste crisis in which Lebanon was engulfed. In a rare and historic public uproar, protesters took it to the streets of major coastal cities—including Beirut—and voiced their deep discontent about the lack of action shown by both the public sector entrusted with such issues and the private subcontractor (Daily Star, 2016).

Briefly reviewing the evolution of solid waste management helps put current crises in context. In most societies that went through large industrialization movements, solid waste management was driven by policy trends in combination with technology options at hand (Durgekar, 2016). The earliest trend is attributed to public health as it is generally accepted that garbage needs to be collected and removed from the vicinity of the household. Followed a second trend related to environmental activism in that the first trend needs to be met without imparting damage of the environment. The third trend is related to treating solid waste as a resource and not as a burden to society—hence, new solutions were introduced whereby MSW is utilized as an energy source (Wilson, 2007). However, the current overarching approach would be to minimize waste from the outset. This mindset addresses the full lifecycle of consumer products by designing them such that they generate the least possible residues (Marshall & Farahbashsh, 2013).

In the 1980s and 1990s, a post-normal approach was explored to address public issues that are characterized by uncertain facts, whose values are in dispute, where societal stakes are high and yet decisions are urgent (Funtowicz & Ravetz, 1993). Issues of solid waste in both developed and developing countries, indeed, face such issues and pose a challenge in the use of traditional science. Various engineering schools of thought came to compete over technological solutions, rather than solve societal problems (Organization for Economic Cooperation and Development [OECD], 2013).

Several developed countries went through the learning curve to establish systems and processes for MSW as we know it today. In the fourteenth century, Europe was plagued due to dumping of organic waste in the streets (Tchobanoglous, Thiesen, & Eliassen, 1978). In the late 1700s, cities in the US suffered public health problems as the practice was that household waste was collected by request of select property owners, and dumped in open sites (Melosi, 1981). Up till the 1890s, large amounts of MSW were a common scene in the streets and it was not until the 1920s that demographic pressures and increased population density in American cities brought progress in the construction of sewage networks and the establishment of public sector processes for solid waste management (Gibson, 1998; Melosi, 2000). It was during that era that ideas such as recycling and reuse of waste material appeared and were applied. For example, the nation's first recycling center was established in New York City in 1898 (Dunson, 1999; Melosi, 1981). Residents were also required to separate waste into three categories—rubbish, comprising paper, glass, wood and metals; garbage, comprising organic food refuse; and other miscellaneous components (Tchobanoglous et al., 1978).

Increased public awareness about the dangers of waste dump sites led to the development of other approaches. Incineration, for example, is an early approach that appeared in Nottingham as far back as 1874, when it was presented as a new technology consisting of cremating large amounts of waste in a government-sponsored, organized manner (Kimball, 1992). Almost a century later, incineration with energy recovery gained attention as it was presented as part of acceptable policies with the 1965 U.S. Solid Waste Disposal Act, amended in 1970 by the Resources Recovery Act (Drobny, Hull, & Testin, 1971; Environmental Protection Agency [EPA], 1988). Within years, the concept moved to state-level regulations for organized solid waste management policies (EPA, 1988). However, what drove state policies for MSW management into local implementation was the 1976 Resource Conservation and Recovery Act, which was amended in 1984 to emphasize standards for sanitary landfills and strict measures for incineration and its impact on air quality.[1]

Over the last few decades, comparative analyses among various methods of MSW management were performed, inclusive of anaerobic digestion and composting (Malinauskaite et al., 2017). Some of the most commonly used still involve landfills and incinerators (Dijkgraaf & Vollebergh, 2004; Finnveden, Bjorklund, Ekvall, & Moberg, 2007; Quina et al., 2018). Relevant models balance waste treatment, as a necessity, with the potential benefit of recovering energy and raw materials generation, thus turning the process into a production operation (Consonni, Giugliano, & Grosso, 2005; Feo, Belgiomo, Rocca, & Napoli, 2003; Joseph, Snellings, Van Den Heede, Matthys, & De Belie, 2017).

Countries of low population density, sparsely inhabited land and comprising acres of unused desert may find landfills to be a possible solution. Not only do landfills provide a fast and inexpensive disposal, but they may be harnessed to generate byproducts related to energy generation and land conditioning (Benson, Barlaz, Lane, & Rawe, 2007; Huang, Zhao, Huang, Wang, & Tseng, 2017). Past research showed that soil fertility is affected positively or negatively by solid waste disposal depending on the ratio of organic material, or on the extent to which other components such as metals may be transmitted from the soil, to plants and into the human body (Fagnano, Adamo, Zampella, & Fiorentino, 2011; Gramatica et al., 2006; Leone, Gilman, & Flower, 1983). Therefore, countries with large surfaces have choices in terms of using landfills to generate energy if the adequate technology for sequestration is put to practice. On the other hand, countries with smaller land surfaces such as Japan realize that landfill site growth is unsustainable, and hence resort to incineration coupled with clean emissions exhaust technologies to curtail the downside of incineration in terms of air pollution (Yolin, EU-Japan Center for Industrial Cooperation, 2015; Zhao, Jiang, & Li, 2016).

2. Intertwined policy and technology options in MSW management

Considering solid waste as raw materials or input in the energy recovery process, a wide range of solutions, technologies and processes exist. Most of these solutions have positive and negative points, and therefore require a comparative study on a case-by-case basis before policy decisions are made (Huang, Ning, Zhang, & Fei, 2015; Tehrani, Karbassi, Ghoddosi, Monavvari, & Mirbagheri, 2009; Wan, Shen, & Choi, 2018). Several parameters have to be studied in parallel including land availability, absorption capacity of sanitary landfills and thermal waste treatment options (Consonni et al., 2005). Practically, MSW solutions often involve more than one type of technology, and rather use a set of integrated systems (Pařízek, Bébar, & Stehlík, 2008). An attractive process would integrate recycling, recovery and energy generation in a waste-to-energy (WTE) conversion. Therefore, one must look at the overall life cycle assessment (LCA) of these processes to determine the cost-benefit of each method (Bjorklund & Finnveden, 2007; Christensen et al., 2007; Hadzic, Voca, & Golubic, 2017; Hupponen, Gronman, & Horttanainen, 2018; Zhou, Tang, Chi, Ni, & Buekens, 2017).

Landfills have long been used as a disposal method, which then evolved to minimize impact on the environment. It relies on breaking down the waste components over a long period of time. Although landfill solutions have been around for a long time, they were used in most less developed countries where many landfills did not have energy conversion mechanisms (Parker, 1983; Pohland, 1991). According to the Environmental Protection Agency in New South Wales (EPA-NSW), "A landfill is an engineered, in-ground facility for the safe and secure disposal of society's waste." However, EPA indicates that landfills may have negative effects on fauna, flora, groundwater and surface water bodies. As such it imposes minimum standards to be met when applying for landfill permits under the Protection of the Environment Operations Act of 1997. No doubt, irrespective of the engineering design details used for landfills, EPA guidelines have policy implications that all stakeholders, especially the end-user in the local community, must have confidence in the overall management of the process. Further, beneficial use or reuse of waste should be central to public management strategies. The excerpt below summarizes the overall EPA objectives attributed to landfills (EPA-NSW, 2015).

> *"... Landfills should be sited, designed, constructed and operated to cause minimum impacts to the environment, human health and amenity.*

The waste mass should be stabilised, the site progressively rehabilitated, and the land returned to productive use as soon as practicable.

Wherever feasible, resources should be extracted from the waste and beneficially reused.

Adequate data and other information should be available about any impacts from the site, and remedial strategies should be put in place when necessary.

All stakeholders should have confidence that appropriately qualified and experienced personnel are involved in the planning, design and construction of landfills to current industry best practice standards..."

Even though EPA published its draft in 2015, landfills were historically considered as an attractive solution in many countries as they were, and still are, convenient and economical provided land areas are available such as in Australia or the US. Technically, refuse is spread in thin layers covered by soil. The layers are typically around 10–12 ft in thickness. Such sites witnessed poor vegetation growth for a variety of reasons particularly the presence of carbon dioxide (CO2) and methane (CH4) caused by anaerobic decomposition (Leone et al., 1983). In addition such sites were regulated against any future construction or the erection of structures. The idea of waste decomposition, whether solid or fluid, evolved in the 1980s and 1990s into an engineered approach that designs and monitors decomposition through anaerobic digestion facilities (Chalhoub, 1992).

As for MSW incineration, it relies on processing waste used as feedstock at very high temperatures, typically exceeding 850°C. Since this process makes air available to combustion, it generates carbon dioxide water, and other material including bottom ash with residual carbon (Zaman, 2010). Ferrous and non-ferrous post-combustion residues are also a source of reusable material (Brunner, 2011). As Brunner puts it:

"... Megacities can produce sufficient amounts of secondary resources for large-scale production of raw materials by urban mining, and cities are always in need of energy. Thus, combining recycling plants for metals such as iron, aluminum, and copper in cities with utilization of waste energy from such plants to fuel the city (heating and cooling, electricity) seems an attractive option for improving the sustainability of cities..."

Pyrolysis-gasification, consisting of incineration in absence of air, is considered an improvement over traditional incineration because it minimizes emissions (Malkow, 2004). In this context, WTE emissions data is an important factor in policy decision-making as it provides a map of the input-output ratios and relationships between emissions and energy production (Cherubini, Silvia Bargigli, & Sergio Ulgiati, 2008; Khoo, 2009).

Clearly, MSW management options are so varied and certain solutions are preferred to others depending on the country and its developmental level. A hierarchy that is commonly acknowledged in literature ranks MSW treatment techniques versus the developmental level of the country. In general, countries that enjoy a higher level of economic development, have a higher level of public awareness and, therefore, have easier time adopting waste avoidance approaches. Countries that are still struggling with developmental issues and whose populations are more focused on short-term gain or survival, resort to less costly low-tech approaches. It is important to clarify here that if the cost-benefit analysis is expanded to include social cost, for instance from environmental degradation or long-term human health hazards, then the results would be different and in favor of avoidance rather than a-posteriori treatment. At the bottom of the treatment hierarchy, you find the landfill approach without energy recovery (Figure 1).

The hierarchy in Figure 1 is not all encompassing. Consider the case of Japan, a country that is very tight on land surfaces and therefore cannot afford the proliferation of landfills. It compensates for that shortage by using incineration as an intermediate step on the way to final disposal. The Japanese practice focuses on optimizing a full cycle starting from collections, transport,

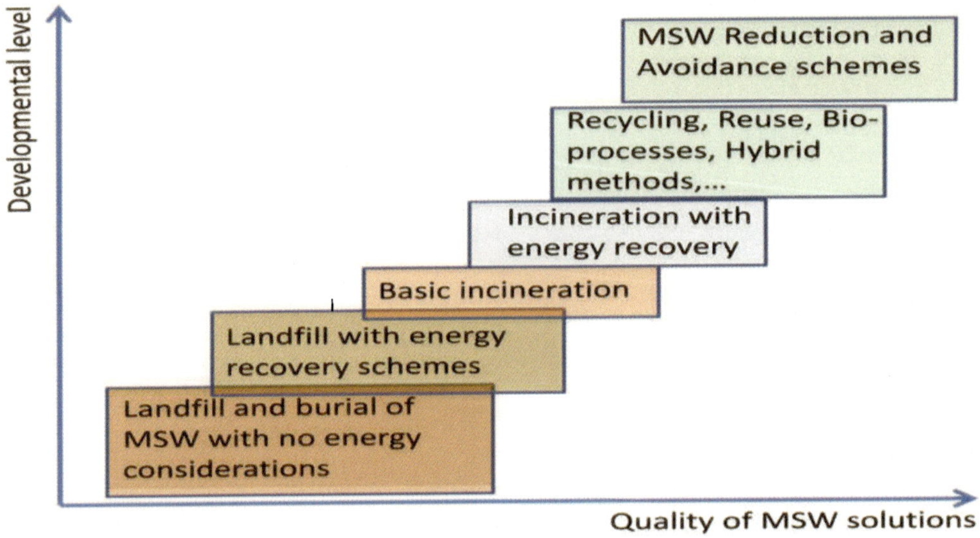

Figure 1. MSW solution versus developmental level (modified from source: UNEP, 2011).

municipal waste incineration technologies, all the way to landfill of a reduced waste mass (MoE of Japan, 2012). The same logic applies to most countries that do have the technology, strict policies, public discipline and awareness, but lack land areas. However, the challenge is two-fold when dealing with countries that have no horizontal space to accommodate landfills, and in addition, lack socio-economic and technological means to apply the same solutions as Japan. Case in point is the East Mediterranean and in particular Lebanon (Chalhoub, 2016).

3. The MSW crisis in Lebanon

Most issues related to MSW are common to less developed countries, but the case of Lebanon is worth mentioning as it reached an all-time high by 2015 year-end, and is still unresolved. Prior to the 1975 civil war, household garbage collectors were full time employees in the public sector who toured door to door on a daily basis, typically starting in the afternoon till early evening.

In the 1960s, it was common in Lebanon to have public sector employees perform door-to-door household garbage collection, trucking to remote sites, dumping and burying. The 1975 civil war brought many discontinuities in public services, and most notably in MSW management. Most public sector services suffered severe setbacks due to lack of safety and security, localized strife at the neighborhood level, public sector budget challenges and demographic movements. Households took it upon their own hands to dispense of their garbage which led to the two most obvious practices; dumping at a remote site or small-scale neighborhood-level incineration.

Between 1975 and 1989 solid waste was not the only sore point because most public services suffered. A combination of security, safety and budget problems crippled whatever was left of government or municipal services. MSW became a matter of localized effort by inhabitants or warring factions to collect and dump without treatment (Nuwayhid, Ayoub, Saba, & Abi-Said, 1996; World Bank. The International Bank for Reconstruction and Development, 1999). In early 1990s, with peace accords and an overall orientation towards privatization of several public sector services, including the reconstruction and rehabilitation of downtown Beirut, a private company was assigned the entire MSW process. Shortly after the Taef accord which received mixed reviews by policy analysts, many public services were put in private hands, including MSW. The MSW contract was described as monopolized by a private company. The new mode of MSW management between 1991 and 1999, and took a branded image with private employees, trucks, signs, branded color codes and regular collection routes. Garbage dumps were identified and the process consisted of collecting MSW from street bins daily and transporting the bulk to remote disposal sites, including a major dump in Burj Hammoud, a densely populated town north of downtown Beirut (Council for Development and Reconstruction [CDR], 2012). In 1998, policymakers decided to interrupt garbage dumps in Burj Hammoud where the dumps reached unacceptable "physical" heights. A mountain of household garbage could be seen and smelled, add to that insects and rodents a stone's throw from homes. So the Naameh landfill, some 18 km south of Beirut, was adopted as a short-term alternative targeting a capacity of about 2.2 million tons of MSW over six consecutive years, i.e. till 2004. The sole private contractor did not process the tonnage progressively so the Naameh landfill remained open well beyond its intended lifetime triggering inhabitants' protests and blockage. As a result of public pressure, it was shutdown in July 2015 (Chalhoub, 2016; Daily Star, 2015a, 2015b; Nature Conservation Center [NCC], 2016).

The sole subcontractor refrained from collections, so households creatively resorted to dumping trash on street corners, under bridge ramps and underpasses, in crevices between buildings, and of course on the Lebanese shore. It was estimated that the Lebanese daily per capita average solid waste production was close to 1 kg. Political debates went in circles, accusations about corruption and kickbacks came in vogue ... while solid waste piled up in the streets. New types of breathing, skin and headache problems appeared causing health hazards (Chalhoub, 2016; Daily Star, 2015c).

A shocking scene came about in October 2015 when heavy rains transported down the streets tons of household trash (Daily Star, 2015d). In the following months, MSW continued to accumulate vertically and resulted in mountains of garbage with decomposition liquids oozing in the streets. A political proposal caught researchers and practitioners by surprise when it declared that an arrangement was being discussed with foreign firms to "export" Lebanese garbage overseas. That was another proposal that faced public opposition as it did not offer a sustainable solution. A simple cost-benefit analysis showed several flaws in the garbage export project. Public pressure rose again on policymakers to seriously support a sustainable local solution for MSW (Chalhoub, 2016; Daily Star, 2016).

The lesson learned was clear. Affluent or not, communities generate solid waste in varying compositions. Although past research efforts were geared towards looking for correlations between waste generation and standards of living, it neither offered a solution to local communities, nor helped design national policies. Countries enjoying communal discipline and a sense of unified environmental purpose at a neighborhood level can realize better results with triage, composting and incineration combined with advanced air pollution technology for pre-sorted solid waste components. When glass, plastics, pulp-based material and metals are separated upstream, incineration becomes more focused, reduces the input mass by orders of magnitude and yields an output that is easier to manage (Chalhoub, 2015).

Reflecting back on the case of Lebanon, it is important to note that household refuse constitutes the majority of its solid waste. Despite the 1997 Environment Emergency Plan, which followed a two-year study, aimed at establishing and implementing sustainable solutions, the country is to-date suffering from waste mismanagement. The greater Beirut area (GBA) was supposed to lead in terms of applying modern practices, but the 1995–1997 study fell short of predicting MSW generation over a ten-year horizon. Comparing 2001 projected to actual data shows that Lebanon generated 1.44 million tons of MSW compared to what was projected for that year back in 1995, which estimated 990,000 tons; a 37% underestimation. Most likely, the reason behind this underestimation is that the study made too many assumptions about public attitudes instead of surveying them. It was over-optimistically assumed that campaigns, public awareness programs and reliance on public managers to curb waste generation and promote recycling, would yield results, while in reality such factors did not materialize as hoped; hence, the wide discrepancy in data (Ministry of Environment, Lebanon, 2001). Further, instead of moving from planning to execution in 2002 and beyond, problems were left unattended until recent crises caused residential streets to be filled with garbage causing serious health hazards, as described earlier in this section. Public unrest built up with a growing feeling that public authorities were not decisive enough in attending to a solution. The Lebanese crisis was a major focusing event that motivated the use of a linear regression analysis to gauge public attitudes towards upstream triage, potential privatization of the whole garbage cycle, issues of quality management, implementation mishaps such as the ones that occurred in both Burj Hammoud and Naameh, and in general, public perception of how reliable public management authorities are.

As for the composition of MSW in Lebanon, it includes a large percentage of organic materials (63%) in GBA and a national average of 52% which did not change over the past four years (Table 1, and Figure 2). On the outset this suggests a favorable condition for composting. But to optimize a composting solution, a consistent waste separation needs to be instilled and practiced at the household, neighborhood, municipal and plant levels, which cannot be guaranteed. Experience has shown that Lebanon had no success with such practices. Case in point, the 2015 crisis came about only a few years after a decision was taken by the Lebanese Council of Ministers in 2010 to develop a proposal for the management of waste in all Lebanese territory, namely adopting thermal decomposition and the conversion of waste into energy in major cities. The objective was to involve the private sector in a turnkey manner from collection to final treatment, and to give incentive to local municipalities to embrace the plan for potential "...thermal treatment and composting plant locations...." However, the public uproar about past practices related to dumping sites such as in Naameh, and open air burning such as in Hbaline, displayed lack of public trust in government promises (CDR (Council for Development and Reconstruction), Lebanon, 2012; Chalhoub, 2015).

As presented in subsequent sections, a multivariate regression analysis is performed taking into consideration variables that are directly relevant to the issues experienced in Lebanon including public attitudes toward public management, privatization and influence of household recycling and triage decisions on final outcomes in MSW management. In particular, waste triage at the household level was included as part of the regression analysis because this step in the overall MSW cycle rests in the hands of the end user prior to the involvement of public authorities or contractors in the collection step of the process.

Table 1. Proportions of solid waste components (source: CDR, 2012)

MSW Component	GBA (%)	Lebanon (%)
Organic	63	51
Paper and cardboard	18	17
Plastic	7	10
Glass	5	9
Textiles	4	3
Metals	3	3
Construction/Demolition	-	5
Other	-	2

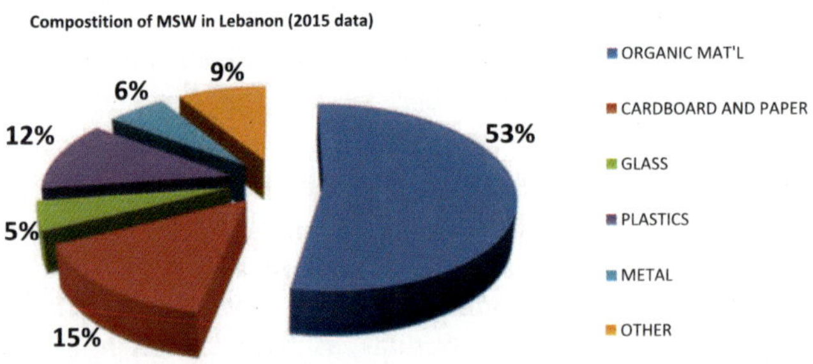

Figure 2. Proportions of solid waste components (source: Nat. Cons. Ctr., 2016).

4. Incineration with clean air technology as an option

If waste is considered as fuel for incineration, there are basic properties that are targeted. Its regular availability around the year and its lower calorific value (LCV) must be above a certain level. Therefore, data cannot be transferred from one region to another. The average LCV is recommended to be at least 6 MJ/kg, although more recent studies favor a higher range, around 7 or even 8 MJ/kg. According to past World Bank studies, a minimum annual provision of 50,000 metric tons of MSW and a weekly plant input variation of less than 20% must be met to run the operation. But such figures cannot be generalized as they depend on local community developmental level, urbanization and growth (Chalhoub, 2015; World Bank, 1999).

Simplified models help determine the appropriateness of solid waste as a fuel in incineration plants without resorting to auxiliary fuel, based on Ash content or ignition residuals, A, combustible fraction, C, and moisture of raw dump, W. Using a Tanner triangle to assess the combustibility of solid waste results in a feasible zone where A is less than 60%, C is greater than 25% and W is less than 50%. Within this context, waste composition is another important factor. The ash and water free calorific value, H_{awf}, greatly depends on the specific component of solid waste. Plastics and polyethylene command a H_{awf} around 45 MJ/kg, paper and cardboard range in between 16 and 19 MJ/kg, and food refuse and vegetables between 15 and 20 MJ/kg. Considering that in Lebanon organic content is 53% and plastics is 12%, paper and cardboard is 15%, while in other countries such ratios may be significantly different, waste as incineration and energy recovery fuel may require differences in design. For instance, surveys from 22 European countries showed a lower mean on organic (23%), on plastics (7%), but higher on paper and cardboard (25%), while in Philippines organic account for 45%, plastics 23% and paper 12% (Chalhoub, 2015; World Bank. The International Bank for Reconstruction and Development, 1999).

On the incineration output side, energy is a benefit and emissions are a liability. Pollutants comprise dust particles, HCl, SO_2, NO_x and HF, among others. A basic emissions control system, for example, through electrostatic precipitators, at minimal investment, would remove particulate matter. More advanced flue gas treatment technologies require significantly higher investments, a more complex operations management expertise, preventive periodic maintenance and monitoring systems. The survey clarified the hazardous effects of air pollution, and provided a simplified description that pollution prevention can be managed through state-of-the-art high investment technology. As such, the environmental engineering treatment for a comparative study in between sub-options of air cleansing technologies in future incineration plants was left outside the scope of this article (Chalhoub, 2015).

5. Empirical analysis of public attitudes toward MSW management options

Our empirical research included two stages; a first stage consisted of focus interviews and a second stage followed with a regression analysis. We started with focus interviews in four communities in Byblos County. Interviews helped identify and confirm issues that the public sees as serious defects in public management, especially in relation to MSW. Lack of transparency, a monopoly on MSW management granted by the government to a private company, extended periods of dumping in the open, were all identified during the first stage as serious issues. There were strong remarks about a low tech incineration site in Hbaline, a small village East of Byblos, which stained the blue sky in an otherwise pristine natural setting. “It is either the smell of rotting garbage, or the smell of incineration smoke, …, we are given no other choices, …” commented a commuter who drives through Hbaline daily.

Upstream triage of garbage was attributed to households whereby small local campaigns spurred inhabitants to sort garbage. Focus interviews showed concerns in that even if households sorts their garbage, once collected by third parties, they end up again in one heap. During the focus interview stage, the majority of participants (65.2%) expressed readiness to start or to continue putting effort in household level triage. Of that percentage, 78.3% viewed household triage as an additional effort that will have no effect on the final outcome; while a meager 9.8% viewed it as effective. " ... No matter what we do at the household or even neighborhood level ... no matter how many bins and colorful dumpsters we implement in our community ... all that garbage is then taken by the contractor and dumped in one heap irrespective of our triage ..." complained a respondent. Another finding from focus interviews is that inhabitants faced a dilemma as to whether to support public sector operations or the intervention of private companies. "If we rely on local municipalities for waste management, we face preferential treatment, but if we hand the matter to a private company, we become totally dependent on its profit-making techniques," commented a respondent who lives in Beirut but spends summers in Byblos County. Most respondents (69.5%) had similar concerns expressed in different ways. Some even voiced their suspicion of corruption in granting potential privatization contracts, especially those who were dissatisfied with the current company holding a monopoly of the MSW entire cycle. Other respondents (34.4%) commented on open incineration and wished there could be a technical solution to avoid air pollution. Although the overall advantages of clean technologies were discussed, specific engineering design methods about how to upgrade an incineration plant into a facility that treats and cleanses exhaust fumes prior to their final release in the air were kept out of scope while interacting with focus interview respondents. The objective was to avoid biasing the discussion toward a mere focus on technical products, and to gauge public attitudes at large by considering the respondent as an end-user and a beneficiary of public management in MSW matters. Majority of respondents (70.2%) were weary of the lack of transparent plans for treatment sites, especially those where incineration was expected, and the weakness of the public sector in project management.

In a second stage, we turned to a quantitative analysis to narrow down the questions into a proposed solution—incineration with clean energy recovery and clean technology. From the preliminary results above we identified five areas to study with a quantitative approach as described in the following sections. A multivariate linear regression analysis was chosen here to study the relationship between a dependent variable and a set of independent variables. The dependent variable was defined as an index representing public attitudes toward handling MSW in an "Incineration with Energy Recovery" mode. The choice by the author of this quantitative analysis was motivated by the fact that any technology solution, whether based on anaerobic digestion or incineration, would have a dim chance to succeed if communities are not fully aware, convinced and engaged in executing the technical solution.

5.1. Multivariate analysis—linear regression analysis model

A multivariate analysis was performed to gauge public attitudes about solid waste management in Lebanon. To help weed through the maze of policy options that were debated in the country for the past 18 months, a simplified IER policy option was described to respondents. Other multivariate statistical analysis methods are also widely used such as the principal component analysis (PCA) method. PCA may be used as a complementary method to a multiple linear regression analysis to confirm the selection of its independent variables. Alternatively, the linear regression expression is developed while observing the significance level of each independent variable in explaining the dependent variable. The author focused on studying the dependence of the regression variable Y on X_1 through X_5. For awareness purposes, the introduction of the survey included a description about two other options; landfill and anaerobic digestion, but then asked the respondent to provide their input regarding IER. It was explained that IER would use incineration with the requisite technology to cleanse exhaust emissions and minimize air pollution to levels significantly below ambient pollution from other sources. Three hundred and seventy-six questionnaires were disseminated, of which 229 included replies to all five items representing the independent variables. A five point likert scale was used for each independent variable with 1 being *least applicable* or *strongly disagree*, and 5 being *most applicable* or *strongly agree*.

5.1.1. The model

The first independent variable, X_1, represents the extent to which public awareness about handling solid waste on the outset, at the household level, would affect the attractiveness and acceptability of IER. The rationale behind this variable is that several studies showed that incineration leaves behind inorganic materials with the bottom ash with metals and other non-perishable components. Whether precious metals or ferrous materials, the recovery from bottom ash is feasible. However, the process would be more efficient if upstream triage is practiced (Grosso, Motta, & Rigamonti, 2010; Muchova, Bakker, & Rem, 2008).

The second independent variable, X_2, represents the affinity with privatizing the MSW cycle, in that private entities would be responsible for public communication, color coding for triage, neighborhood collection, transportation, incineration and energy redemption. As seen with many public services in less developed countries, privatization has been long debated as it offers advantages and disadvantages (Fahmi, 2005). Cost reduction is often stated as an intended objective of privatization. However, it has been shown that cost savings are not systematic in the privatization of solid waste management (Bel & Warner, 2008). Studies by OECD found that municipal services offered through private entities are more expensive than the municipal service provision (OECD, 2000). In 1965, empirical studies in the US for over 24 municipalities in St. Louis County, Missouri, showed that there is no significant cost difference between public and private solid waste service provision (Hirsch, 1965). In 1974, similar results were reached in a statewide study in Montana, USA (Pier, Vernon, & Wicks, 1974). More recent studies show that there are some savings from privatizing the solid waste management cycle but that this effect erodes over the subsequent few years (Dijkgraaf & Gradus, 2008). Still, it is expected that privatization would improve efficiency through technology and private sector work productivity. One of the main points gleaned during focus interviews is that the privatization process itself may suffer from corruption and kickbacks. In Lebanon, a country known for its liberal and open economic practices, privatization has seen a tumultuous past.

The third independent variable, X_3, represents the requisite quality management in daily operations. As seen in previous sections, operating an incineration plant is broader than just the technical burning process but rather encompasses a full cycle. A panoramic view is critical to assess supply of waste as fuel, demand on energy and pollution control in a fully coordinated cycle. Concerns voiced during focus interviews revolved around short-sightedness of management style in the region and Lebanon in particular, whether in the public or private sector. In addition, since air quality is a public good, the issue opens a whole range of questions traditionally related to the tragedy of the commons and the free rider effect.

The fourth independent variable, X_4, represents challenges in implementation. A major concern that transpired during focus interviews is that the speeches and even the written plans may be completely different from the actual execution of the idea. Several respondents asked about potential locations of such facilities. However, specific location selection was kept outside the scope of the questions and replies to maintain an overall view on the issue, rather than trigger reactions based on the respondent's address. Nevertheless, such issue must be dissected in great detail in a follow-up study where location should be of paramount importance in the survey. This would be part of a study that puts forth a tangible execution plan with a back-up study on all parameters affecting emissions and dispersion of stack exhaust with ambient physical properties such as wind direction, speed, neighboring communities and their existing levels of industrialization, and physical geographic coordinates.

The fifth independent variable, X_5, addresses the reliability of public management and authorities from an end-user perspective. This particular point triggered a reply in light of recent national history known for its lack of success in public management including the 20-year civil war, to political deals and settlements such as the Taef agreement, to the Normandy garbage mountain and land-grab into Mediterranean sea, to the telecom bidding crisis, and more recently the neighborhood garbage crisis of 2015, and to-date in 2016. All things considered, it was important to include it because any technology option regarding MSW management is unavoidably linked to potential public policy analysis and public management. The simplified mathematical model is expressed as:

$$Y = \beta o + \sum_{i=1}^{5} \beta_i X_i \quad (1)$$

5.1.2. Regression analysis results

Analysis results from the multivariate linear regression analysis were interpreted using a 5% significance level. The coefficient of determination, R^2, for the overall equation is 74% showing an acceptable comparison between the estimated *Y*-value and actual *Y*-value. The R^2 value measures how close the data are to the fitted regression line. A coefficient of determination of 74% was found to be satisfactory in that the selected model explains most of the variability of response data around its mean. The F-statistic is 128 and the degrees of freedom measure, d_f, is 223. The regression sum of squares SS_{reg} is 176 while the residual sum of squares SS_{resid} is 61. The multivariate linear regression model computation resulted in the following relationship between the dependent and the independent variables:

$$Y = 0.955 + 0.0099X_1 - 0.0468X_2 + 0.5695X_3 + 0.2787X_4 - 0.1096X_5 \quad (2)$$

Having checked the overall validity of the resulting linear regression expression, we turn to the coefficients and their standard error values. All *p*-values or se_i values were found to be less than 0.05 with the exception of se_1 which is equal to 0.052, i.e. still very close to 0.05. The large R^2 value combined with small se_i values make potential collinearity among independent variables of a lesser concern and hence there was no major effort extended on reducing the equation dimensions any further then the five independent variables with the use of complementary methods such as PCA. In terms of explaining the dependent variable *Y*, it is found that there is a statistically significant and positive correlation with the independent variable X_3 representing requisite quality management in daily operations with a standard error $se_3 = 0.050$ ($\leq\alpha = 0.05$). This result is somewhat expected. The Lebanese end users at large are cognizant of modern management techniques and the importance of their application to achieve expected results. There is also enough awareness in Lebanon that strict managerial quality techniques are not observed locally. It is found that there is a statistically significant and positive correlation with the independent variable X_4 representing requisite technological upfront investment that poses challenges in implementation with a standard error $se_4 = 0.048$ ($\leq\alpha = 0.05$). Clearly, public attitudes show awareness about the importance of emitting clean air and a reluctance to go for a tradeoff between getting garbage off the streets while returning air that causes health hazards.

It is found that there is a statistically significant and negative correlation with the independent variable X_5, representing the reliability of public management with a standard error $se_5 = 0.032$ ($\leq\alpha = 0.05$). There have been past public management failures in other sectors such as transportation, water and power, all of which are intimately related to the MSW cycle. Therefore, it is most likely that any solution related to MSW management with public sector involvement or responsibility would find a negative correlation with acceptance by end users. It is also found that there is a statistically significant and negative correlation with the independent variable X_2 representing agreement with privatization with a standard error $se_2 = 0.028$ ($\leq\alpha = 0.05$). As for the independent variable X_1, representing public awareness about upstream household level triage, no statistically significant correlation was found. The regression coefficient is positive but the standard error is $se_1 = 0.052$ ($\geq\alpha = 0.05$). Note that the error is very small to the 5% significance level, and that X_1 would be considered to be significant at a 10% level. Keeping with the 5% significance level, this last result may be partially related to the low expectations about the impact of household actions on public policy outcomes. Household-level triage is marginalized by the garbage collection and dumping process used by the sole subcontractor who does not discriminate in between MSW components but rather collects and dumps mixed bags. This is another possible reason why the public may perceive itself to be of little relevance to MSW collection, transport and dumping process.

6. Discussion and interpretation

Empirical results from this field work are overall in good agreement with results from other countries. Our results related to household triage, although important, yielded a statistically insignificant relationship, most likely due to how the full cycle is managed by the MSW collections company. In a MSW study on Thailand, and another one in Ghana, the involvement of household was also highlighted and the need for better communication and coordination between residents and local municipalities was identified as an opportunity for improvement (Boateng, Amoako, Appiah, Poku, & Garsonu, 2016; Yukalang, Clarke, & Ross, 2017). Regarding privatization, there are both positive and negative attributions to handing MSW management control to private parties. Our results yielded a statistically significant and negative correlation. In a study on MSW privatization conducted in Bostwana, Egypt and a number of other countries also led to conclusions that capacity building in eth public sector is preferred to privatization. The same studies emphasize the importance of capacity building rather than subcontracting or privatizing MSW, and the perception of respondents that such management should reside in the hands of the public sector despite the advantages of privatization at large in other sectors. These finding are consistent with ours in relation to privatization and the requisite quality in service of a public entity to succeed (Bolaane & Isaac, 2015; Fahmi, 2005).

Our results showed that upfront technology investment is required in order to make IER solutions viable. Comparative studies to seek corroboration with other countries would be somewhat challenging for this particular independent variable because technology investments depend on several parameters. Those parameters are very different from country to country, and are distributed all along the value chain. Nevertheless, upfront technology planning and investment is described as one key success factor (Elagroudy, Warith, & El Zayat, 2016; Schubeler, Wehrle, & Christen, 1996; WEC—World Energy Council, 2016).

7. Future research recommendations

The inquiries during the focus interview phase, the preliminary results from qualitative data, and the results from the quantitative regression analysis pointed out that future research may include other independent variables. Using the present regression analysis as a pilot project, several other factors could be explored by collecting data from other counties such as Kesrouane County or Batroun County, Lebanon. Such broader data collection allows for the investigation of data samples from areas that are at different stages of rural development, education, employment rate, commercialization, public transport networks that may ease MSW collection and processing, maturity of municipal services, and urbanization. Upon expanding the dimension of potential independent variables, other multivariate methods such as PCA would be used upstream to reduce the total number of independent variables into the principal factors affecting the analysis outcome, prior to performing a linear regression analysis.

8. Conclusions

Municipal solid waste issues are of paramount importance as they bear directly on public health and the economy. Further, MSW requires a multidisciplinary convergence between technology or engineered solutions and public policy design, taking into consideration socio-economic factors. MSW management evolved as countries became industrialized and was influenced by environmental activism and a pro-active effort to identify alternative sources of energy. Mismanagement of MSW is more exacerbated in less developed countries. Across a range of solutions, the most desirable ones comprise waste avoidance, and the least desirable ones revolve around traditional landfills. A correlation between MSW public policies and the developmental level of the country was presented in literature. However, this cannot be generalized because hybrid solutions are developed on a case-by-case basis depending on the country's limitations in land area or in technological sophistication.

A set of focus interviews was conducted to identify areas of concerns from the end-user perspective. Local community engagement, transparency, adoption of cleaner technologies and better planning and implementation were pointed out by most respondents. Following the qualitative interview stage, a multivariate linear regression analysis was used. The majority of respondents in the focus interviews stage expressed concern about the current state of affairs in MSW management in Lebanon. These concerns were taken as a basis for the selection of regression analysis variables to gauge public attitudes. Incineration with energy recovery and clean exhaust technology was explored as a solution and its attractiveness to the public was tested. The dependent variable is an index quantifying the extent to which respondents agree with IER, while the independent variables address awareness building about upstream triage, privatization, quality management in daily operations, technology and investment challenges in implementation, and reliability of public management authorities. Results showed a statistically significant and positive correlation with quality in daily operations and in technological upfront investment, but showed a statistically significant and negative correlation with the reliability of public management, and privatization. Upstream household level triage, although important in future community orientation, was not found statistically significant. There are various explanations to this latter counter-intuitive result related to respondents' low expectations and simmering socio-economic problems. Overall, IER is perceived as a viable option but requires strict public policies.

Future research should include other factors such as the location of potential incinerators. This was kept out of scope during this phase of the study for simplicity. Other factors such as gender, age and occupation, would also be an interesting follow-up to see how respondents may be predisposed for or against a certain MSW solution. It is recommended that follow-up research be focused on upfront triage and the role of household-level waste management, involvement of local authorities such as municipalities, and collaboration schemes with contractors and investors. In fact, some of the recent proposals in Lebanon consider imposing fines on those who do not abide by certain triage rules. However, highly regulated approaches may not offer sustainable solutions as culture- and incentive-based approaches would.

Funding
The author received no direct funding for this research.

Competing Interests
The author declares no competing interests.

Author details
Michel Soto Chalhoub[1]
E-mail: mchalhoub@live.com
[1] Department of Civil and Environmental Engineering, Notre Dame University, Louaize, Lebanon.

References

Aye, L., & Widjaya, E. R. (2006). Environmental and economic analyses of waste disposal options for traditional markets in Indonesia. *Waste Management, 26* 1180–1191. doi:10.1016/j.wasman.2005.09.010

Bel, G., & Warner, M. (2008). Does privatization of solid waste and water services reduce cost? A review of empirical studies. *Resources, Conservation and Recycling, 52*, 1337–1348. doi:10.1016/j.resconrec.2008.07.014

Benson, C. H., Barlaz, M. A., Lane, D. T., & Rawe, J. M. (2007). Practice review of five bioreactors/recirculation landfills. *Waste Management, 27*(1), 13–29. doi:10.1016/j.wasman.2006.04.005

Bjorklund, A., & Finnveden, G. (2007). Life cycle assessment of a national policy proposal – The case of a Swedish waste incineration tax. *Waste Management, 27*(8), 1046–1058. doi:10.1016/j.wasman.2007.02.027

Boateng, S., Amoako, P., Appiah, D. O., Poku, A. A., & Garsonu, E. K. 2016. Comparative analysis of households solid waste management in urban and rural Ghana. *Journal of Environmental and Public Health, 2016*, 1–10. Art. ID 5780258. doi:10.1155/2016/5780258.

Bolaane, B., & Isaac, E. (2015, June). Privatization of solid waste collection services: Lessons from Gaborone. *Waste Management, 40*, 14–21. doi:10.1016/j.wasman.2015.03.004

Brunner, P. H. (2011). Urban Mining: A Contribution to Reindustrializing the City. *Material Flow Analysis, Journal of Industrial Ecology, 15*(3), 339–341. doi:10.1111/j.1530-9290.2011.00345.x

CDR (Council for Development and Reconstruction), Lebanon. (2012, October). *Basic services report, ch. on solid waste.* Beirut: CDR.

Chalhoub, M. S., (1992), "Anaerobic Digestion from Engineering Design to Operations Management: Opportunities and Challenges at the Los Angeles County Hyperion Treatment Plant," Rep. No. 92-021-ES, R. M. Parsons Co., Pasadena, CA.

Chalhoub, M. S., (2015, February), Solid waste as fuel for renewable energy with an optimization for air pollution impact, *Working paper*, NDU.

Chalhoub, M. S., (2016, January), Solid waste field reconnaissance and environmental implications in Lebanon, *Working paper*, NDU.

Cheng, S., Chan, C. W., & Huang, G. H. (2002). Using multiple criteria decision analysis for supporting decisions of solid waste management. *Journal of Environmental Science & Health Part A: Toxic/ Hazardous Substances and Environmental Engineering, 37*, 975–990. doi:10.1081/ESE-120004517

Cherubini, F., Silvia Bargigli, S., & Sergio Ulgiati, S. (2008). Life Cycle Assessment (LCA) of waste management strategies: Landfilling, sorting plant and incineration. *Energy, 34*(12), 2116–2123. doi:10.1016/j.energy.2008.08.023

Christensen, T. H., Bhander, G., Lindvall, H., Larsen, A. W., Fruergaard, T., Anders, D., ... Hauschild, M. (2007). Experience with the use of LCA modeling (EASEWASTE) in waste management. *Waste Management Research, 25*(3), 257–262. doi:10.1177/0734242X07079184

Consonni, S., Giugliano, M., & Grosso, M. (2005). Alternative strategies for energy recovery from municipal solid waste, part A: Mass and energy balances. *Waste Management, 25*(2), 123–135. doi:10.1016/j.wasman.2004.09.007

Daily Star. (2015a). *Naameh residents in uproar following delay of landfill closure*. Beirut: Lebanon News. January 14, 2015.

Daily Star (2015b). Protesters enforce naameh dump closure. *Lebanon News*, July 17, 2015, Retrieved from July 18, 2015.

Daily Star. (2015c, July 21). *Garbage Collection Halted as Naameh Stays Shut*. Beirut: Lebanon News, 2015.

Daily Star. (2015d). *Beirut trash washed away by heavy rain*. Beirut: Lebanon News. October 25, 2015.

Daily Star. (2016). *Lebanon scraps trash export deal with British firm chinook*. Beirut: Lebanon News. February 19, 2016.

De Feo, G., & De Gisi, S. (2010a). Using an innovative criteria weighing tool for stakeholders involvement to rank MSW facility sites with the AHP. *Waste Management, 30*, 958–976. doi:10.1016/j.wasman.2010.02.019

Dijkgraaf, E., & Gradus, R. (2008). Institutional developments in the dutch waste collection market. *Environment and Planning C: Government & Policy, 26* (1), 110–126. doi:10.1068/c0636

Dijkgraaf, E., & Vollebergh, H. R. J. (2004). Burn or bury? A social cost comparison of final waste disposal methods. *Ecological Economics Journal, 50*, 233–247. doi:10.1016/j.ecolecon.2004.03.029

Drobny, N. L., Hull, H. E., & Testin, R. F. (1971). *Recovery and utilization of municipal solid waste*. (pp. SW–127). Washigton, DC: Wiley.

Dunson, C. (1999, December). Waste and wealth. *Waste Age*. Retrieved from https://www.waste360.com/mag/waste_waste_wealth

Durgekar, V. (2016). Towards sustainable waste management through technological innovations, effective policy, supply chain integration & participation. *Procedia Environmental Sciences, 35*, 140–149. doi:10.1016/j.proenv.2016.07.061

Elagroudy, S., Warith, M. A., & El Zayat, M. (2016). *Municipal solid waste management and green economy*. (M. Gardner, Ed). Berlin, Germany: Pub. Global Young Academy.

Environmental Protection Agency (1988). *Report to congress: Solid waste disposal in the U.S.*, Vol *(ii)*, EPA Report No. 530-SW 88-011B, EPA, Washington DC.

Environmental Protection Agency. (2015). *Draft environmental guidelines: SOLID waste landfills* (2nd ed.). Washington, DC: New South Wales.

Fagnano, M., Adamo, P., Zampella, M., & Fiorentino, N. (2011). Environmental and agronomic impact of fertilization with composted organic fraction from municipal solid waste: A case study in the region of Naples, Italy. *Agriculture, Ecosystems & Environment, 141*(1–2), 100–107. doi:10.1016/j.agee.2011.02.019

Fahmi, W. S. (2005, October). The impact of privatization of solid waste management on the Zabaleen garbage collectors of Cairo. *Environment & Urbanization, 17*(2), doi:10.1630/095624705774481740

Feo, G. D., Belgiomo, V., Rocca, C. D., & Napoli, R. M. A. (2003). Energy from gasification of solid wastes. *Waste Management, 23*(1), 1–15. doi:10.1016/S0956-053X(02)00149-6

Finnveden, G., Bjorklund, A., Ekvall, T., & Moberg, A. (2007). Environmental and economic assessment methods for waste management decision-support: Possibilities and limitations. *Waste Management Research, 25*, 263–269. doi:10.1177/0734242X07079156

Funtowicz, S. O., & Ravetz, J. R. (1993). Science for the post-normal age. *Futures Sep, 1993*, 739–755.

Gibson, L. (1998). Institutionalizong regional science. *Annals Reg Sciences, 32*, 459–467. doi:10.1007/s001680050083

Gramatica, P., Battaini, F., Giani, E., Papa, E., Jones, R. J. O., Preatoni, D., & Cenci, R. M. (2006). Analysis of mosses and soils for quantifying heavy metal concentrations in Sicily: A multivariate and spatial analytical approach. *Environmental Science and Pollution Research, 13*(1), 28–36.

Grosso, M., Motta, A., & Rigamonti, L. (2010). Efficiency of energy recovery from waste incineration in the light of the new waste directive. *Waste Management, 30*, 1238–1243. doi:10.1016/j.wasman.2010.02.036

Guerrero, L. A., Maas, G., & Hogland, W. (2013). Solid waste management challenges for cities in developing countries. *Waste Management, 33*(1), 220–232. doi:10.1016/j.wasman.2012.09.008

Hadzic, A., Voca, N., & Golubic, S. (2017). Life-cycle assessment of solid-waste management in city of Zaghreb, Croatia. *Journal of Material Cycles and Waste Management, 20*(2), 1286–1298. doi:10.1007/s10163-017-0693-2

Hirsch, W. Z. (1965). Cost Functions of an Urban Government Service: Refuse Collection. *Review of Economics and Statistics, 47*(1), 87–92. doi:10.2307/1924127

Huang, J., Zhao, R., Huang, T., Wang, X., & Tseng, M. L. (2017). Sustainable municipal solid waste disposal in the belt and road initiative: A preliminary proposal for Chengdu city. *Sustainability, Pub. MDPI, 10*(1147), 1–15.

Huang, Y., Ning, Y., Zhang, T., & Fei, Y. (2015). Public acceptance of waste incineration power plants in China: Comparative case studies. *Habitat International, 47*, 11–19. doi:10.1016/j.habitatint.2014.12.008

Hupponen, M., Gronman, K., & Horttanainen, M. (2018). Areas on which to focus when seeking to reduce the greenhouse gas emissions of commercial waste management: A case study of a hypermarket, Finaland. *Waste Management, 76*, 1–18. doi:10.1016/j.wasman.2018.03.024

Joseph, A. M., Snellings, R., Van Den Heede, P., Matthys, S., & De Belie, N. (2017). "The use of municipal solid waste incineration ash in various building materials: A Belgian point of view," Materials, pub. *Mdpi, 11*(141), 1–30.

Khoo, H. H. (2009). Life cycle impact assessment of various waste conversion technologies. *Waste Management, 29*(6), 1892–1900. doi:10.1016/j.wasman.2008.12.020

Kimball, D. (1992). *Recycling in America*. Santa Barbara, CA: ABC-CLIO Inc.

Leone, I. A., Gilman, E. F., & Flower, F. B. (1983). Growing trees on sanitary landfills. *Arboricultural Journal, 7*, 247–252. doi:10.1080/03071375.1983.9746626

Malinauskaite, J., Jouhara, H., Czajczynska, D., Stanchev, P., Katsou, E., Rostkowski, P., ... Spencer, N. (2017, Dec 15). Municipal solid waste management and waste-to-energy in the context of a curcular economy and energy recycling in Europe. *Energy, Pub. Elsevier, 141*, 2013–2044.

Malkow, T. (2004). Novel and innovative pyrolysis and gasification technologies for energy efficient and environmentally sound MSW disposal. *Waste Management, 24* (1), 53–79. doi:10.1016/S0956-053X(03)00038-2

Marshall, R. E., & Farahbashsh, K. (2013). Systems approaches to integrated solid waste management in developing countries. *Waste Management, 33*, 988–1003. doi:10.1016/j.wasman.2012.12.023

Melosi, M. (1981). *Garbage in the cities: Refuse reform and the environment 1880–1980*. Texas: Texas A&M Press.

Melosi, M. (2000). *The Sanitary City*. Baltimore, MD: Johns Hopkins University Press.

Ministry of Environment, Lebanon. (2001). *Lebanon state of the environment report, ch.14, solid waste management*. Beirut: Lebanese Environment and Development Observatory (LEDO).

Ministry of the Environment, Japan. (2012, February). *Solid waste management and recycling technology of Japan – Toward a sustainable society*. Tokyo, Japan: Minister's Secretariat, Waste Management and Recycling Department, Policy Planning Division.

Muchova, L., Bakker, E., & Rem, P. (2008). Precious metals in municipal solid waste incineration bottom ash. *Water Air Soil Pollution: Focus, 9*(1), 107-116. Ruine Antoniasub.

NCC (Nature Conservation Center), Beirut. (2016)*Waste crisis. Human & health*. (Vol. 34). Winter. Beirut, Lebanon.

Nuwayhid, R. Y., Ayoub, G. M., Saba, E. F., & Abi-Said, S. (1996). The solid waste management scene in Greater Beirut. *Waste Management & Research, 14*, 171–187. doi:10.1177/0734242X9601400207

OECD, Organization for Economic Cooperation and Development. (2000). *Environmental performance reviews*. Paris, France.

OECD, Organization for Economic Cooperation and Development, (2013). *Policy Roundtables – Waste Management Services*, OECD, Report DAF/COMP(2013) 26, pp. 17–32.

Pařízek, T., Bébar, L., & Stehlík, P. (2008). Persistent pollutants emission abatement in waste-to-energy systems. *Clean Technologies and Environmental Policy, 10*(2), 147–153. doi:10.1007/s10098-007-0135-2

Parker, A. (1983). Behaviour of wastes in landfill-leachates, ch. 7; behaviour of wastes in landfill-methane generation, ch. 8. In J. R. Holmes (Eds.), *Practical waste management*. England: John Wiley & Sons, Chister.

Pier, W. J., Vernon, R. B., & Wicks, J. H. (1974). An empirical comparison of government and private production efficiency. *National Tax Journal, 27*(4), 653–656.

Pohland, F. G., (1991), Fundamental principles and management strategies for landfill codisposal practices. *Proceedings Sardinia 91, 3rd International Landfill Symposium*, (Vol. 2, pp. 1445–1460), Grafiche Galeati, Imola, Italy.

Quina, M. J., Bontempi, E., Bogush, A., Schlumberger, S., Weibel, G., Braga, R., ... Lederer, J. (2018, Sep). Technologies for the management of msw incineration ashes from gas cleaning: New perspectives on recovery of secondary raw materials and circular economy. *Science of the Total Environment, Pub. Elsevier, 635*, 526–542. doi:10.1016/j.scitotenv.2018.04.150

Schubeler, P., Wehrle, K., & Christen, J. (August 1996), Conceptual framework for solid waste management in low-income countries, *UNDP/UNCHS (Habitat)/World Bank/SDC Collaborative Programme - Urban Management and Infrastructure, Working Paper No. 9*, p. 53.

Tchobanoglous, G., Thiesen, H., & Eliassen, R. (1978). *Solid waste: Engineering principles and management issues*. New York, NY: McGraw Hill.

Tehrani, S. M., Karbassi, A. R., Ghoddosi, J., Monavvari, S. M., & Mirbagheri, S. A. (2009). Prediction of energy consumption and urban air pollution reduction in e-shopping adoption. *International Journal of Food, Agriculture and the Environment*, 7(3–4), 898–903.

UNEP. (2011). *United nations environment programme*. Towards a Green Economy: Pathways to Sustainable Development and Poverty Eradication. New York, NY. ISBN 978-92-807-3143-9.

Wan, C., Shen, G. Q., & Choi, S. (2018). Differential public support for waste management policy: The case of Hong Kong. *Journal Clean Products, 175*, 477–488. doi:10.1016/j.jclepro.2017.12.060

WEC – World Energy Council. (2016). *World energy resources: Waste to energy*. London.

Wilson, D. C. (2007). Development drivers for waste management. *Waste Management & Research, 25*, 198–207. doi:10.1177/0734242X07079149

World Bank. The International Bank for Reconstruction and Development. (1999). *Municipal solid waste incineration*. Washington DC: World Bank Technical Guidance Report.

Yolin, C. (2015, September). *Waste management and recycling in Japan: Opportunities for European companies with SME focus*. Tokyo: European Union-Japan Center for Industrial Cooperation.

Yukalang, N., Clarke, B., & Ross, K. (2017). Barriers to effective municipal solid waste management in a rapidly urbanizing area in Thailand. *International Journal of Environmental Research and Public Health, 14*(9), 1–23. doi:10.3390/ijerph14091013

Zaman, A. U. (2010). Comparative study of municipal solid waste treatment technologies using life cycle assessment. *International Journal of Environmental Science and Technology, 7*(2), 225–234. doi:10.1007/BF03326132

Zhao, X., Jiang, G., & Li, A. (2016). Technology, cost, a performance of waste-to-energy incineration industry in China. *Renewable Sustainable Energy Review, 55*, 115–130. doi:10.1016/j.rser.2015.10.137

Zhou, Z., Tang, Y., Chi, Y., Ni, M., & Buekens, A. (2017). Waste-to-energy: A review of life cycle assessment and its extension methods. *Waste Management and Research, 36*(1), 3–16. doi:10.1177/0734242X17730137

3

Expected impacts of active mining on the distribution of heavy metals in soils around Iludun-Oro and its environs, Southwestern Nigeria

Abiola Oyebamiji[1,2,3]*, Amobichukwu Amanambu[2,4], Tehseen Zafar[1,2], Adeniyi JohnPaul Adewumi[5] and Damilare Stephen Akinyemi[6,7]

*Corresponding author: Abiola Oyebamiji, Department of Science Laboratory Technology, Ekiti State University, Ado-Ekiti, Nigeria
E-mail: abiola.oyebamiji@eksu.edu.ng

Reviewing editor: Peter Fantke, Technical University of Denmark (DTU), Denmark

Abstract: This present study is aimed at evaluating the level of environmental degradation and potential ecological risk in soils around an active (tantalum-niobium) mining site of Iludun-Oro, southwestern Nigeria. A total of 18 topsoil samples were randomly collected from different parts of the study area and analysed using the inductively coupled plasma-mass spectrophotometry. The concentrations of Pb, Zn, Fe, Cu, Cr and Ni varied between 0–678, 7–222, 1.64–8.87, 8.8–145, 1.4–81 and 1.6–93.1 ppm, respectively, with an average concentration of 107.72, 48.17, 5.17, 45.28, 14.76 and 21.53 ppm, respectively. The order of abundance of these heavy metals analysed in the soils is as follows: Pb > Zn> Cu> Ni> Cr> Fe. The concentration values of Pb, Fe, Cu, Ba and Sr are higher than the background values indicating that mining activities have an adverse impact on the study area.

On accessing the soil quality in the study area, the results of the geo-accumulation, contamination factor (CF) and pollution load indices (PLIs) indicated high concentrations of Pb in the soils. Likewise, most of the Igeo values reported uncontaminated to moderately contaminated for all elements except for Pb. This study has further provided the background information for a vulnerability assessment of the study area as well as for future impact assessment and proper planning

ABOUT THE AUTHOR

The authors are a group of multidisciplinary scientific researchers, actively involved in the study of the ecological and geochemical study of the environments, most importantly the distribution and monitoring of elevated levels of heavy metals in soils, rocks, waters and vegetation. Their current study focuses on the assessment of heavy metals around an active mining site, its potential effects on human health using multiple indices. It provides background information for a vulnerability evaluation of the study area and recommends remediation plans for the control.

PUBLIC INTEREST STATEMENT

The impacts of mining activities through the expulsion of mine wastes pose a serious threat to human life, vegetation and animals across the globe. This perspective research describes some of the distribution of these heavy metals, their pollution level, ecological threat and potential sources of these hazardous metals around an active mining site in southwestern Nigeria. It was found that these metals range from moderately contaminated to contaminated for most of the elements. The presence of these toxic elements has adverse effects on the immediate environment. Understanding these effects can provide useful information for appropriate environmental laws and mitigate the influence of pollution on the environment in this study area and other similar places.

of the mining site. These findings have provided useful information for making appropriate environmental management strategies to mitigate the influence of heavy metal pollution generated from an active mining site in Nigeria and other similar areas.

Subjects: Environment & Agriculture; Earth Sciences; Environmental Studies & Management; Physical Sciences

Keywords: ecological risk; heavy metals; PCA; Southwestern Nigeria; topsoil

1. Introduction

The discoveries and exploitation of natural and mineral resources in Nigeria and Africa as a whole have led to the development of the most sectors of their economy and their people's livelihood (Boakye, Dessus, Foday, & Oppong, 2012; Fayiga, Ipinmoroti, & Chirenje, 2017; Wilson, Wang, Kabenge, & Qi, 2017). The exploitation of these resources (gold, copper, tin, tantalum, columbite, niobium, natural oil and gas) has negatively impacted the environment. Several factors such as population growth, urbanization, mining, agricultural activities, and industrialization have contributed to the present pollution in the world. These global trends in environmental pollution possess a challenge, but the levels of these contaminations vary in different parts of the world, especially in developing economies.

The environmental impacts of mining activities have similar effects worldwide. These effects on the environment are determined by the chemical attributes of the ore, the method of extraction and the environmental conditions (Adriano, 2001; Higueras et al., 2004; Hsu et al., 2016; Khan, Yusoff, Bakar, Bakar, & Alias, 2016; Lupankwa, Love, Mapani, Mseka, & Meck, 2006; Mierzwa-Hersztek, Gondek, Klimkowicz-Pawlas, Baran, & Bajda, 2018; Oelofse, 2009; Potra, Dodd, & Ruhl, 2017; Yan, Liu, Zhong, Guo, & Wu, 2018). The mobility and bioavailability of these elevated heavy metals in the environment primarily depend on some physicochemical characteristics of the depositional environment such as pH, textural characteristics, organic matter content, speciation or chemical form, and electrical conductivity (Adewuyi & Osobamiro, 2016; Chukwu, 2011; Equeenuddin, Tripathy, Sahoo, & Panigrahi, 2013; Lu, Liu, Wang, & Wu, 2017; Sarapulova, Dampilova, Bardamova, Doroshkevich, & Smirnova, 2017; Zhang, You, & Hu, 2016). Higher concentrations of heavy metals in soils have been documented at a variety of sites associated with Pb–Zn mining in the southeast Missouri, USA (Potra et al., 2017). These have adverse effects on the floristic quality where no other human activities are obvious. Trace elements' (As, Cu, Zn and Pb) concentration has also been documented around a gold mine in Ghana (Antwi-Agyei, Hogarh, & Foli, 2009) with As concentrations in soils significantly higher than the Netherlands intervention value of 55 $mgkg^{-1}$. An anomalous abundance of rare earth elements in surface soils has also been reported in Bayan Obo Mine that was influenced by the exploitation of these minerals around the mining site (Wang & Liang, 2016).

The natural soil is not only a geochemical habitat for the contaminants, but also serves as a natural buffer for transportation of elements in the air, water, and biomass (Benhaddya & Hadjel, 2014; Nsenga Kumwimba et al., 2017; Wang, Deng, Yin, & Tang, 2018). Due to the high elevated values of heavy metals in soils, they have posed an adverse effect on human health and aquatic environment, which can be easily inhaled into human bodies from suspended dust or by uptake in plants (Benson, Anake, Adedapo, Fred-Ahmadu, & Ayejuyo, 2017; Jan et al., 2015; Madrid, Díaz-Barrientos, & Madrid, 2002; Mortazavi, Mortazavi, & Paknahad, 2016; Wang & Yong, 2005). Mining activities, as the hazardous sources of toxic elements, relatively affect localized areas. The tailings, gangue minerals and subsequent smelting operations have caused noticeably significant environmental concerns (Mwesigye, Young, Bailey, & Tumwebaze, 2016; Okang'Odumo et al., 2014). The fine-grained particles of mine tailings or the ore may be blown or diffused into adjacent areas. Acid-mine drainage contamination has been reported in the Abuja leather mine in Komu, southwest Nigeria (Oyebamiji, Odebunmi, Ruizhong, & Rasool, 2017), which has led to the contamination of nearby aquatic habitats.

Previous studies on the distribution of these heavy metals in our immediate environments have not only become a local phenomenon, but also a global one as well (Harikrishnan et al., 2017). Hence, geochemical studies of the mining sites using sample media like soils, stream sediments and waters are fundamental in determining the dispersion pattern of these trace elements.

In order to better understand the pollution level, ecological risk level and potential sources of heavy metals in soils of Iludun-Oro mining district, the main objectives of this study were to: (i) determine the spatial distribution of heavy metals in soils; (ii) identify their sources by using the principal component analysis (PCA); (iii) assess the contamination level using various risk assessment indices such as; CF, geoaccumulation index (Igeo), degree of contamination (CD), PLI, ecological risk index (ERI) and potential ecological risk index (PERI).

2. Materials and methods

2.1. Study area

The Iludun-Oro mining district is located in the southwestern part of Nigeria, lies between longitude 08°09'00"N to 08°16'00"N and latitude 004°49'00"E to 004°54'00"E and extends 12.5 km from east to west and 15.8 km from south to north. The study area is surrounded by several towns namely Okerimi-Oro, Agberun, Ajasse, Sanmora, Agbe-Ola, Ibode-Oro and Igbonla (Figure 1). This district is well accessible with the presence of major and minor roads created by miners and farmers during their quest to find means of livelihood. The study area is located within the tropical rainforest belt characterized by an alternation of dry and wet seasons. The rainy season runs from April through October when the effect of the southwest monsoon wind overwhelms that of the northeast trade wind, which commonly causes dry seasons from November to April. The wet season is characterized by heavy rainfall with

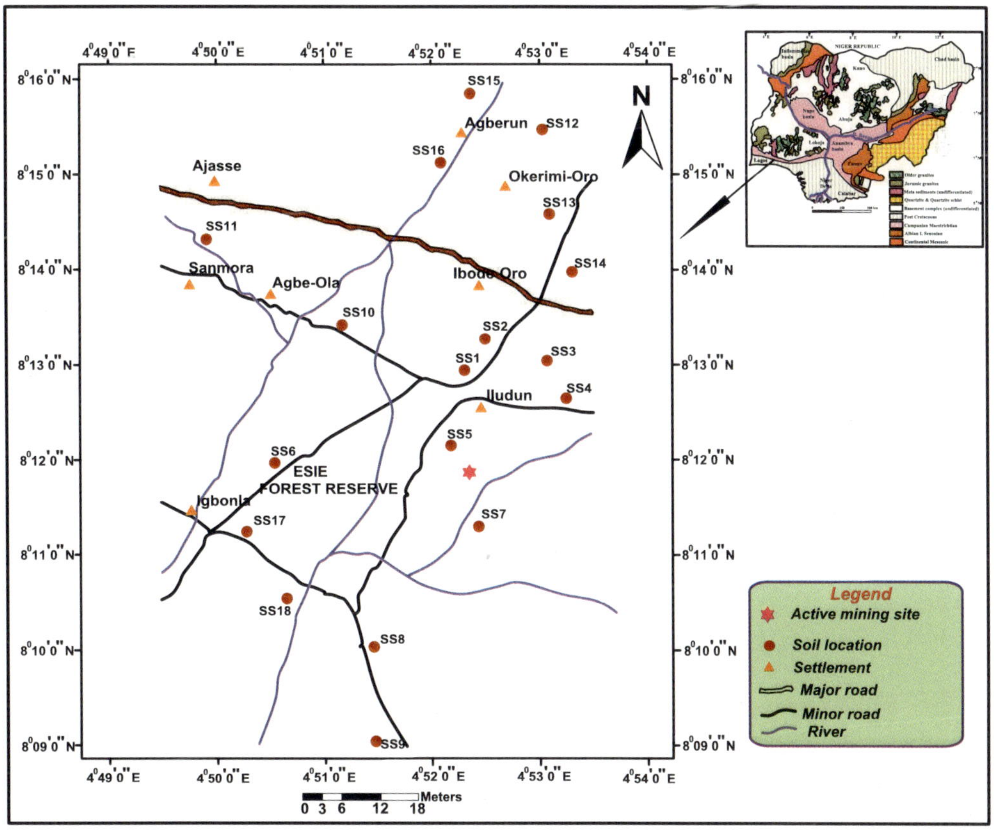

Figure 1. Location map of the study area showing sampling points.

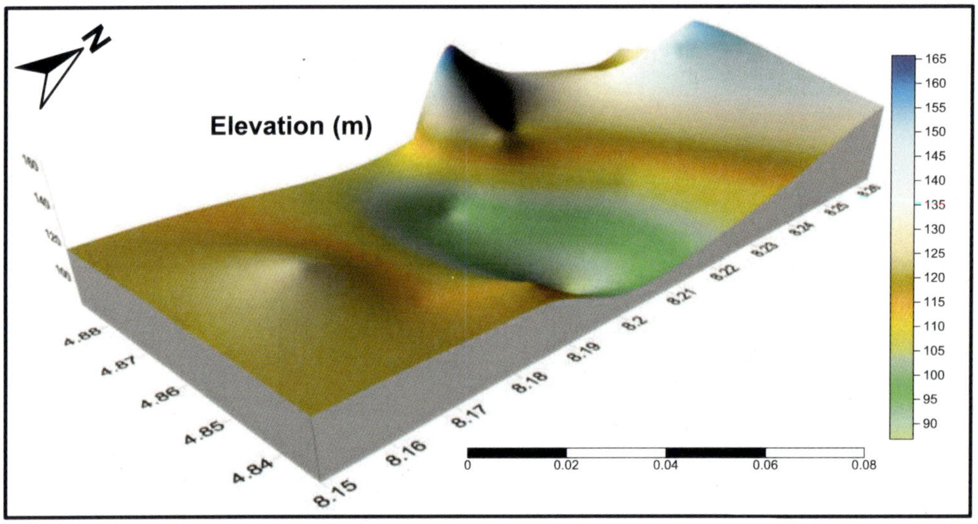

Figure 2. Geomorphological view of the study area showing the peaks and nadirs above sea level.

corresponding low temperatures whilst the dry season comes with high temperature and little or no rain. Annual rainfall varies from 1150 mm to 2000 mm in the northern and southern areas, respectively. Annual mean temperature ranges from 21 to 29°C whilst the relative humidity ranges from 27.5 to 98.2% (Ogunrayi, Akinseye, Goldberg, & Bernhofer, 2016). The soil types mainly include reddish-brown and brown-black loam. Land use is predominantly controlled by agriculture and mining activities. The elevation of the study area varies from 90 to 165 m above sea level as shown in Figure 2. The peak of the elevation is located on the north-west flank of the study area. The area is drained by several tributaries of river running in the north-south direction. The geology of the study area is underlain by the Precambrian basement complex rocks, younger granite of Jurassic age, and the Cretaceous to Recent sediments (Obaje, 2009). The study area is one of the major production sites for tantalite, tin, niobium, columbite and gemstones, which are mined from mineralized pegmatite veins hosted within the quartz-mica schist and granite gneiss (Figure 3). These pegmatites occur as coarse-grained rocks and mainly trend in the NNW-SSE direction.

2.2. Sample collection and preparation

In this study, a total of 18 topsoil samples (0–20 cm) were randomly collected from the different parts of the study area. Composite samples were also collected in dug pits at a depth interval of 1 m, 2 m, and 3 m. These samples were carefully collected; freshly accumulated soils were avoided. A representative soil sample about 1 kg was subjected to pulverization, air-dried and later passed through a 200-mesh nylon sieve to remove unwanted materials. Proper delineation of the mining sites was done during the reconnaissance survey, which was followed by a systematic sampling pattern. The coordinates of the sample location were recorded with the aid of a GPS. All collected samples were stored in a clean polythene bags to avoid contamination before the laboratory analysis.

2.3. Sample analysis procedure and accuracy

The dried fine-grained soil particles were digested in aqua regia using a mixture of 25% HNO_3 and 75% HCl. After a near-total absorption by hydrofluoric-perchloric acid, the major and trace elements were analysed using the inductively coupled plasma-mass spectrophotometry (ICP-MS) in a certified commercial laboratory in Canada (Bureau Veritas Mineral Laboratories). The accuracy of the analytical methods was monitored by the repeated analysis of a standard reference material done together with a batch of soil samples according to the technique of Dulski (2001). These data give satisfactory results, with accuracy values within ± 1–10% for various elements. The recovery rate for the analysis ranges from 95 to 110%. The minimum

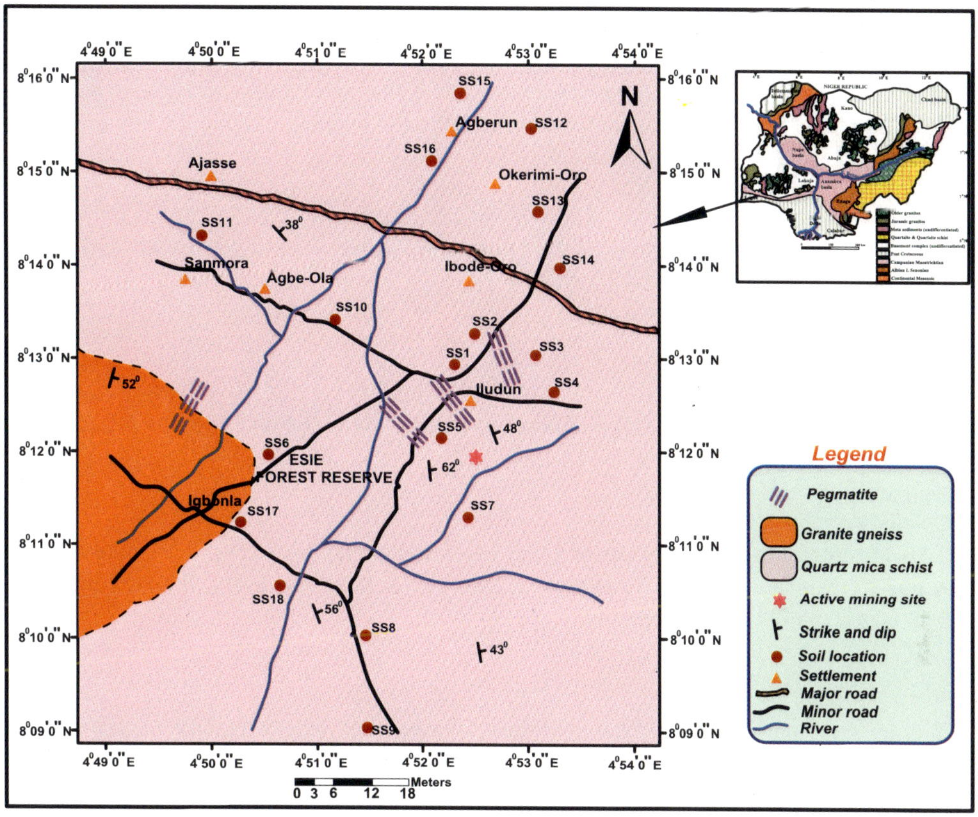

Figure 3. Location map showing sampling pattern superimposed on the geologic map.

detection limits for Fe, Pb, Zn, Cu, Cr, Ni, Ba, Sr, Th, Nb, Zr, Y, Sc and Mn were 0.04, 0.1, 1.0, 0.1, 0.1, 0.1, 1.0, 0.5, 0.2, 0.1, 0.1, 0.1, 1.0 and 0.01, respectively.

2.4. Statistical analysis and spatial distribution maps

PCA was used to distinguish the different groups of heavy metals and classify heavy metals from different sources based on similarities in their chemistry. The location and spatial distribution maps of heavy metals were produced with the method of kriging interpolation technique using the software package SURFER version 15.0 (Golden Software Inc., CO, USA).

2.5. Methods of heavy metal pollution assessment

2.5.1. Contamination factor (CF)

CF is a quantitative evaluation of the level of contamination and sources of pollution. The CF makes use of conservative elements as the reference element (Cabrera, Clemente, Barrientos, López, & Murillo, 1999). CF is expressed as follows:

$$C_f^i = C_i / C_{ri} \tag{i}$$

where C_i is the content of metal *i* instead of mean content from at least five sample sites; C_{ri} is the reference value, baseline level of metal *i*. The degree of heavy metal pollution of CF is divided into four classes.

2.5.2. Geoaccumulation index (Igeo)

Igeo index as depicted by Eq. (ii) is used to classify identification of environmental influences by anthropogenic activities, also used to quantify the extent of heavy metal contamination associated with soils. Igeo is expressed as follows (Vural, 2015; Zhao, Feng, Yang, Niu, & Shen, 2012):

$$I_{geo} = log_2 \frac{Cn}{1.5*Bn} \quad \text{(ii)}$$

where *Cn* is the measured concentration of metal *n, Bn* is the geochemical background value and constant 1.5 is used to represent the natural fluctuations and minor anthropogenic influences. The Igeo was distinguished into seven classes.

2.5.3. Contamination degree (CD)

This is expressed by the summation of all the CFs for each metal in the sequence. CD is expressed as follows:

$$\mathrm{CD} = \sum_{(i=1)}^{n} Cf \quad \text{(iii)}$$

where *n* is the number of metals, *i* is the baseline level of the metal and *Cf* is the CF. The CD is divided into four classes.

2.5.4. Pollution load index (PLI)

PLI is used to determine the magnitude of heavy metal contamination in soils. It is a potent tool in heavy metal pollution evaluation (Tomlinson, Wilson, Harris, & Jeffrey, 1980). The PLIs are calculated for all the metals. PLI is expressed as follows:

$$\mathrm{PLI} = (Cf_1 \times Cf_2 \times Cf_3 \ldots\ldots \times Cf_n)^{1/n} \quad \text{(iv)}$$

where Cf_1 is the CF and *n* is the number of metals. The degree of heavy metal risk assessed by PLI is mainly divided into three classes (Chakravarty & Patgiri, 2009).

2.5.5. Ecological risk index (ERI)

The ERI is used to evaluate heavy metal pollution in soil to associate ecological and environmental effects with their toxicology and the toxic-response factor (T_{ri}) of Cu, Zn, Cd, Cr, Ni and Pb is 5, 1, 30, 2, 5 and 5 (μg/g), respectively (Hakanson, 1980). ERI is expressed as follows:

$$E_r^i = T_r^i \times Cf_i \quad \text{(v)}$$

where T_r^i is the toxic-response factor for a given substance and Cf_i is the CF. The degree of heavy metal risk assessed by ERI is mainly divided into five classes.

2.5.6. Potential ecological risk index (PERI)

PERI is employed to determine the semi-quantitative evaluation of regional pollution level. It can be expressed as follows (Wang et al., 2015):

$$PERI = \sum_{i=1}^{n} E_r^i \quad \text{(vi)}$$

where E_r^i is the ecological risk factor of element *i*. The degree of heavy metal assessment by PERI is categorized into four classes.

3. Results and discussion

3.1. Heavy metals concentration in soils

The statistics of heavy metals are presented in Table 1. Concentrations of Pb, Zn, Cu, Cr, Ni, Ba, Sr, Th, Zr, Y, Sc and Mn in the soils, as well as the background values of heavy and trace metals, are presented. The concentrations of Pb, Zn, Fe, Cu, Cr and Ni varied between 0–678, 7–222, 1.64–8.87, 8.8–145, 1.4–81 and 1.6–93.1 ppm, respectively, with average concentrations of 107.72, 48.17, 5.17, 45.28, 14.76 and 21.53 ppm, respectively. The order of abundance of these heavy metals analysed in the soils followed this pattern Pb > Zn> Cu> Ni> Cr> Fe. The values of Pb, Fe, Cu, Ba and Sr are higher than the background values (samples collected 3.5 km away from the mining site)

Table 1. The summary of chemical parameters of the soil samples of the study area

Elements	Minimum	Maximum	Mean	Standard deviation	Coefficients of variation (%)	Background value
Fe (%)	1.64	8.87	5.17	2.22	42.9	4.06
Pb (ppm)	0	678	107.72	174.12	161.64	19
Zn (ppm)	7	222	48.17	63.09	130.97	72
Cu (ppm)	8.8	145	45.28	39.01	86.15	29
Cr (ppm)	1.4	81	14.76	18.62	126.15	18
Ni (ppm)	1.6	93.1	21.53	27.31	126.85	24
Ba (ppm)	63	3022	626.19	916.57	146.37	573
Sr (ppm)	18.1	570.2	107.74	168.54	156.43	63
Th (ppm)	3.3	16.7	11.29	4.77	42.25	21
Nb (ppm)	23.76	972.05	119.33	215.41	180.52	125
Zr (ppm)	60.2	488.2	255.71	125.13	48.93	409
Y (ppm)	7.8	71.2	27.53	18.63	67.67	33
Sc (ppm)	6	36	15.33	9.09	59.30	13
Mn (ppm)	0.01	0.11	0.05	0.04	80	127.6

indicating that mining activities had influence in the study area. The concentrations of Pb, Cu, Ba, Sr, Nb and Zr varied greatly across the study area. Mn is relatively low and slightly homogenous in all the soil samples analysed. The coefficient of variation (CV) is a measure of relative variability and used to compare the degree of variation from one data series to another. Fe, Th and Zr had the lowest CV of 42.9, 42.25, and 48.93%, respectively, followed by Cu, Mn and Y having CVs of 86.15, 80 and 67.67%, respectively. Pb, Zn, Cr, Ni, Ba, Sr and Nb had their CV more than 90% (161.64, 130.97, 126.15, 146.37, 156.43 and 180.52%, respectively). This indicates that the (CV < 90%) values of heavy metals are dominated by anthropogenic sources are generally high compared to (CV > 90%) dominated by geogenic sources. Several other factors could be responsible for the large coefficient variation across the study area and these might include the following: sampling methods, sample preparation methods and analytical techniques. However, these variations could be further separated into direct and indirect contributions to the total variation with the use of PCA. Therefore, Pb, Zn, Cr, Ni, Ba, Sr and Nb concentrations in the soil samples tend to be high and introduced by anthropogenic activities whilst Fe, Th and Zr are associated with natural sources. The high concentration of Pb, Cu and Sr (Table 2) in the soisls posed a serious threat to vegetation and animal rearing and ultimately affecting the overall quality of human health through consumption. Moderately high concentrations of Fe, Zn, Cu and Cr indicate an evidence of leaching of mine waste under acidic conditions. In general, different sample points showed varied soil heavy metal concentrations, indicating influence of human activities from mineral exploitation and processing.

3.2. Spatial distribution of heavy metals

The spatial distribution of heavy metals concentrations is a veritable tool used to access the possible sources of enrichment and identify likely hotspot regions with high metal concentration (Li, Lu, Zhang, Liu, & Hou, 2008). In this study, the concentration variability maps of Zn, Pb, Ni, Mn, Fe, Cu, Cr and Sr were made with the method of kriging interpolation using the Surfer (version 15.0) software package. The spatial distribution maps of the study area are presented in Figure 4a-h. The distribution patterns of Zn, Ni, Mn, Fe, Cu, Cr and Sr were decreasing in concentrations away from the active mining site to the adjacent non-active mining region. They peaked close to the mining site.

The spatial distribution of Zn, Ni, Mn and Cu vary slightly from other metals. They peaked at two points located in the northwestern part of the study area, associated with increased agricultural

Table 2. Geochemical data of selected elements in the soil samples of the study area

	Elements (all the elements are in ppm except Fe in %)													
Sample ID	Pb	Zn	Cu	Cr	Ni	Co	Ba	Sr	Th	Nb	Zr	Y	Sc	Fe	Mn
SS1	24	24	32.2	12.3	12.3	0.1	232	71.5	16.7	92.39	488.2	28.1	19	6.01	0.09
SS2	34	14	20	8.4	8.3	BDL	166	42.4	16.1	138.42	278	13.3	11	6.43	0.02
SS3	56	129	98.5	21.5	55.7	0.1	656	30	10.1	127.86	195.5	46.4	26	7.13	0.08
SS4	71	8	13.8	4.3	3.6	0.1	101	39.6	7.7	40.1	148.1	9.7	7	3.37	0.02
SS5	34	7	8.8	1.4	1.6	0.1	92	92.9	3.3	972.05	401	14.4	7	1.64	0.02
SS6	78	10	13.4	2.4	2.7	BDL	63	18.1	4.1	26.89	64.2	7.8	6	2.85	0.01
SS7	31	20	30.8	12.3	13.1	0.1	182	45.6	14.8	70.61	288.2	19.5	16	5.30	0.06
SS8	134	145	107.1	24.2	65.3	BDL	762	30.3	10.1	58.24	184.1	55.3	30	7.44	0.09
SS9	456	30	34.7	12	15.3	0.1	258	69	16	70.28	388.9	26.5	18	6.01	0.09
SS10	678	10	12.4	2.1	2.9	0.1	63	18.6	4.3	43.06	60.2	7.9	6	2.92	0.01
SS11	89	128	92.7	20.6	59.4	0.1	634	31.5	9.4	64.18	221.3	49	27	6.98	0.09
SS12	67	14	24	81	7.7	BDL	217	84.9	15	76.06	296.6	15.2	11	6.78	0.06
SS13	51	13	18	7.4	8.2	0.1	166	48.4	15.1	111.2	236.6	12.2	11	6.37	0.02
SS14	34	52	45.4	13.9	24.5	0.1	379	67.1	14	90.24	265.8	26.8	17	6.77	0.07
SS15	56	21	28.8	6.6	7.6	0.1	137	96.5	10.2	73.25	206.8	13.3	12	4.29	0.04
SS16	34	10	45.7	2.3	3.1	0.1	3022	570.2	15.3	34.32	397.6	39.7	8	1.88	0.02
SS17	12	10	43.8	2.4	3.1	0.1	3012.5	561.7	16.1	35.14	399.2	39.3	8	1.90	0.02
SS18	0	222	145	30.5	93.1	0.1	1129	21.1	5	23.76	82.4	71.2	36	8.87	0.11

*BDL = Below Detection Limit

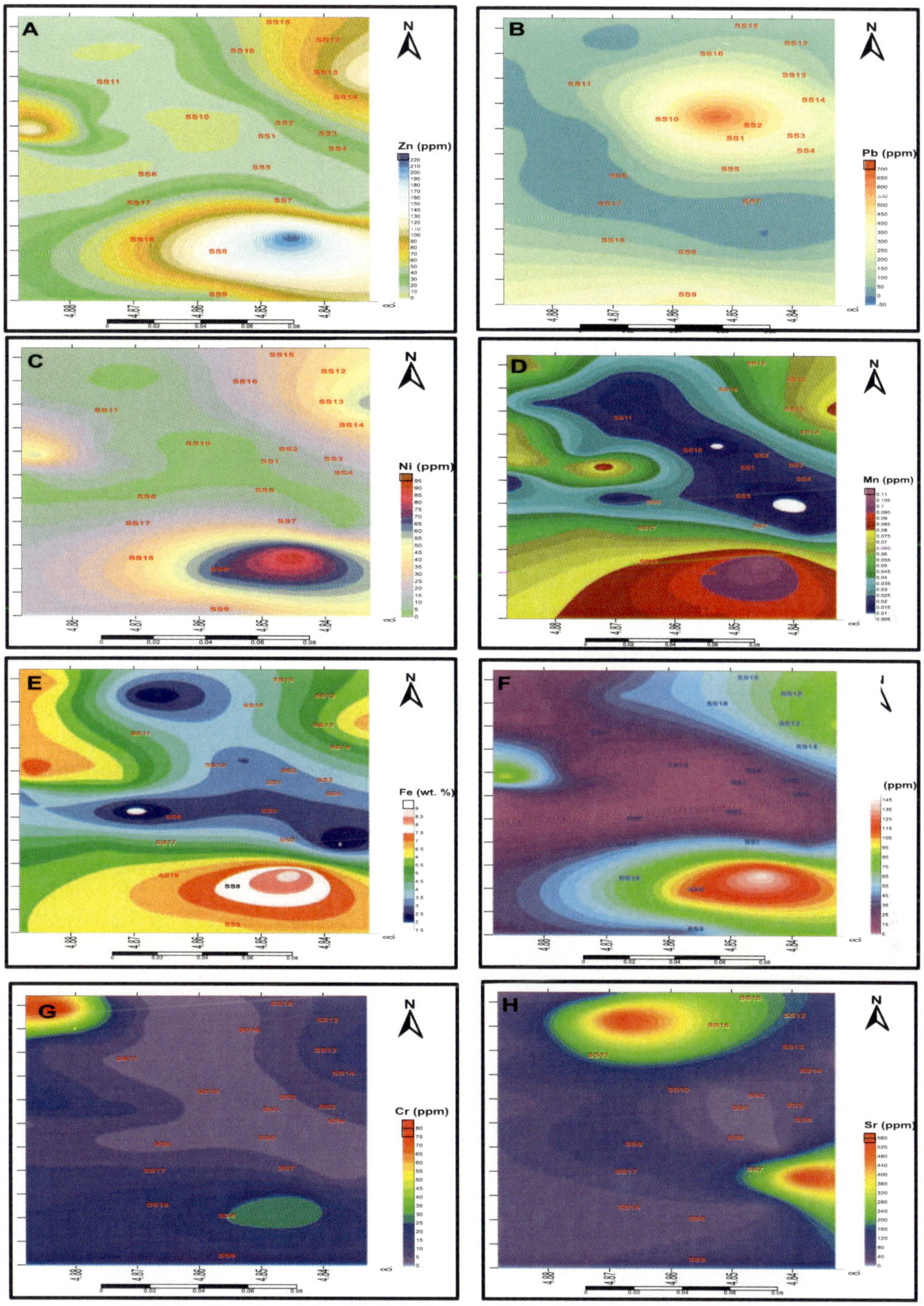

Figure 4. a-h: Spatial distribu-tion maps of Zn, Pb, Ni, Mn, Fe, Cu, Cr and Sr in study area.

activities and high traffic density. The second peak is located in the southeastern part of the study area, related to the presence of active mining of tantalum-niobium concentrates. Pb distribution across the study area showed high spatial variability. Only one hotspot is visible, and it is located close to the central area of the study area around the mines. The spatial distribution of Fe is quite different from other heavy metals. A single peak around the mine with relatively low spatial variability is shown. The mean concentration is comparable to the background value, suggesting that Fe concentrations in the soils have not been altered by anthropogenic activities. The spatial distribution of Cr showed relatively low spatial variability across the study area. A single hotspot is shown at the extreme northwestern region indicating the presence of intense agricultural activities. Cr may also be inputted into the soil by the parent material.

Also, the spatial distribution of Sr concentrations showed very high spatial variability with two distinct peaks around the mining area and the agricultural land space. The mean concentrations of Sr are higher compared with the background values, suggesting that Sr concentrations in the soils of the study area have been affected by anthropogenic activities or derived mainly due to the input of weathered parent material.

3.3. Indices of pollution assessment

3.3.1. Assessment using CF, CD, Igeo and PLI

Based on the monitoring data of soil quality in the study area, a quantitative analysis of heavy metal pollution assessment around the tantalum-niobium mining area was undertaken to determine the environmental effects. The evaluation was done using the CF and geoaccumulation index as presented in Tables 3 and 4.

The CF was calculated by comparing the observed heavy metal concentration with the background values for soil taken in an undisturbed location about 3.5 km away from the mining site. The CF values followed the descending sequence: Pb > Sr> Cu> Fe> Sc> Ba> Nb> Ni> Y> Cr> Zn> Zr> Th (Figure 5). Pb and Sr had the higher CF values of 35.68 ± 0.1 and 9.05 ± 0.05, indicating that the soils were considerably contaminated to highly contaminated. The values of the CF of various trace and heavy metals in the soils are summarized in Table 3 is as follows; Zn (0.10–3.08), Cu (0.43–5.00), Cr (0.08–4.50), Ni (0.07–3.88), Ba (0.11–5.27), Th (0.16–0.80), Nb (0.19–1.11), Zr (0.15–1.19), Y (0.24–2.16), Sc (0.46–2.77) and Fe (0.40–2.18). The values derived from these metals indicate that their mode of occurrence was a natural process (geogenic). Zn, Cu, Cr, Ni and Ba showed low-to-considerable CF ($Cf < 1: 3 \leq Cf < 6$), whilst Th, Nb, Zr, Y, Sc and Fe showed low-to-moderate CF ($Cf < 1: 1 \leq Cf < 3$) as presented in Table 5.

The Igeo values for each heavy metals are Pb (−1.25 to 4.57), Zn (−3.32 to 1.62), Cu (−2.32 to 2.59), Cr (−4.32 to 1.58), Ni (−4.64 to 1,37), Ba (−3.84 to 1.82), Sr (−2.40 to 2.59), Th (−3.32 to −0.92), Nb (−2.94 to 2.37), Zr (−3.32 to −0.32), Y (−2.64 to 0.53), Sc (−1.69 to 0.89) and Fe (−1.69 to 0.55). The Igeo values followed the descending order: Pb > Cu> Fe> Sc> Sr> Y> Nb> Zr> Ba> Zn> Th> Cr> Ni (Figure 6). Most of the values indicate uncontaminated to moderately contaminated, except for Pb that indicates uncontaminated to heavily contaminated. Based on the average values, the soils of the study area are mostly slightly contaminated by almost all elements and more pronounced for Pb. Zn, Cu, Cr, Ni and Ba showed uncontaminated (Igeo < 0) to moderately contaminated (1 < Igeo < 2) of the class 1 order. Also, Th, Nb, Zr, Y, Sc and Fe showed uncontaminated (Igeo < 0) to moderately contaminated (1 < Igeo< 2) of the class 2 order (Table 5).

The degree of contamination, which is expressed by the summation of all the CFs for each metal, varies from moderate degree of contamination to very high degree of contamination; indicating an alarming anthropogenic influence. Zn, Cr, Ni, Ba, Nb, Y, Sc and Fe showed a considerable degree of contamination for all the soil samples, whilst Pb, Cu and Sr indicate a very high degree of contamination depicting alarming anthropogenic input due to mining activities and increased agricultural activities.

The values of the PLI in Table 5 were found to be generally (<1) in most of the sampled points except for three sites which is slightly higher than 1. The PLI provides a simple, comparative means to determine the magnitude of heavy metals in a site. A value of PLI = 0 indicates perfection, PLI = 1 indicates only baseline levels of pollutants, whilst PLI > 1 indicates progressive deterioration of site (Chakravarty & Patgiri, 2009).

Table 3. Contamination factor for metals in the soil samples of the study area

Sample ID	Pb	Zn	Cu	Cr	Ni	Ba	Sr	Th	Nb	Zr	Y	Sc	Fe
SS1	1.26	0.3	1.11	0.68	0.51	0.40	1.13	0.80	0.74	1.19	0.85	1.46	1.48
SS2	1.79	0.19	0.69	0.47	0.35	0.29	0.67	0.77	1.11	0.68	0.40	0.85	1.58
SS3	2.95	1.79	3.40	1.19	2.32	1.14	0.48	0.48	1.02	0.48	1.41	2.00	1.76
SS4	3.74	0.11	0.48	0.24	0.15	0.18	0.63	0.37	0.32	0.36	0.29	0.54	0.83
SS5	1.79	0.10	0.30	0.08	0.07	0.16	1.47	0.16	7.78	0.98	0.44	0.54	0.40
SS6	4.11	0.14	0.46	0.13	0.11	0.11	0.29	0.20	0.22	0.16	0.24	0.46	0.70
SS7	1.63	0.28	1.06	0.68	0.55	0.32	0.72	0.70	0.56	0.70	0.59	1.23	1.31
SS8	7.05	2.01	3.69	1.34	2.72	1.33	0.48	0.48	0.47	0.45	1.68	2.31	1.83
SS9	24.00	0.42	1.20	0.67	0.64	0.45	1.10	0.76	0.56	0.95	0.80	1.38	1.48
SS10	35.68	0.14	0.43	0.12	0.12	0.11	0.30	0.20	0.34	0.15	0.24	0.46	0.72
SS11	4.68	1.78	3.20	1.14	2.48	1.11	0.50	0.45	0.51	0.54	1.48	2.08	1.72
SS12	3.53	0.19	0.83	4.50	0.32	0.38	1.35	0.71	0.61	0.73	0.46	0.85	1.67
SS13	2.68	0.18	0.62	0.41	0.34	0.29	0.77	0.72	0.89	0.58	0.37	0.85	1.57
SS14	1.79	0.72	1.57	0.77	1.02	0.66	1.07	0.67	0.72	0.65	0.81	1.31	1.67
SS15	2.95	0.29	0.99	0.37	0.32	0.24	1.53	0.49	0.59	0.51	0.40	0.92	1.06
SS16	1.79	0.14	1.58	0.13	0.13	5.27	9.05	0.73	0.27	0.97	1.20	0.62	0.46
SS17	0.63	0.14	1.51	0.13	0.13	5.26	8.92	0.77	0.28	0.98	1.19	0.62	0.47
SS18	0	3.08	5.00	1.69	3.88	1.97	0.33	0.24	0.19	0.20	2.16	2.77	2.18

Table 4. Geoaccumulation index for metals in the soil samples of the study area

Sample ID	Pb	Zn	Cu	Cr	Ni	Ba	Sr	Th	Nb	Zr	Y	Sc	Fe
SS1	−0.25	−1.60	−0.43	−1.12	−1.56	−1.89	−0.40	−0.92	−1.03	−0.32	−0.81	−0.04	−0.01
SS2	0.25	−2.40	−1.12	−1.69	−2.12	−2.40	−1.15	−0.97	−0.43	−1.15	−1.89	−0.84	0.08
SS3	0.97	0.84	1.18	−0.32	0.63	−0.40	−1.64	−1.64	−0.56	−1.64	−0.09	0.41	0.23
SS4	1.32	−3.18	−1.64	−2.64	−3.32	−3.06	−1.25	−2.06	−2.25	−2.06	−2.32	−1.47	−0.86
SS5	0.25	−3.32	−2.32	−4.32	−4.64	−3.18	−0.03	−3.32	2.37	−0.62	−1.79	−1.47	−1.89
SS6	1.45	−2.84	−1.69	−3.47	−3.64	−3.84	−2.40	−2.94	−2.84	−3.32	−2.64	−1.69	−1.09
SS7	0.12	−1.84	−0.49	−1.12	−1.47	−2.25	−1.06	−1.09	−1.36	−1.09	−1.36	−0.29	−0.20
SS8	2.23	1.01	1.30	−0.15	0.86	−0.17	−1.64	−1.64	−1.69	−1.74	0.16	0.62	0.29
SS9	4.00	−1.25	−0.32	−1.18	−1.22	−1.74	−0.45	−0.97	−1.43	−0.67	−0.89	−0.12	−0.01
SS10	4.57	−2.84	−1.79	−3.64	−3.64	−3.84	−2.32	−2.84	−2.12	−3.32	−2.64	−1.69	−1.06
SS11	1.64	0.83	1.09	−0.40	0.72	−0.43	−1.60	−1.74	−1.56	−1.47	−0.01	0.46	0.20
SS12	1.23	−2.40	−0.86	1.58	−2.25	−2.00	−0.15	−1.06	−1.29	−1.06	−1.69	−0.84	0.15
SS13	0.84	−2.47	−1.29	−1.89	−2.12	−2.40	−0.97	−1.06	−0.76	−1.36	−2.00	−0.84	0.07
SS14	0.25	−0.47	0.06	−0.97	−0.56	−1.18	−0.49	−1.18	−1.06	−1.22	−0.89	−0.20	0.15
SS15	0.97	−1.79	−0.60	−2.06	−2.25	−2.64	0.03	−1.64	−1.36	−1.56	−1.89	−0.69	−0.49
SS16	0.25	−2.84	0.07	−3.47	−3.47	1.82	2.59	−1.03	−2.47	−0.62	−0.32	−1.29	−1.69
SS17	−1.25	−2.84	0.01	−3.47	−3.47	1.81	2.57	−0.97	−2.40	−0.62	−0.34	−1.29	−1.69
SS18	0.00	1.62	1.74	0.18	1.37	0.39	−2.18	−2.64	−2.94	−2.94	0.53	0.89	0.55

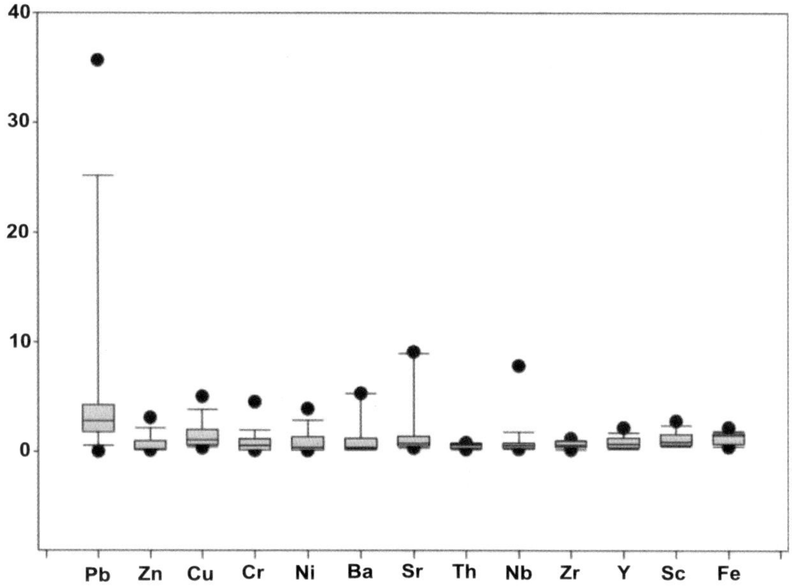

Figure 5. Indices of CF for metals in soil samples of the study area.

3.3.2. Assessment using potential ecological risk

The PERI was introduced to determine the semi-quantitative evaluation of regional pollution level (Wang et al., 2015), according to the toxicity of heavy metals and the response of the environment. Based on the standardized heavy metal toxic factor expressed by Hakanson (1980), the toxic response factor (T_{ri}) of Cu, Zn, Cd, Cr, Ni and Pb is 5, 1, 30, 2, 5 and 5 (μg/g), respectively. The ecological risk indices of Pb (0–178.4), Zn (0.1–3.08), Cu (1.5–18.45), Ni (0.6–13.6) and Cr (0.16–9.0) with an average of 28.35, 0.67, 7.81, 4.49 and 1.64, respectively, in all the sampling sites indicate low risk ($E_r^i < 30$). The highest ecological risk can be found in two sites with values (120, 178.4) indicating high risk. The main donation of the potential ecological risk comes from Pb, which indicates a significantly high risk with a value greater than 440.

3.4. Multivariate analysis

3.4.1. Pearson's correlation analysis

Pearson's correlation coefficient of heavy metals in the soil samples is presented in Table 6. There are varying levels of correlation between the metals in the soils. The inter-elemental relationship in the soils suggests different geochemical pathways are responsible for the metal concentration. A positive correlation between heavy metals suggests that these metals have common sources, mutual dependence and identical behaviour during transport. On the other hand, negative correlations suggest that they do not share the previously mentioned metal traits with each other. The results showed a strong correlation between Cu–Zn, Ni–Zn, Ni–Cu, Sr–Ba, Y–Cu, Sc–Zn, Sc–Cu, Sc–Ni, Sc–Y, Fe–Zn, Fe–Cu, Fe–Ni, Fe–Sc, Mn–Zn, Mn–Cu, Mn–Ni, Mn–Y, Mn–Sc and Mn–Fe. This may indicate that elements have similar sources of input or are precipitated in the same media. Pb, Th and Nb correlate negatively with other elements indicating that they originated from different or mixed sources. These diverse sources can be; gangue minerals, atmospheric deposition, community sewage system, use of pesticide and fertilizers amongst others.

3.4.2. Identification of source

Using the multivariate statistical analysis, PCA can be used to identify the sources of pollution (Zheng-Qi, Shi-Jun, Xian-Guo, & Cheng-Jiang, 2008). The results of PCA for heavy metals concentration are presented in Table 7. Two rotated principal components (PCs) with eigenvalues of 6.401 and 2.859 accounted for 66.14% of the total variance (Figure 7). These results showed two component models for all the data variation. PC1 accounted for 45.72% of the

Table 5. Summary of the pollution indices calculated for the study area

Indices	Equation	Classification	Pb	Zn	Cu	Cr	Ni	Ba	Sr	Th	Nb	Zr	Y	Sc	Fe
Contamination factor (Cf)	$Cf = \frac{C_n}{B_n}$	Cf < 1 (low Cf)	2	14	8	13	13	12	10	18	15	17	12	10	6
		1 ≤ Cf < 3 (moderate Cf)	9	3	6	4	4	4	6		2	1	6	8	12
		3 ≤ Cf < 6 (considerable Cf)	4	1	4	1	1	2							
		Cf ≥ 6 (very high Cf)	3						2		1				
Geoaccumulation Index (Igeo)	$Igeo = log_2 \frac{Cn}{1.5 * Bn}$	Igeo < 0 Class 0 (uncontaminated)	2	14	11	16	14	15	15	18	17	18	16	14	10
		0 < Igeo < 1 Class 1 (uncontaminated to moderately contaminated)	9	2	3	1	3	1	1				2	4	8
		1 < Igeo < 2 Class 2 (moderately contaminated)	4	2	4	1	1	2							
		2 < Igeo < 3 Class 3 (moderately to heavily contaminated)	1						2		1				
		3 < Igeo < 4 Class 4 (heavily contaminated)	1												
		4 < Igeo < 5 Class 5 (heavily to extremely contaminated)	1												
		Igeo ≥ 5 Class 6 (extremely contaminated)													
Contamination degree (CD)	$CD = \sum_{i=1}^{n} Cf$	CD < 6 (low CD)													
		6 ≤ CD < 12 (moderate CD)	2												
		12 ≤ CD < 24 (considerable CD)	8												
		CD ≥ 24 (very high CD) indicating alarming anthropogenic contamination	3												

(Continued)

Table 5. (Continued)

Indices	Equation	Classification	Pb	Zn	Cu	Cr	Ni	Ba	Sr	Th	Nb	Zr	Y	Sc	Fe
Pollution Load Index (PLI)	$PLI = (Cf_1 \times Cf_2 \times Cf_{3....} \times Cf_n)^{1/n}$	PLI = 0 (perfection)	10												
		PLI = 1 (baseline levels of pollutants present)													
		PLI > 1 (progressive deterioration of site)	3												
Ecological Risk (Er)	$E_r^i = T_r^i \times Cf_i$	E_r^i < 30 (low risk)	15	18	18	18	18								
		E_r^i: 30–60 (moderate risk)	1												
		E_r^i: 60–120 (considerable risk)													
		E_r^i: 120–240 (high risk)	2												
		E_r^i : > 240 (significantly high risk)													
Potential Ecological Risk (PERI)	$PERI = \sum_{i=1}^{n} E_r^i$	PERI < 110 (low risk)		1		1	1								
		PERI: 110–220 (moderate risk)			1										
		PERI: 220–440 (high risk)													
		PERI > 440 (significantly high risk)	1												

Cf = Contamination Factor; Igeo = Geo-Accumulation Index; CD = Contamination Degree; PLI = Pollution Load Index; Er = Ecological Risk; PERI = Potential Ecological Risk Index; Cn = Observed Element Concentration; Bn = Background Concentration of the Element.

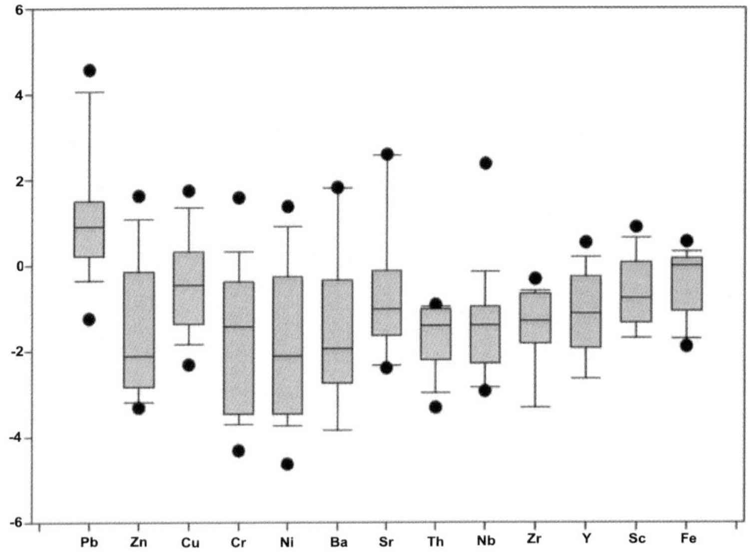

Figure 6. Indices of geoaccumu-lation for metals in soil samples of the study area.

total variance with strong loadings of Zn and Cu (0.781 and 0.873, respectively), supported by the strong correlations between Zn and Cu (Table 6) and their low Igeo and CF values. Hence, they may be contributed by geogenic sources. Coefficients of PC1 are supported by moderate loadings of Ni, Y, Sc, Fe and Mn (0.586, 0.635, 0.687, 0.629 and 0.643, respectively). These metals are siderophilic (having a strong affinity for metallic phase and usually depleted in the silicate portion of the earth) and were correlated well with one another.

The second principal component (PC2) accounted for 20.42% of the total variance with moderate loadings of Ba and Sr. This is further supported by weak correlation of Ba and Sr with other elements. Pb was equivocally different in the two-rotated component matrix; this is supported by the negative correlation of Pb with other elements. Pb average was 5.6-fold higher than the background value, indicating that it was inputted by anthropogenic activities. This was further supported by its high Igeo and CF values.

4. Conclusions

Different veritable tools, methods and indices were employed for the contamination assessment of soils heavy metals and trace elements in a typical active mining site of Iludun-Oro and its environs, southwestern Nigeria. This study was aimed at determining the spatial variances of heavy metals and identifying the pollution level, ecological risk level and potential sources of heavy metals in the soils. A total of 18 topsoil samples were randomly collected across different part of the study area and analysed using the ICP-MS. The concentrations of Pb, Zn, Fe, Cu, Cr and Ni varied between 0–678, 7–222, 1.64–8.87, 8.8–145, 1.4–81 and 1.6–93.1 ppm, respectively, with an average concentration of 107.72, 48.17, 5.17, 45.28. 14.76 and 21.53 ppm, respectively. The sequence of occurrence of these heavy metals analysed in the soils followed this pattern Pb > Zn> Cu> Ni> Cr> Fe. The values of Pb, Fe, Cu, Ba and Sr are higher than the background values indicating that mining activities influenced the study area. Pb, Zn, Cr, Ni, Ba, Sr and Nb had their CV more than 90% (161.64, 130.97, 126.15, 146.37, 156.43 and 180.52%, respectively). This indicates that the (CV < 90%) values of heavy metals are dominated by anthropogenic sources are high compared to (CV > 90%) dominated by geogenic sources.

Results of the spatial distribution maps and multivariate statistical analysis of the heavy metals showed that mining activities, atmospheric deposition, poor community sewage system, and the use of pesticide and fertilizers represented the most important sources of heavy metal pollution. Accessing the soil quality in the study area, the results of the geoaccumulation, CF and PLIs

Table 6. Results of Pearson's correlation matrix for soil samples

Element	Pb	Zn	Cu	Cr	Ni	Ba	Sr	Th	Nb	Zr	Y	Sc	Fe	Mn
Pb	1	-.143	-.199	-.130	-.146	-.237	-.205	-.213	-.130	-.249	-.235	-.154	-.136	-.084
Zn		1	**.970****	.308	**.997****	.096	-.283	-.289	-.174	-.384	**.872****	**.938****	**.696****	**.753****
Cu			1	.293	**.967****	.307	-.067	-.143	-.249	-.254	**.954****	**.922****	**.631****	**.746****
Cr				1	.325	-.126	-.215	.160	-.160	-.037	.218	.347	**.567***	**.470***
Ni					1	.070	-.307	-.244	-.186	-.362	**.867****	**.956****	**.733****	**.780****
Ba						1	**.920****	.280	-.205	.322	**.523***	.009	-.303	-.077
Sr							1	.382	-.073	**.507***	.181	-.333	**-.554***	-.328
Th								1	-.343	**.671****	-.004	-.039	.207	.142
Nb									1	.326	-.204	-.212	-.323	-.205
Zr										1	-.012	-.168	-.193	.072
Y											1	**.845****	**.489***	**.708****
Sc												1	**.816****	**.910****
Fe													1	**.795****
Mn														1

**. Correlation is significant at the 0.01 level (2-tailed).

*. Correlation is significant at the 0.05 level (2-tailed).

Bold correlation is significant at the 0.05 level.

Table 7. The result of principal component analysis for soils

	Component		
Component matrix	**Coefficients of PC1**	**Coefficients of PC2**	**Communalities**
Pb	−0.065	−0.248	0.066
Zn	**0.781**	−0.015	0.610
Cu	**0.873**	0.118	0.776
Cr	0.179	−0.032	0.046
Ni	**0.586**	−0.017	0.627
Ba	0.013	**0.533**	0.284
Sr	−0.129	**0.528**	0.295
Th	−0.034	0.352	0.125
Nb	−0.111	−0.084	0.019
Zr	−0.111	0.382	0.158
Y	**0.635**	0.265	0.474
Sc	**0.687**	0.014	0.472
Fe	**0.629**	−0.110	0.506
Mn	**0.643**	0.019	0.432
Initial Eigenvalues	6.401	2.859	
Percentage of Variance (%)	45.72	20.42	
Cumulative percentage (%)	45.72	66.14	

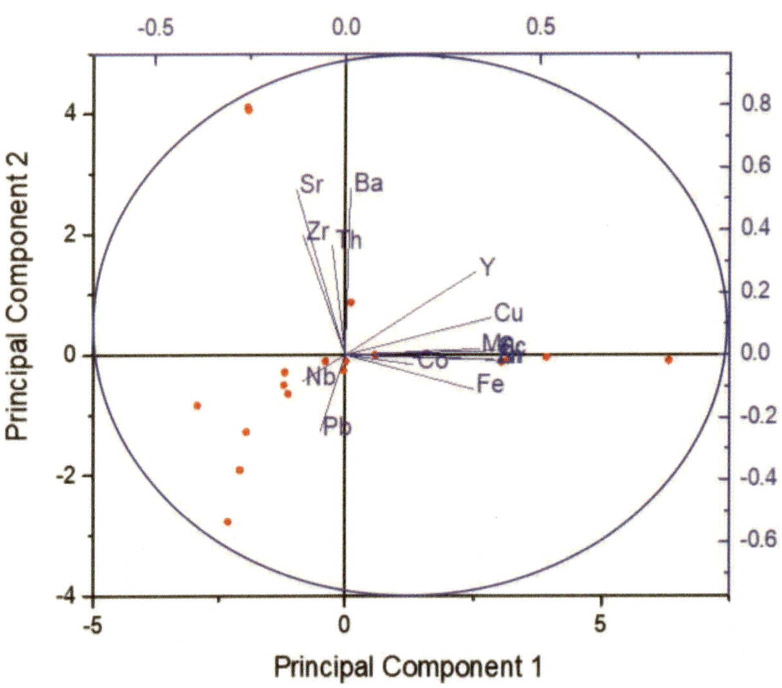

Figure 7. Rotated principal components binary plot for PC1 and PC2.

indicated high concentrations of Pb in the soils of Iludun-Oro. Also, most of the values indicated uncontaminated to moderately contaminated for all elements except for Pb. This research suggested that the lessening of anthropogenic sources in the study area should be considered in further future control measures. Our study highpoints that multiple indices provide more reasonable assessments, which will be considered in the future studies on soils with heavy metals.

Conclusively, this paper has provided background information for a vulnerability assessment of the study area as well as for future impact assessment and proper planning of the mining site. Besides, this study would recommend a remediation plan for analysis of the quality of drinking water and soil quality for crop production in the study area. To sum up, there is the need for implementation and enforcement of environmental laws, which may be an aid to pollution control around the mining sites.

Funding
The authors received no direct funding for this research.

Competing Interest
The authors declare no competing interests.

Author details
Abiola Oyebamiji[1,2,3]
E-mail: abiola.oyebamiji@eksu.edu.ng
ORCID ID: http://orcid.org/0000-0002-8776-8690
Amobichukwu Amanambu[2,4]
E-mail: jonesamanambu@yahoo.com
Tehseen Zafar[1,2]
E-mail: tehseenabbas11@yahoo.com
Adeniyi JohnPaul Adewumi[5]
E-mail: adewumiadeniyi27@yahoo.com
ORCID ID: http://orcid.org/0000-0003-1091-743X
Damilare Stephen Akinyemi[6,7]
E-mail: sakinyemi@oauife.edu.ng
[1] State Key Laboratory of Ore Deposit Geochemistry, Institute of Geochemistry, Chinese Academy of Sciences, Guiyang, China.
[2] University of Chinese Academy of Sciences, Beijing, China.
[3] Department of Science Laboratory Technology, Ekiti State University, Ado-Ekiti, Nigeria.
[4] State Key Laboratory of Desert and Oasis Ecology, Xinjiang Institute of Ecology and Geography, Chinese Academy of Sciences, Urumqi, China.
[5] Department of Geological Sciences, Achievers University, Owo, Nigeria.
[6] State Key Laboratory of Vegetation and Environmental Change, Institute of Botany, Chinese Academy of Sciences, Beijing, China.
[7] Department of Botany, Obafemi Awolowo University, Ile Ife, Nigeria.

Authors' contributions
All the authors contributed equally to the preparation of this manuscript. All authors read and approved the final manuscript.

Compliance with ethical standards
Yes

References

Adewuyi, G. O., & Osobamiro, M. T. (2016). Chemical speciation and potential mobility of some toxic metals in tropical agricultural soil. *Research Journal of Environmental Toxicology, 10*, 159–165. doi:10.3923/rjet.2016.159.165

Adriano, D. C. (2001). *Trace elements in terrestrial environments biogeochemistry bioavailability and risk of metals* (Vol. 2, 2nd ed., pp. 61–89). NewYork: Springer.

Antwi-Agyei, P., Hogarh, J. N., & Foli, G. (2009). Trace elements contamination of soils around gold mine tailings dams at Obuasi, Ghana. *African Journal of Environmental Science and Technology, 3*(11), 353-359.

Benhaddya, M. L., & Hadjel, M. (2014). Spatial distribution and contamination assessment of heavy metals in surface soils of Hassi Messaoud, Algeria. *Environmental Earth Sciences, 71*(3), 1473–1486. doi:10.1007/s12665-013-2552-3

Benson, N. U., Anake, W. U., Adedapo, A. E., Fred-Ahmadu, O. H., & Ayejuyo, O. O. (2017). Toxic metals in cigarettes and human health risk assessment associated with inhalation exposure. *Environmental Monitoring and Assessment, 189*(12), 619. doi:10.1007/s10661-017-6348-x

Boakye, D., Dessus, S., Foday, Y., & Oppong, F. (2012). *Investing mineral wealth in development assets*. Ghana: Liberia and Sierra Leone.

Cabrera, F., Clemente, L., Barrientos, E. D., López, R., & Murillo, J. M. (1999). Heavy metal pollution of soils affected by the Guadiamar toxic flood. *Science of the Total Environment, 242*(1–3), 117–129.

Chakravarty, M., & Patgiri, A. D. (2009). Metal pollution assessment in sediments of the Dikrong River, NE India. *Journal of Human Ecology, 27*(1), 63–67. doi:10.1080/09709274.2009.11906193

Chukwu, E. (2011). Small scale mining and heavy metals pollution of agricultural soils: The case of Ishiagu Mining District, South Eastern Nigeria. *Journal of Geology and Mining Research, 3*(4), 87–104.

Dulski, P. (2001). Reference materials for geochemical studies: New analytical data by ICP-MS and critical discussion of reference values. *Geostandards and Geoanalytical Research, 25*(1), 87–125. doi:10.1111/ggr.2001.25.issue-1

Equeenuddin, S. M., Tripathy, S., Sahoo, P. K., & Panigrahi, M. K. (2013). Metal behavior in sediment associated with acid mine drainage stream: Role of pH. *Journal of Geochemical Exploration, 124*, 230–237. doi:10.1016/j.gexplo.2012.10.010

Fayiga, A. O., Ipinmoroti, M. O., & Chirenje, T. (2017). Environmental pollution in Africa. *Environment, Development and Sustainability*, 20(1), 1–33.

Hakanson, L. (1980). An ecological risk index for aquatic pollution control. A sedimentological approach. *Water Research, 14*(8), 975–1001. doi:10.1016/00431354(80)90143-8

Harikrishnan, N., Ravisankar, R., Chandrasekaran, A., Gandhi, M. S., Kanagasabapathy, K. V., Prasad, M. V. R., & Satapathy, K. K. (2017). Assessment of heavy metal contamination in marine sediments of east coast of Tamil Nadu affected by different pollution sources. *Marine Pollution Bulletin, 121*(1–2), 418–424. doi:10.1016/j.marpolbul.2017.05.047

Higueras, P., Oyarzun, R., Oyarzún, J., Maturana, H., Lillo, J., & Morata, D. (2004). Environmental assessment of copper-gold-Mercury mining in the Andacollo and Punitaqui districts, northern Chile. *Applied Geochemistry, 19*(11), 1855–1864. doi:10.1016/j.apgeochem.2004.04.001

Hsu, L. C., Huang, C. Y., Chuang, Y. H., Chen, H. W., Chan, Y. T., Teah, H. Y., ... Tzou, Y. M. (2016). Accumulation of heavy metals and trace elements in fluvial sediments received effluents from traditional and semiconductor industries. *Scientific Reports*, *6*, 34250. doi:10.1038/srep3425

Jan, A. T., Azam, M., Siddiqui, K., Ali, A., Choi, I., & Haq, Q. M. R. (2015). Heavy metals and human health: Mechanistic insight into toxicity and counter defense system of antioxidants. *International Journal of Molecular Sciences*, *16*(12), 29592–29630. doi:10.3390/ijms161226183

Khan, A. M., Yusoff, I., Bakar, N. K. A., Bakar, A. F. A., & Alias, Y. (2016). Assessing anthropogenic levels, speciation, and potential mobility of rare earth elements (REEs) in ex-tin mining area. *Environmental Science and Pollution Research*, *23*(24), 25039–25055. doi:10.1007/s11356-016-7641-x

Li, C., Lu, F. Y., Zhang, Y., Liu, T. W., & Hou, W. (2008). Spatial distribution characteristics of heavy metals in street dust in Shenyang city. *Ecology Environment*, *17* (2), 560–564.

Lu, L., Liu, G., Wang, J., & Wu, Y. (2017). Bioavailability and mobility of heavy metals in soil in vicinity of a coal mine from Huaibei, China. *Human and Ecological Risk Assessment: an International Journal*, *23*(5), 1164–1177. doi:10.1080/10807039.2017.1308816

Lupankwa, K., Love, D., Mapani, B., Mseka, S., & Meck, M. (2006). Influence of the Trojan Nickel Mine on surface water quality, Mazowe valley, Zimbabwe: Runoff chemistry and acid generation potential of waste rock. *Physics and Chemistry of the Earth, Parts A/B/C*, *31*(15–16), 789–796. doi:10.1016/j.pce.2006.08.030

Madrid, L., Díaz-Barrientos, E., & Madrid, F. (2002). Distribution of heavy metal contents of urban soils in parks of Seville. *Chemosphere*, *49*(10), 1301–1308.

Mierzwa-Hersztek, M., Gondek, K., Klimkowicz-Pawlas, A., Baran, A., & Bajda, T. (2018). Sewage sludge biochars management—Ecotoxicity, mobility of heavy metals, and soil microbial biomass. *Environmental Toxicology and Chemistry*, *37*(4), 1197–1207. doi:10.1002/etc.4045

Mortazavi, S. M. J., Mortazavi, G., & Paknahad, M. (2016). Levels of arsenic, mercury, cadmium, copper, lead, zinc, and manganese in serum and whole blood of resident adults from mining and non-mining communities in Ghana. *Environmental Science and Pollution Research International*, *23*(21), 22220. doi:10.1007/s11356-016-7697-7

Mwesigye, A. R., Young, S. D., Bailey, E. H., & Tumwebaze, S. B. (2016). Population exposure to trace elements in the Kilembe copper mine area, Western Uganda: A pilot study. *Science of the Total Environment*, *573*, 366–375. doi:10.1016/j.scitotenv.2016.08.125

Nsenga Kumwimba, M., Zhu, B., Dong, Z., Tang, J., Wang, T., Xiao, L., & Kavidia Muyembe, D. (2017). Assessing nutrient, biomass, and sediment transport of drainage ditches in the Three Gorges Reservoir Area. *CLEAN–Soil, Air, Water*, *45*(1). doi:10.1002/clean.201501012

Obaje, N. G. (2009). *Geology and mineral resources of Nigeria* (Vol. 120). Springer. doi:10.1007/978-3-540-92685-6_1

Oelofse, S. H. (2009). *Mine water pollution-acid mine decant, effluent and treatment: A consideration of key emerging issues that may impact the state of the environment*. Department of Environmental Affairs & Tourism. The Icfai University Press.

Ogunrayi, O. A., Akinseye, F. M., Goldberg, V., & Bernhofer, C. (2016). Descriptive analysis of rainfall and temperature trends over Akure, Nigeria. *Journal of Geography and Regional Planning*, *9*(11), 195–202. doi:10.5897/JGRP2016.0583

Okang'Odumo, B., Carbonell, G., Angeyo, H. K., Patel, J. P., Torrijos, M., & Martín, J. A. R. (2014). Impact of gold mining associated with mercury contamination in soil, biota sediments and tailings in Kenya. *Environmental Science and Pollution Research*, *21*(21), 12426–12435. doi:10.1007/s11356-014-3190-3

Oyebamiji, A., Odebunmi, A., Ruizhong, H., & Rasool, A. (2017). Assessment of trace metals contamination in stream sediments and soils in Abuja leather mining, southwestern Nigeria. *Acta Geochimica*, 1–22. doi:10.1007/s11631-017-0256-1

Potra, A., Dodd, J. W., & Ruhl, L. S. (2017). Distribution of trace elements and Pb isotopes in stream sediments of the Tri-State mining district (Oklahoma, Kansas, and Missouri), USA. *Applied Geochemistry*, *82*, 25–37. doi:10.1016/j.apgeochem.2017.05.005

Sarapulova, A., Dampilova, B. V., Bardamova, I., Doroshkevich, S. G., & Smirnova, O. (2017). Heavy metals mobility associated with the molybdenum mining-concentration complex in the Buryatia Republic, Germany. *Environmental Science and Pollution Research*, *24*(12), 11090–11100. doi:10.1007/s11356-016-8105-z

Tomlinson, D. L., Wilson, J. G., Harris, C. R., & Jeffrey, D. W. (1980). Problems in the assessment of heavy-metal levels in estuaries and the formation of a pollution index. *Helgoländer meeresuntersuchungen*, *33*(1), 566. doi:10.1007/BF02414780

Vural, A. (2015). Contamination assessment of heavy metals associated with an alteration area: Demirören Gumushane, NE Turkey. *Journal of the Geological Society of India*, *86*(2), 215–222. doi:10.1007/s12594-015-0301-9

Wang, L., & Liang, T. (2016). Anomalous abundance and redistribution patterns of rare earth elements in soils of a mining area in Inner Mongolia, China. *Environmental Science and Pollution Research*, *23*(11), 11330–11338. doi:10.1007/s11356-016-6351-8

Wang, X., Deng, C., Yin, J., & Tang, X. (2018). Toxic heavy metal contamination assessment and speciation in sugarcane soil. *IOP Conference Series: Earth and Environmental Science*, *108*(4), 042059. IOP Publishing. doi:10.1088/1755-1315/108/4/042059

Wang, X. S., & Yong, Q. I. N. (2005). Correlation between magnetic susceptibility and heavy metals in urban topsoil: A case study from the city of Xuzhou, China. *Environmental Geology*, *49*(1), 10–18. doi:10.1007/s00254-005-0015-1

Wang, Z., Wang, Y., Chen, L., Yan, C., Yan, Y., & Chi, Q. (2015). Assessment of metal contamination in coastal sediments of the Maluan Bay (China) using geochemical indices and multivariate statistical approaches. *Marine Pollution Bulletin*, *99* (1–2), 43–53. doi:10.1016/j.marpolbul.2015.07.064

Wilson, S. T., Wang, H., Kabenge, M., & Qi, X. (2017). The mining sector of Liberia: Current practices and environmental challenges. *Environmental Science and Pollution Research*, *24*(23), 18711–18720. doi:10.1007/s11356-017-9647-4

Yan, X., Liu, M., Zhong, J., Guo, J., & Wu, W. (2018). How human activities affect heavy metal contamination of soil and sediment in a long-term reclaimed area of the Liaohe River Delta, North China. *Sustainability, 10* (2), 338. doi:10.3390/su10020338

Zhang, W., You, M., & Hu, Y. (2016). The distribution and accumulation characteristics of heavy metals in soil and plant from Huainan coalfield, China. *Environmental Progress & Sustainable Energy, 35*(4), 1098–1104. doi:10.1002/ep.12336

Zhao, S., Feng, C., Yang, Y., Niu, J., & Shen, Z. (2012). Risk assessment of sedimentary metals in the Yangtze Estuary: New evidence of the relationships between two typical index methods. *Journal of Hazardous Materials, 241*, 164–172. doi:10.1016/j.jhazmat.2012.09.023

Zheng-Qi, X., Shi-Jun, N., Xian-Guo, T., & Cheng-Jiang, Z. (2008). Calculation of heavy metals' toxicity coefficient in the evaluation of potential ecological risk index [J]. *Environmental Science & Technology, 2*(8), 31.

4

Gender stereotyping: Evidence from gender differentiated household vulnerability to climate change in Eastern Uganda

Kenneth Balikoowa [1]*, Gorettie Nabanoga[2] and David. M. Tumusiime[1]

*Corresponding author: Kenneth Balikoowa, School of Forestry, Environment and Geographical Sciences, College of Agricultural and Environmental Sciences, Makerere University, P.O. Box 7062 Kampala, Uganda
E-mail: balikenyy@gmail.com

Reviewing editor: Conor Buggy, University College Dublin, Ireland

Abstract: A narrative linking women and vulnerability has been overarching in vulnerability literature. However, this narrative has not been empirically proven aside from generalizations based on poverty-gender linkages. This study used primary data collected in 2016 from 735 randomly selected households in Eastern Uganda to construct a gender vulnerability index based on the livelihood vulnerability framework to determine the most vulnerable gender differentiated households. The results showed that single male-headed households were the most vulnerable to climate change, even worse than single female-headed households. Mixed gender households showed lowest vulnerability. Presence of a female spouse reduced the vulnerability of the traditionally "male headed" households. The results suggest that disparity in access to requisite resources mediates vulnerability among groups of households. It is concluded that the straight assumption that vulnerability is associated with female-headed households is misleading and could disenfranchise some male-headed households that are more vulnerable.

Subjects: Agriculture & Environmental Sciences; Climate Change; Environment & Gender

Keywords: vulnerability; adaptive capacity; gender; climate change; Eastern Uganda

ABOUT THE AUTHOR

Kenneth Balikoowa is a PhD candidate whose research interest is natural resources management. He conducts research on gender and climate change adaptation, forestry and environmental management in Uganda. This work is one of the outputs from the Regional Capacity Building for Sustainable Resource Management and Agricultural Productivity under Changing Climate (CAPSNAC) project covering three countries whose main goal is to strengthen research capabilities in climate change and to generate knowledge for policy formulation.

PUBLIC INTEREST STATEMENT

Identifying the most vulnerable sections of society is important in informing climate change adaptation policy. An assertion that women are more vulnerable to climate change has often been alluded to in climate change adaptation literature. In this paper, we interrogate the assertion that women are more vulnerable to climate change and that female-headed households are particularly disadvantaged using data from Eastern Uganda, our research demonstrates that over emphasizing that women are more vulnerable to climate change may disenfranchise some men or could fail to recognize the moderate status of certain women. Our results suggest that gender stereotypes can be counterproductive when identifying the most vulnerable people for adaptation planning.

1. Introduction

Even the best efforts at mitigating climate change will fall short of stemming climate change due to historical emissions (Pielke, Prins, Rayner, & Sarewitz, 2007). This realization has shifted emphasis toward enabling societies to cope with the new biophysical conditions caused by climate change. The need to identify who will be most affected by climate change in planning for climate change responses cannot be overemphasized (Holand, 2014; Holand, Lujala, & Rød, 2011; O'Brien, Eriksen, Schjolden, & Nygaard, 2004). Those individuals, households, communities or countries likely to be most affected are considered vulnerable. Vulnerability here implies the degree of susceptibility to harm from a natural hazard (Lewis, 1999). But to identify the most vulnerable entities requires that vulnerability is measured in ways that make comparison of different contextual groups possible.

Different conceptualizations of vulnerability have been suggested (see Ribot, 2010). Following the integrative approach, vulnerability depends on both biophysical and human elements of a system. Society mediates climate vulnerability by deliberate action or unavoidable inaction to adjust their livelihoods. The resultant vulnerability is thus a function of not only the level of environmental stimuli but also the pre-existing economic, cultural and social situation that would make some sections of society fail to make the necessary adjustments to shield themselves from the climate stress. In fact according to Holand et al. (2011), vulnerability is inherent to a system even before a potential threat. The integrative approach captures the connection between people and the biophysical environment by highlighting the fact that observed vulnerability is ultimately the interaction between the physical phenomenon and the human context within which it occurs (Vincent & Cull, 2010). Like Kousky and Cooke (2009) noted, climate change does not occur in a vacuum. It is an additional constraint to already existing risks to livelihoods like poverty and gender inequality (Cannon, 2002). Climate change will lead to even more livelihood vulnerability as it places additional hardship on people who aspire to maintain their livelihoods.

Social vulnerability expresses the totality of factors that make people more susceptible to harm from specific threats. These factors can be external and/or internal characteristics of a person or group that influence their capacity to anticipate, cope with, resist and recover from the impact of a natural hazard (Blaikie, Cannon, Davis, & Wisner, 1994). Social vulnerability intricately connects to entitlements and endowment of people, by focusing on the need to have the requisite resources to cope with climate-related stress as a major driver for vulnerability rather than the magnitude of the stress itself. Social vulnerability in a common-sensical way highlights the role of gender in determining vulnerability and specifically suggests that women are more vulnerable since they are poorer than men (Arora-Jonsson, 2011; Rahman, 2013) because of low economic opportunities (Johnsson-Latham, 2007; O'Sullivan, Rao, Banerjee, Gulati, & Vinez, 2014). Since vulnerability to climate change is transformed into differentiated outcomes via social structure (Ribot, 2010), it is expected to reflect differently on female and male-headed households due to disparity of heads of households in access to resources needed to safeguard their livelihoods from climate-related stress (Sharmin & Islam, 2013).

This study investigates the a priori determination that female-headed households are by implication more vulnerable to climate change (see Waite, 2000) and assesses whether blanket statements about gender and vulnerability may disenfranchise even more vulnerable groups of people. The assertion that women are more vulnerable to climate change may be less definitive due to the heterogeneity of women. For instance, the marital status of a woman can have mediating effects on access to resources which would increase adaptive capacity. Misconstruing vulnerability as being linked with women has the undesirable consequence that vulnerability is thought to be confined to female-headed households and overlooks the situation of certain kinds of male-headed households. It also obfuscates the agency, knowledge and resilience of women (Dankelman, 2010), as they become the subject of study instead of contributors to knowledge. The objective of this study was to interrogate the gender stereotypes in climate change vulnerability particularly regarding the assertion that women are more vulnerable to climate change using gender differentiated household data collected from four districts in Eastern Uganda. It is anticipated that such a research is imperative in identifying the most vulnerable households to climate change, but could help to illuminate the merit in generalizations regarding gender and vulnerability.

2. Methodology

2.1. Description of the study area

The study was conducted in four districts in Eastern Uganda (Figure 1). Bududa and Manafwa (elevation of >1300 m above sea level) have wet tropical climate (Vlaeminck et al., 2016) with temperatures ranging from 18°C to 22°C (BDPU, 2012). They receive bimodal rainfall regime of approximately 1500–2000 mm pa with peaks in April–May and August–September (Kitutu, Poesen, & Deckers, 2011; Mbogga, 2012). Iganga and Namutumba are located at lower elevation (<1100 m above sea level) and have warm humid climate with mean temperature of 23°C and annual precipitation of 900–1200 mm over two rain seasons in March–May and August–October. Bududa and Manafwa have experienced heavy rainfall events in the last 10 years the worst being in 2010 which caused the infamous Bududa landslide (Staudt, Kuosmanen, Babirye, & Lugaizi, 2014). Majority of farmers in Bududa and Manafwa practice a predominantly coffee-banana system but also grow other bi-annuals and annuals like beans, maize and sweet potatoes. In Namutumba and Iganga, the major crops grown include maize, cassava and beans and sweet potatoes (UBOS, 2010). Like other parts of Uganda, the major livelihood activity is farming for both cash and subsistence using low input practices that make people very sensitive to climatic changes even over a short period of time.

2.2. Research design

The study followed a cross-sectional research design. A descriptive cross-sectional household survey of 735 households was conducted during the first season of 2016 cropping year. The survey

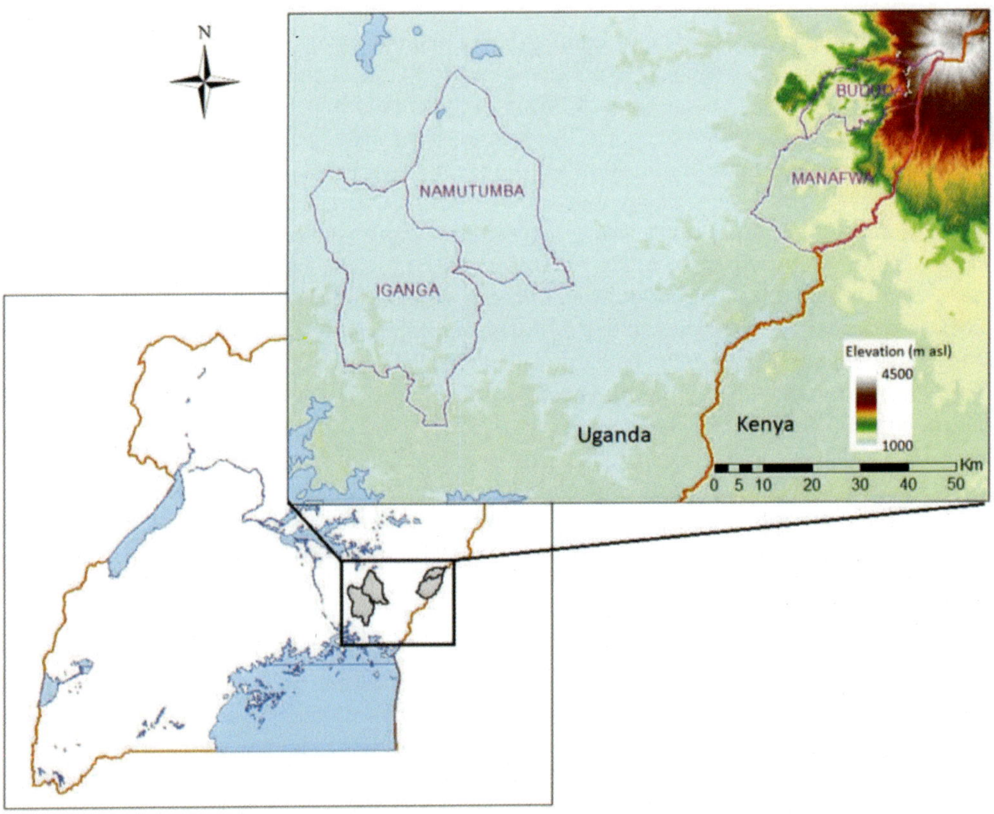

Figure 1. Map of Uganda show-ing the location of the four dis-tricts in Eastern Uganda.

employed a 21-page pre-tested questionnaire administered in the local languages (*Lusoga* in Iganga and Namutumba or *Lumasaba* in Bududa and Manafwa) through a local translator. In the survey, farmers were asked questions about their perceptions on long-term (5–30 years) climatic changes based on temperature and precipitation patterns in the area. The household survey also collected data on socio demographics, asset ownership, livelihood activities, crops grown and the observed climate change impacts.

2.3. Sampling and subject selection

The districts were purposely selected to cover areas that have reported heavy rainfall and drought associated with climate change. Selection of respondents was done using multi-stage sampling with subcounty and household selection at different stages. At the first stage, a list of sub counties was acquired from a contact person at each of the four districts. From this list, eight sub counties were randomly selected (two from each District). The sub counties selected were Bukaye and Namungalwe in Iganga, Namutumba and Magada in Namutumba, Nalondo and Khabutoore in Manafwa and Bushika and Nakatisi in Bududa. For each of the sub counties selected, a local contact person was identified with the help of the district production office. The local person generated a list of households in any two adjacent villages. Respondents were selected from adjacent villages for logistical purposes to reduce the travel time for the respondents. Each household was given a number. The household list comprised the sampling frame from which a random sample was drawn by use of a table of random numbers. Where a sampled household could not be located, the nearest house was chosen to replace it. About 50 farm households were sampled from each village.

3. Constructing an index of vulnerability to climate change

Vulnerability to climate change comprises of both internal and external factors to a system. Accordingly, mere presence of a risk (e.g. high intensity rainfall and drought) does not imply vulnerability of the system but rather the potential for perturbation. Vulnerability occurs when the system cannot cope with the new set of biophysical conditions to maintain a given prior standard. The vulnerability of a system or group is ultimately driven by the intensity of the hazard (exposure), factors that predispose the group to climate hazards (sensitivity) and the capabilities available that mediate the group's ability to adjust relative to the climate event (adaptive capacity). Exposure and sensitivity determine the potential impacts to a system but the ability to adjust determines its vulnerability. For this reason, vulnerability is intensified among the poor, marginalized and underrepresented people because of lack of access to requisite resources to make them more resilient to harm from climate change related vagaries (Ribot, 2010).

Vulnerability assessments need to answer the question on who is most vulnerable and how to assist them (Ribot, 2010). One way to do this is by constructing vulnerability indices. A vulnerability index integrates several indicators (Hahn, Riederer, & Foster, 2009) into a single numerical variable that can be used to compare contextual groups of households. The theoretical relationship between the indicators and vulnerability can be based on understanding of relationships (Wiréhn, Danielsson, & Neset, 2015) in an inductive or deductive manner (Adger, Brooks, Bentham, Agnew, & Eriksen, 2004; Hinkel, 2011), guided by literature review.

Each of the three dimensions of vulnerability, i.e. exposure, sensitivity and adaptive capacity was constructed using a set of indicators (Table 1). The conceptual relationship between indicators and any of the dimensions of vulnerability was used to classify indicators as proxies of either adaptive capacity, sensitivity or exposure. All indicators of adaptive capacity show capabilities available to the household. This may be in terms of social networks, technology or livelihood options. Adaptive capacity indicators were all hypothesized to reduce vulnerability of the household to climate change. Sensitivity indicators describe aspects of the household that make it susceptible to even short-term variability. Such indicators include level to which the household is dependent on season crops and the lack of climate neutral livelihood options. In this study, exposure was constructed in terms of actual change realized within a household to the climate hazard (see Ionescu, Klein, Hinkel, Kumar, & Klein, 2009).

Table 1. Vulnerability indicators used to calculate values of major components for the GVI-IPCC index

GVI-IPCC Vulnerability index								
Major component	**Adaptive capacity**				**Sensitivity**			**Exposure**
Sub-component	Human resource capacity	Technology	Livelihood strategy	Social networks	Non-farm income	Food security	Land characteristics	Natural disaster
Vulnerability indicators	Hire labor (%)	Access to extension (%)	Income index	Social group (%)	No salary income (%)	Dependence on seasonal crops index	Perceive land as infertile (%)	Death of persons and/or livestock (%)
	Educ level of HH head	ICT diversity index	Cash crop income index	Government assistance (%)	No remittance income (%)	Don't store food (%) Months of food shortage	No access to wetland (%) No agroforestry (%)	Temporarily migrated (%)
	Age of HH head	Inorganic fertilizers (%)	Livestock asset index					Diseases to persons and livestock (%)
	HH size	Irrigation (%)	Livestock cash income index					Property destroyed (%)
		Use improved seed (%)	Farm size (acres)					Exposure index
		Weather information (%)						
		Use animal traction (%)						

4. Gender vulnerability index specification

In order to test the assertion on gender and climate change vulnerability, households' data were categorized based on the gender of the household head. Households were grouped as being their single male headed (male headed), single female headed (female headed) or mixed gender households (where household head enjoys shared responsibility with a spouse).

Exposure, sensitivity and adaptive capacity are each constructed from *n* sub-components. The sub-components were first standardized using the min–max normalization method (Nardo et al., 2005).

The standardized values for each sub-component were given by

$$I_g = \frac{x_g - \text{min}_g(x_g)}{\text{max}_g(x_g) - \text{min}_g(x_g)}$$

where x_g, is the value of the indicator sub-component for each gender differentiated household,

and

$\text{min}_g(x_g)$ and $\text{max}_g(x_g)$ are the minimum and maximum values, respectively, for each indicator sub-component determined using combined data for the whole sample.

Hahn et al. (2009) illustrates how the index is calculated.

Diversity of crops, livelihood strategies, livestock assets, crop and livestock cash income and technology options were calculated following Shah, Dulal, Johnson, and Baptiste (2013). The resulting index was subtracted from one such that a household with more crop choices had a higher diversity index.

The standardized sub-components were then averaged using the equation below

$$M_g = \frac{\sum_{i=1}^{n} I_g}{n}$$

where M_g = one of the eight major components for gender g_i [Human capacity (HC), Livelihood diversity (LD), Social Networks (SN), Land/use characteristics (LC), Food security (FS), Technology (T), Non-farm Income (NFI) or Climate change impact (CI)], $Index_{sg^i}$ represents the sub-components, indexed by *i*, that make up each major component, and *n* is the number of sub-components in each component.

The three components were segregated following the Intergovernmental Panel on Climate Change (IPCC) dimensions of vulnerability and each calculated as an index containing several sub-components.

The index for Addptive capacity (*Adpt*), exposure (*Exp*), and sensitivity (*Sen*) were calculated as follows:

The index for adaptive capacity (*Adpt*) is determined by

$$Adpt = \frac{w_{HC}HC_i + w_{LD}LD_i + w_{SN}SN_i + w_T T_i}{w_{HC} + w_{LD} + w_{SN} + w_T}$$

where:

W_{HC},W_{LD}, W_{SN} and W_T is the weight for human capacity, livelihood diversity, social networks and technology calculated from the number of indicators under the sub-components.

The index for exposure (*Exp*) was determined as follows:

$$Exp = \frac{w_{CI}CI_i}{w_{CI}}$$

where:

W_{CI} the weight for climate change impact calculated from the number of indicators under the sub-component.

The index for sensitivity (*Sen*) was determined as follows:

$$Sen = \frac{w_{LC}LC_i + w_{NFI}NFI_i + w_{FC}FS_i}{w_{LC} + w_{NFI} + w_{FS}}$$

W_{LC}, W_{NFI} and W_{FS} is the weight for land characteristics, non-farm income and food security respectively calculated from the number of indicators under the sub-components

All sub-components contribute equally to the overall vulnerability index as they were given equal weights (Nardo et al., 2005).

The three contributing factors are combined to derive the GVI-IPCC index based on the IPCC framework by the IPCC (Parry et al., 2007) that encapsulates vulnerability for each gender differentiated household into a single value. The model specification for GVI-IPCC index is as shown in equation below

$$\text{GVI} - \text{IPCC} = (Exp_{g_i} - Adpt_{g_i})^{*}Sen_{g_i} \tag{1}$$

where GVI-IPCC is the Gender Vulnerability Index for gender differentiated household category g_i expressed using the IPCC framework, Exp_{g_i} is the calculated exposure score (equivalent to the Natural Disaster component), $Adpt_{g_i}$ is the calculated adaptive capacity score (weighted average of the human capacity, livelihood diversity, social networks and technology components) and Sen_{g_i} is the calculated sensitivity score (weighted average of the Food security, Land/use characteristics and Non-farm income sub-components).

An independent sample *t*-test was used to numerical sub-components and the Pearson chi square at 5% level of significance to compare non-numerical sub-components for gender differentiated household categories.

5. Results

5.1. Respondents' socio-economic and demographic characteristics

The respondents were mainly from two ethnic groups i.e. *basoga* in Iganga and Namutumba and *bagishu* in Bududa and Manafwa. Majority (79%) of household heads were male. A sizeable number (23%) of households were female headed, these were mainly where the man was either deceased or continuously absent from the home for long periods of the year. The age of the respondents ranged between 18 and 91 years of age with an average at 48 years old. Most of the household heads had attained education mainly up to secondary level (13 years of school). Average farm size was 2.8 acres, but small parcels were mainly found in the high elevation districts with most of the respondents using all their land for production continuously without fallow.

5.2. Vulnerability indicator scores

There were more differences among household categories across adaptive capacity indicators than those of sensitivity and exposure. Differences in adaptive capacity were more between mixed gender households and female-headed households with 15 out of 18 indicators being significant. Mixed gender households differed from single male-headed households in 4 of 18 indicators of adaptive capacity, while single male and female-headed households only differed significantly in two of the 18 indicators. Differences between mixed gender and female-headed households were found in two of the eight indicators of sensitivity. Single male and female-headed households only differed significantly in one indicator of sensitivity. The GVI-IPCC contributing factors calculations are shown in Table 2 and the individual indicator results are presented below.

5.2.1. Adaptive capacity

There was disparity in many of the indicators of adaptive capacity among the three household categories (Figure 2). Mixed gender households had the largest family size of 7.64 individuals followed by 6.82 in single male-headed households and only 5.57 individuals in single female-headed households. Family size in mixed gender households was significantly larger than both single male (t = 2.92, p = 0.004) and single female-headed households (t = 2.56, p = 0.011). Single female and single male-headed households did not differ with respect to family size. Compared to female-headed households, the heads in mixed gender households had attained 5.6 years of school which was not different from 5.58 years of school attained in single male-headed households but significantly higher (t = 2.099, p = 0.036) than the 4.9 years attained in single female-headed households. The oldest household heads were found in single female-headed households while the youngest were in mixed gender households (44.9). Age of the household head was significantly different between female headed and mixed gender households (t = 2.07, p = 0.039) but not different from single male-headed households (p = 0.895). Mixed gender and single male-headed households did not differ with respect to age of the household head. Thirty-nine percent of mixed gender households could hire labor for agricultural activities. This proportion was significantly higher (t = 8.616, p = 0.003) than 26.6% of single female-headed households but not different from 28% in single male-headed households.

Mixed gender households had a higher index of information and communications technology than both single female-headed households (t = 5.646, p < 0.00) and single male-headed households (t = 0.2.435, p = 0.0.021). Single male and single female-headed households were not significantly different with respect to access to ICTs (t = 0.182, p = 0.856). Almost 75% of mixed gender households had access to weather forecast information which was significantly higher than the 61.5% of single female-headed households (χ^2 = 10.644, df = 1, p = 0.001), but no difference with the single male-headed households (67.9%). While 59.8% of mixed gender households used improved seed, only 46.4% and 45% of male-headed and female-headed households, respectively did the same. Mixed gender households were only different from single female-headed households in the use of improved seed (χ^2 = 11.497, p = 0.001).

More than half of mixed gender households (53.3%) had a member belonging to a local social group which was significantly higher than the 40.2% of single female-headed households (χ^2 = 8.84, p = 0.003) but no difference with single male-headed households. Access to assistance from government in the event of climate change-related disaster was highest among single male-headed households (44%) followed by mixed gender households (41.3%) and single female-headed households (20.3%). Female-headed households performed significantly poorer than the other two household types with respect to access to government assistance for climate related stresses. Government assistance was reportedly in form of dry rations, agricultural seeds and clothing materials for affected families.

Mixed gender households owned 3.08 acres of agricultural land which was not different from 1.68 acres owned by single male-headed households but significantly different from 2.19 acres owned by single female-headed households (t = 3.587, p < 0.05). Single male and female-headed households did not differ from each other in the amount of land owned (t = 1.641, p = 0.108).

The diversity of income sources was similar between mixed gender and male-headed households but higher than for female-headed households (t = 5.316, p < 0.05). Single male and single

Table 2. GVI-IPCC contributing factors calculations for mixed, male and female-headed households in Eastern Uganda

Major component	Subcomponent	Major component values based on household headship			Number of sub-components	Contributing factor values based on household headship			GVI-IPCC index based on household headship		
		Mixed gender	Single male	Single female		Mixed gender	Single male	Single female	Mixed gender	Single male	Single female
Exposure	Natural disaster	0.151	0.199	0.133	5	0.15	0.20	0.13	−0.156	−0.099	−0.121
Adaptive capacity	Social networks	0.473	0.419	0.303	2	0.44	0.38	0.35			
	Livelihood diversity	0.541	0.479	0.446	5						
	Technology	0.409	0.337	0.305	7						
	Human capacity	0.359	0.319	0.317	4						
Sensitivity	Non-farm income	0.790	0.750	0.793	2	0.53	0.54	0.56			
	Food security	0.539	0.550	0.533	4	0.53	0.54	0.56			
	Land characteristics	0.505	0.563	0.556	3						

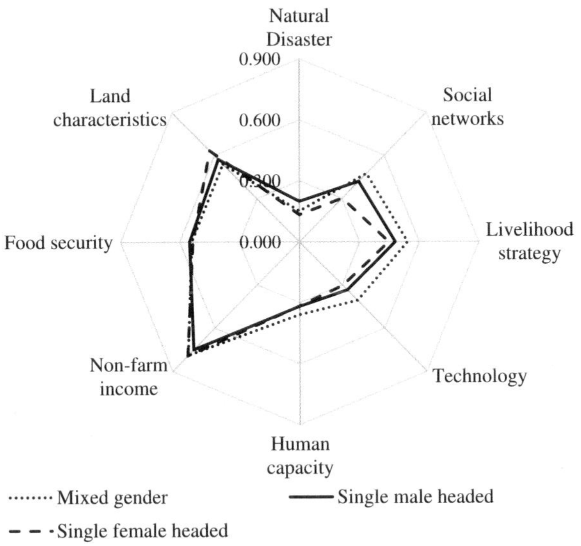

Figure 2. Vulnerability spider diagram of major components for the gender vulnerability index for mixed gender, male and female-headed households in Eastern Uganda.

female-headed households did not differ in diversity of income sources. The diversity of livestock types sold for cash was higher than for mixed gender households than for both single male headed ($t = 2.283$, $p = 0.023$) and single female-headed households ($t = 4.565$, $p < 0.05$). There was no significant difference in the diversity of income, crops sold for cash and livestock between mixed gender and single male-headed households.

5.2.2. Gender differentiated sensitivity and exposure to climate change

There was no significant difference among the different households across most of the sensitivity indicators considered in the analysis. However, a significantly larger proportion of single female-headed households did not practice agroforestry compared to mixed gender households ($\chi^2 = 23.566$, $p = 0.000$).

More households headed by single men reported having temporarily migrated due to climate change-related stress. Ten percent of single male-headed households had sought temporary migration which was significantly higher than 1.5% among mixed gender households ($\chi^2 = 11.891$, $p = 0.001$) and 1.8% among female-headed households ($\chi^2 = 6.5$, $p = 0.011$). Thirty-five percent of single male-headed households reported experiencing diseases to persons and livestock which was significantly higher than 15.2% of mixed gender ($\chi^2 = 5.130$, $p = 0.024$) and 16% of female-headed households ($\chi^2 = 6.135$, $p = 0.013$).

Overall the GVI-IPCC vulnerability index showed that single male-headed households were more vulnerable to climate change than either mixed gender or single female-headed households (Figure 3).

6. Discussion

The heads of households were mainly male due to a patriarchal system followed in the study area. Women were household heads only when the man is away for long periods of the year or when widowed. The unexpected high number of female-headed households highlights an increasing trend where men move away from home seeking for other income opportunities. This probably explains why the percentage of households receiving remittance income was also high. Due to the large proportion of female-headed households, it is important to assess whether these households are disadvantaged on account of the gender of the household head. Many household heads had attained an education possibly due to opportunities for free primary and secondary education provided by the government of Uganda. High membership to social groups is characteristic of rural behavior where social groups are avenues for livelihoods support (Lahiri-Dutt & Samanta, 2006;

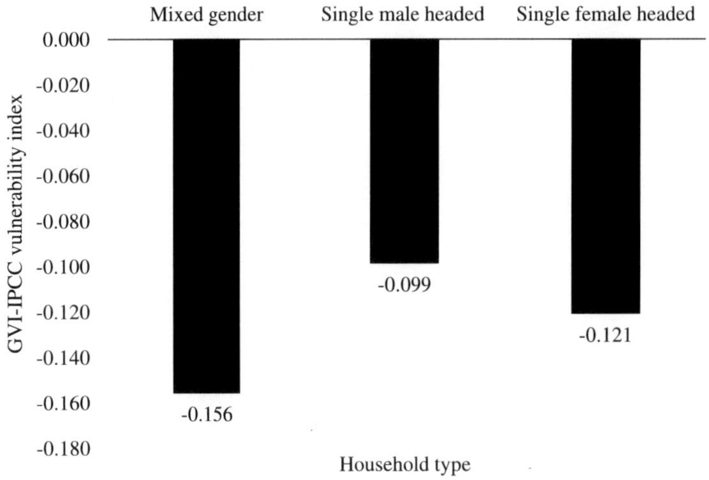

Figure 3. GVI-IPCC Vulnerability index for mixed gender, male and female-headed households in Eastern Uganda.

Narayan & Pritchett, 1999). Socio-economic characteristics corroborate other studies in the area (e.g. Mubangizi, Kyazze, & Mukwaya, 2018) showing high potential for impact on local livelihoods given the high dependence on agriculture. But vulnerability of livelihood is not expected to be uniform across all households due to pre-existing conditions that affect their adaptive capacity (Holand et al., 2011).

The assertion on women vulnerability may be misleading especially if the conclusion is drawn from household surveys using the usual gender dichotomy. Comparing female and male-headed households is in effect comparing single and mixed gender households. This is because most male-headed households will also have a female adult, while female-headed households are comprised mostly by widows who are too old to remarry after the loss of their spouses. By further decomposing household headship, the results contest the a priori determination placed on women vulnerability and show that instead women are positive contributors to reducing vulnerability in their households. It is probably the reason why men without women (in single male-headed households) performed worse than women without men (in single female-headed households). In fact the results agree with Davids and Van Driel (2001) that female household heads may benefit from being without a dominating male. The autonomy and ability to make independent decisions leads to better outcomes for female-headed households and compensate for the lack of an adult male.

The overall GVI-IPCC vulnerability index favored mixed gender households when compared with either single male or female-headed ones probably due to gender synergism within such households. Households with couples benefit from division of responsibilities between adult members. The men undertake off farm income generating activities as the women are entrusted in securing the food security of the household. This would increase the adaptive capacity of such households.

Few female-headed households hired additional labor probably because of cultural constraints hindering females from taping into the predominately male-dominated agricultural labour market (O'Sullivan et al., 2014). Culturally, women find it more comfortable to negotiate and hire fellow women to offer agricultural labor. Because few women work full-time as hired agricultural laborers, women have a smaller pool of hired labour to choose from. This could reduce ability of female-headed households to manage their cropping activities especially when the growing season is shortened by late onset or early cessation of rainfall as was reported.

The number of years of schooling attained by the household head was lowest among female-headed households probably because of gender bias that female children face in attaining education (Kabeer, 2005). Subsequently, these households have limited opportunities to engage in gainful employment or even to make good decisions regarding how to cope with or safeguard their livelihoods against climate change.

The small family size of single male-headed households could be due to the fact that these were mainly elderly men whose adult children had left to live independently. Single male-headed households therefore experience more acute labour shortages. Additionally, the absence of a spouse in the lone male/female-headed households deprived them of the contribution that the partner could have made toward household welfare in terms of labour of earned income.

All indicators of technology show that single female-headed households were particularly disadvantaged. For instance, use of irrigation requires manual power which may not be readily available in a household without a male adult. Use of agricultural inputs like improved cultivars and inorganic fertilizers require readily available cash which may disadvantage female-headed households because women usually concentrate on subsistence crop production (Palacios-Lopez, Christiaensen, & Kilic, 2017) and seldom have the money required to purchase agricultural inputs.

The livelihood diversity indicators showed mixed gender households outperforming both single male and female-headed households. Having two adults enables the household to diversify its livelihood options due to division of tasks (Boserup, 2007; Simiyu & Foeken, 2014). Mixed gender households in our study have an advantage of benefiting from the enterprise of a female partner, which explains why their income diversity index was higher than for lone male/female-headed households. In mixed households the division of tasks between male and female members probably creates labour idling which according to Ellis (2000) forces members to venture into other income generating activities.

The big portfolio of livestock among mixed gender households could be as a result of "gendering livestock" whereby men tend to concentrate on large commercially more valuable livestock (Bravo-Baumann, 2000) while the women maintain the smaller livestock like goats and poultry (Kechero, 2008; Njuki & Sanginga, 2013). Moreover, different kinds of livestock serve different purposes, sale of small livestock easily caters for short-term needs like purchase of inputs, while large livestock are reserved for big shortfalls in household income. This array of possibilities improves the adaptive capacity of the households and reduces their sensitivity to disturbances related to climate change.

Low social networks among female-headed households echoes what Nelson, Meadows, Cannon, Morton, and Martin (2002) reported that even government programs fail to adequately target female-headed households. Women from female-headed households are constrained by lack of time due to many responsibilities of looking after their homes to participate in social groups (Chant, 2004). Mixed gender households can afford to have one of the adults attend social group events while the other attends to the household chores. Issuance of government provided relief tends to be a male dominated activity which would greatly disadvantage those households without a male adult.

The high impact suffered by single male-headed households could be explained by these households having the lowest resource rich individuals. Men who could not get married are the poorest of the poor which explains why they were the more likely to take up temporary migration. In addition, male-headed households which had the smallest family size would find it logistically easier to temporarily migrate in the event of a climate stress.

7. Conclusion

A definite assertion that women and therefore female-headed households are more vulnerable to climate change without empirical proof could fail the process of determining those to be prioritized for adaptation assistance. The results show that the straight assumption that vulnerability is always associated with female-headed households is wrong and misleading. It is wrong because single male-headed households were the most vulnerable to climate change in Eastern Uganda. It is misleading because it fails to query the situation of women in male-headed households and obscures vulnerability often experienced by men in single male-headed households. Moreover, portraying women as the most vulnerable obscures the virtuousness of women and draws a veil over their contribution in mixed gender households. While denying that women are more vulnerable to climate change can potentially deprive them of adaptation assistance, emphasizing their vulnerability blurs over the moderate status of women in male-headed households or their contribution in reducing vulnerability in their households.

The results imply that vulnerability is not sex linked, but rather one of the symptoms of the lack of gender synergism. Gender synergism or the lack of creates big differences even within one gender category. There were differences in vulnerability among both men and women based on their marital status. As such, any vulnerability reducing intervention that would target women would not only disenfranchise a more vulnerable group of single men but also fail to identify the more vulnerable among women.

The main contribution to understanding vulnerability is that whereas the results do not dispel the common notion that women are more vulnerable to climate change, they show that the sex dichotomy of gender is not sufficient to identify the most vulnerable members of society, as there were differences within gender categories. In conclusion, it is imperative to sway the climate change vulnerability narrative away from stereotyping women as the most vulnerable (cf. Dankelman, 2010) and recommend that more research is done to reveal the inter household socio-economic characteristics other than gender which reveal greatest differences in vulnerability to climate change.

Funding
This study was funded by the Regional Capacity Building for Sustainable Resource Management and Agricultural Productivity under Changing Climate (CAPSNAC) project with generous support from the Norwegian Programme for Capacity Development in Higher Education and Research for Development (NORHED).

Author details
Kenneth Balikoowa [1]
E-mail: balikenyy@gmail.com
ORCID ID: http://orcid.org/0000-0001-7358-608X
Gorettie Nabanoga[2]
E-mail: goret69@gmail.com
ORCID ID: http://orcid.org/0000-0002-1243-2703
David. M. Tumusiime[1]
E-mail: tumusid@yahoo.com
[1] School of Forestry, Environment and Geographical Sciences, College of Agricultural and Environmental Sciences, Makerere University, P.O. Box 7062 Kampala, Uganda.
[2] School of Agricultural Sciences, College of Agricultural and Environmental Sciences, Makerere University, P.O. Box 7062 Kampala, Uganda.

References
Adger, W. N., Brooks, N., Bentham, G., Agnew, M., & Eriksen, S. (2004). *New indicators of vulnerability and adaptive capacity* (Vol. 122). Norwich: Tyndall Centre for Climate Change Research Norwich.
Arora-Jonsson, S. (2011). Virtue and vulnerability: Discourses on women, gender and climate change. *Global Environmental Change*, *21*(2), 744–751. doi:10.1016/j.gloenvcha.2011.01.005
BDPU. (2012). *Bududa district statistical abstract*. Bududa: Bududa District Local Government.
Blaikie, P., Cannon, T., Davis, I., & Wisner, B. (1994). *At risk: Natural hazards, people's vulnerability and disasters* London: Routledge.
Boserup, E. (2007). *Woman's role in economic development*. London: Earthscan.
Bravo-Baumann, H. (2000). *Livestock and gender: A winning pair: Capitalization of experiences on the contribution of livestock projects to gender issues: Implications of gender aspects for project design*. Bern, Switzerland: Swiss Agency for Development and Cooperation.
Cannon, T. (2002). Gender and climate hazards in Bangladesh. *Gender & Development*, *10*(2), 45–50. doi:10.1080/13552070215906
Chant, S. (2004). Dangerous equations? How female-headed households became the poorest of the poor: Causes, consequences and cautions. *IDS Bulletin*, *35*(4), 19–26. doi:10.1111/idsb.2004.35.issue-4
Dankelman, I. (2010). *Gender and climate change: An introduction*. London: Routledge.
Davids, T., & Van Driel, F. (2001). Globalisation and gender: Beyond dichotomies. *Globalisation and Development Studies Challenges for the 21st Century*, *1*, 153–175.
Ellis, F. (2000). The determinants of rural livelihood diversification in developing countries. *Journal of Agricultural Economics*, *51*(2), 289–302. doi:10.1111/j.1477-9552.2000.tb01229.x
Hahn, M. B., Riederer, A. M., & Foster, S. O. (2009). The livelihood vulnerability index: A pragmatic approach to assessing risks from climate variability and change—A case study in Mozambique. *Global Environmental Change*, *19*(1), 74–88. doi:10.1016/j.gloenvcha.2008.11.002
Hinkel, J. (2011). "Indicators of vulnerability and adaptive capacity": Towards a clarification of the science–Policy interface. *Global Environmental Change*, *21*(1), 198–208. doi:10.1016/j.gloenvcha.2010.08.002
Holand, I. S. (2014). Lifeline issue in social vulnerability indexing: A review of indicators and discussion of indicator application. *Natural Hazards Reviews*, *16*(3), 04014026. doi:10.1061/(ASCE)NH.1527-6996.0000148

Holand, I. S., Lujala, P., & Rød, J. K. (2011). Social vulnerability assessment for Norway: A quantitative approach. *Norsk Geografisk Tidsskrift-Norwegian Journal of Geography, 65*(1), 1–17. doi:10.1080/00291951.2010.550167

Ionescu, C., Klein, R. J., Hinkel, J., Kumar, K. K., & Klein, R. (2009). Towards a formal framework of vulnerability to climate change. *Environmental Modeling & Assessment, 14*(1), 1–16. doi:10.1007/s10666-008-9179-x

Johnsson-Latham, G. (2007). *A study on gender equality as a prerequisite for sustainable development.* Retrieved from http://www.uft.oekologie.uni-bremen.de/hartmutkoehler_fuer_studierende/MEC/09-MEC-reading/gender%202007%20EAC%20rapport_engelska.pdf

Kabeer, N. (2005). Gender equality and women's empowerment: A critical analysis of the third millennium development goal 1. *Gender & Development, 13*(1), 13–24. doi:10.1080/13552070512331332273

Kechero, Y. (2008). Gender responsibility in smallholder mixed crop-Livestock production systems of Jimma zone, South West Ethiopia. *Livestock Research for Rural Development, 20*(11), 12.

Kitutu, M., Poesen, A. M. J., & Deckers, J. (2011). Farmer's perception on landslide occurrences in Bududa District, Eastern Uganda. *African Journal of Agricultural Research, 6*(1), 7–18.

Kousky, C., & Cooke, R. M. (2009). Climate change and risk management: Challenges for insurance, adaptation, and loss estimation. *SSRN Electronic Journal.* doi:10.2139/ssrn.1346387

Lahiri-Dutt, K., & Samanta, G. (2006). Constructing social capital: Self-help groups and rural womens development in India. *Geographical Research, 44*, 285–295. doi:10.1111/ages.2006.44.issue-3

Lewis, J. (1999). *Development in disaster-prone places: Studies of vulnerability.* London: ITDG Publishing.

Mbogga, M. (2012). *Climate profiles and climate change vulnerability assessment for the Mbale Region of Uganda.* Retrieved from http://www.adaptation-undp.org/resources/assessments-and-background-documents/tacc-uganda-%E2%80%93-climate-profiles-and-climate-change

Mubangizi, N., Kyazze, F. B., & Mukwaya, P. I. (2018). Smallholder Farmers' perception and adaptation to rainfall variability in Mt. Elgon Region, Eastern Uganda. *International Journal of Agricultural Extension, 5*(3), 103–117.

Narayan, D., & Pritchett, L. (1999). Cents and sociability: Household income and social capital in rural Tanzania. *Economic Development and Cultural Change, 47*(4), 871–897. doi:10.1086/452436

Nardo, M., Saisana, M., Saltelli, A., Tarantola, S., Hoffman, A., & Giovannini, E. (2005). *Handbook on constructing composite indicators.* Paris: OECD Publishing.

Nelson, V., Meadows, K., Cannon, T., Morton, J., & Martin, A. (2002). Uncertain predictions, invisible impacts, and the need to mainstream gender in climate change adaptations. *Gender & Development, 10*(2), 51–59. doi:10.2307/4030574

Njuki, J., & Sanginga, P. C. (2013). *Women, livestock ownership and markets: Bridging the gender gap in eastern and southern Africa.* New York: Routledge.

O'Brien, K., Eriksen, S. E., Schjolden, A., & Nygaard, L. P. (2004). *What's in a word? Conflicting interpretations of vulnerability in climate change research.* Retrieved from

O'Sullivan, M., Rao, A., Banerjee, R., Gulati, K., & Vinez, M. (2014). *Levelling the field: Improving opportunities for women farmers in Africa.* Washington DC.

Palacios-Lopez, A., Christiaensen, L., & Kilic, T. (2017). How much of the labor in African agriculture is provided by women? *Food Policy, 67*, 52–63. doi:10.1016/j.foodpol.2016.09.017

Parry, M. L., Canziani, O., Palutikof, J., Van der Linden, P., & Hanson, C. (2007). *Climate change 2007-impacts, adaptation and vulnerability: Working group II contribution to the fourth assessment report of the IPCC.* Cambridge, UK: Cambridge University Press.

Pielke, R., Prins, G., Rayner, S., & Sarewitz, D. (2007). Climate change 2007: Lifting the taboo on adaptation. *Nature, 445*(7128), 597–598. doi:10.1038/445597a

Rahman, M. S. (2013). Climate change, disaster and gender vulnerability: A study on two divisions of Bangladesh. *American Journal of Human Ecology, 2* (2), 72–82. doi:10.11634/216796221302315

Ribot, J. (2010). Vulnerability does not fall from the sky: Toward multiscale, pro-poor climate policy. In R. Mearns & A. Norton (Eds.), *Social dimensions of climate change: Equity and vulnerability in a warming world* (Vol. 2, pp. 47–74). Washington, DC. The World Bank

Shah, K. U., Dulal, H. B., Johnson, C., & Baptiste, A. (2013). Understanding livelihood vulnerability to climate change: Applying the livelihood vulnerability index in Trinidad and Tobago. *Geoforum, 47*, 125–137. doi:10.1016/j.geoforum.2013.04.004

Sharmin, Z., & Islam, M. (2013). Consequences of climate change and gender vulnerability: Bangladesh perspective. *SSRN Electronic Journal.* doi:10.2139/ssrn.2200116

Simiyu, R., & Foeken, D. (2014). Gendered divisions of labour in urban crop cultivation in a Kenyan town: Implications for livelihood outcomes. *Gender, Place & Culture, 21*(6), 768–784. doi:10.1080/0966369X.2013.810602

Staudt, M., Kuosmanen, E., Babirye, P., & Lugaizi, I. (2014). *The Bududa landslide of 1 March 2010. Geological Survey of Finland (Special Paper 56), 373-384*

UBOS. (2010, December). *Uganda census of agriculture 2008/2009. Volume III: Agricultural household and holding characteristics report.* Kampala, Uganda: Uganda Bureau of Statistics.

Vincent, K., & Cull, T. (2010). *A Household Social Vulnerability Index (HSVI) for evaluating adaptation projects in developing countries.* Paper presented at the PEGNet Conference 2010: Policies to foster and sustain equitable development in times of crises: Midrand, 2-3rd September 2010, Midrand, South Africa.

Vlaeminck, P., Maertens, M., Isabirye, M., Vanderhoydonks, F., Poesen, J., Deckers, S., & Vranken, L. (2016). Coping with landslide risk through preventive resettlement. Designing optimal strategies through choice experiments for the Mount Elgon region, Uganda. *Land Use Policy, 51*, 301–311. doi:10.1016/j.landusepol.2015.11.023

Waite, L. (2000). How is household vulnerability gendered? Female-headed households in the collectives of Suleimaniyah, Iraqi Kurdistan. *Disasters, 24*(2), 153–172.

Wiréhn, L., Danielsson, Å., & Neset, T.-S.-S. (2015). Assessment of composite index methods for agricultural vulnerability to climate change. *Journal of Environmental Management, 156*, 70–80. doi:10.1016/j.jenvman.2015.03.020

5

Pollution status of shooting range soils from Cd, Cu, Mn, Ni and Zn found in ammunition

Pogisego Dinake[1*], Onneetse Maphane[1], Karabo Sebogisi[2] and Obed Kamwi[3]

*Corresponding author: Pogisego Dinake, Department of Chemical and Forensic Sciences, Botswana International University of Science and Technology, Biust, Plot 10071, Boseja-Khurumela Private Bag 16, Palapye, Botswana
E-mail: dinakep@biust.ac.bw

Reviewing editor: Keng Yuen Foo, River Engineering and Urban Drainage Research Centre, Universiti Sains Malaysia - Kampus Kejuruteraan Seri Ampangan, Malaysia

Abstract: Evaluation of concentration of heavy metals found in shooting range soil is important in assessing the pollution risk posed to biota. Soil samples from five shooting ranges found in Botswana were used in this study. All the five shooting ranges accumulated high concentration of Cu ranging from 67.4 ± 0.05 mg/kg to 1569 ± 13 mg/kg followed by Mn (25.9 ± 0.1–953.8 ± 2.8 mg/kg). Pollution risk indices were used to quantify the environmental pollution risk posed by the different heavy metals studied. It was established that even though all the five shooting ranges recorded low concentrations of Cd, this metal still posed the highest pollution risk than any other metal with S/P shooting range recording the highest potential ecological risk index *(peri)* of 8141 (C_{Cd} ~ 3.6 ± 0.03 mg/kg) and TAB at *PERI* of 3507 (C_{Cd} ~ 4.9 ± 0.02 mg/kg). Similarly, contamination factor *(CF)* value of 271 for Cd was measured at S/P shooting range indicating high contamination from Cd. Pollution risk indices were able to establish that even though concentration of Cd accumulated in the soil was low this heavy metal still posed highest pollution risk to biota. Continuous assessment of the pollution status of these shooting ranges should be carried out in order to establish appropriate best shooting range management practices and remedial strategies.

ABOUT THE AUTHOR

Pogisego Dinake works as a lecturer in the department of Chemical and Forensic Sciences, Botswana International University of Science & Technology (BIUST), Botswana. He obtained his PhD in Chemistry in 2012 at Howard University in the United States of America (USA). His research interests are in Environmental Analytical Chemistry. He investigates environmental pollution of air, soil, water and living organisms from trace metals and organic pollutants. He also studies; chemical speciation of heavy metals in water, soil and biological systems and chemical remediation of toxic trace metal contaminated water and soil. He also applies supramolecular chemistry in the design, synthesis and application of macro ligands for environmental pollution monitoring such as molecular sieves for waste water purification. He has collaboration with researchers in Botswana, South Africa and USA. He has published in international journals. Recently he has established his own research laboratory and working on various environmental pollution projects.

PUBLIC INTEREST STATEMENT

This study is the first in Botswana in assessing pollution in shooting range soils from ammunition. Shooting ranges are a source of environmental pollution since toxic heavy metals are major constituents of bullets. Concentrations of toxic metals in the five military shooting ranges that have been in operation for over 30 years exceeded the world health organization (WHO) set limits. These toxic metals can be absorbed by plants endangering humans and animals that consume them. Adverse health effects such as respiratory and kidney problems can result from exposure to these hazardous metals. Most of the shooting ranges studied are not fenced making them accessible as grazing fields. Surface and underground water pollution is possible resulting in these toxic metals finding their way into the food chain. Therefore remedial measures such as cleaning of contaminated soils and fencing of shooting ranges should be carried out as soon as possible.

Subjects: Environment & agriculture; Earth sciences; Environmental studies & management; Physical sciences

Keywords: ammunition; shooting range; pollution risk index; environmental pollution; Botswana

1. Introduction

Heavy metals such as arsenic (As), antimony (Sb), cadmium (Cd) and zinc (Zn) present multi-element detrimental effects to biota (Seshadri et al., 2017). Shooting ranges have been found to be among the main sources of heavy metal pollution of the environment. Shooting ranges are the main source of environmental lead (Pb) pollution after the battery industry (Sehube et al., 2017) and most if not all heavy metal pollution of shooting range soil has focused on Pb (Mariussen, Johnsen, & Strømseng, 2017). This is due, in part, to the fact that ammunition contains an excess of 80% Pb slug and small traces of As, Sb, Cd, Cu, Ni, Mn and Zn whose environmental impact is rarely studied (Hockmann et al., 2015). Ammunition shots and bullets have been found to contain 95–97% metallic Pb by weight, 0.4–2.0% Sb, 0.2–0.8% As and mean concentrations of Sn, Se, Mn, Cd, Cr, Cu and Ni greater than 30 mg/kg (Fayiga & Saha, 2016). Pb is mainly found in the core of bullets and shots while Cu, Zn and Sb are constituents of bullet ore heads and cartridges (Mariussen et al., 2017; Urrutia-Goyes, Argyraki, & Ornelas-Soto, 2017).

Copper, zinc and nickel are essential elements in plants. Copper is a constituent of enzymes and plays a significant role in photosynthesis (Lange et al., 2017). Zinc Nickel is a component of the enzyme urease which plays an important function in seed germination (de Queiroz Barcelo et al., 2017). However, increased concentrations of these biologically important trace elements can have detrimental effects to biota. Accumulation of Zn in plants inhibits the uptake of iron and copper leading to pale yellow interveinal chlorosis on the leaves (Broadley, White, Hammond, Zelko, & Lux, 2007). Exposure to antimony causes respiratory problems such as chronic bronchitis, chronic emphysema, inactive tuberculosis, pleural adhesions and respiratory irritation characterized by chronic coughing, wheezing and upper airway inflammation (Sundar & Chakravarty, 2010). The adverse health effects associated with exposure to As include replacing other elements in plant enzymes leading to plant stunted growth and eventual death (Nagajyoti, Lee, & Sreekanth, 2010). Adverse health effects associated with exposure to arsenic (As) also include cardiovascular, diabetes and cancer as usually reaches the blood stream through ingestion, inhalation and skin contact(Alexander, Gulledge, & Han, 2016). Cardiovascular effects due to exposure to Sb in humans include increased blood pressure and alteration of electrocardiography (ECG). Abdominal pain, diarrhoea, vomiting, and ulcers are some of the gastrointestinal effects associated with exposure to Sb (Okkenhaug et al., 2013). High level of Cd causes decrease in seed germination, stunted growth in plants, disrupt enzyme activities and makes plants prone to fungal invasion (Laghlimi, Baghdad, El Hadi, & Bouabdli, 2015). Its compounds relatively have high solubility in the soil making Cd more mobile and bioavailable (Sharma & Sachdeva, 2015).

Copper (Cu) is an essential element to plants, however, high levels of copper can lead to disruption of plant photosynthesis and reproduction processes (Laghlimi et al., 2015). High concentrations of Cu have adverse health effects to the gastrointestinal tract and the liver. Pulmonary fibrosis and increased vascularity have also been reported in people exposed to high levels of copper (Laghlimi et al., 2015). Low concentration of Cu has been reported to cause brain damage in animals (Department of Water Affairs and Forestry, 1996). Nickel is a known epigenetic carcinogen and can cause genetic modifications (Brocato & Costa, 2017). Physicochemical processes in the soil such as physical weathering, oxidation, sulfonation, dissolution and precipitation transform these heavy metals into the mobile and bioavailable forms that are detrimental to the biota (Kelebemang et al., 2017). Antimony is the second largest shooting range pollutant after Pb comprising 7% of bullet shot and bullet (Rooney, McLaren, & Cresswell, 1999). Sb concentration of 13,800 mg/kg has been found in a Swiss shooting range (Johnson, Moench, Wersin, Kugler, &

Wenger, 2005). Elevated concentrations of 5200 mg/kg, 1100 mg/kg and 830 mg/kg corresponding to Cu, Zn and Sb have been established at a Norwegian shooting range (Mariussen et al., 2017). These high concentrations of heavy metals were able to find their way into the water sources with concentrations of Zn, Cu and Sb in surface water measuring 1.6, 0.9 and 0.15 mg/l respectively (Mariussen et al., 2017). Other such high pollution from heavy metals such as Cu, Cd, Zn, Sb, As and Ni other than Pb emanating from ammunition have been found in Canada (Lafond, Blais, Martel, & Mercier, 2013), Spain (Rodríguez-Seijo, Alfaya, Andrade, & Vega, 2016), Switzerland (Evangelou, Hockmann, Pokharel, Jakob, & Schulin, 2012), South Korea (Lee et al., 2002) and the United States of America (Sehube et al., 2017). Most of these shooting range soils accumulated heavy metals concentrations far exceeding the USEPA critical limits. Heavy metal contamination of shooting range soils also poses risk of surface and underground water pollution (Wersin, Johnson, & Furrer, 2002). Concentrations of Sb as high as 1.59 mg/L have been reported in water found in close proximity to a shooting range (Okkenhaug et al., 2013). The accumulation of heavy metals into the soil which in turn may bind to the soil organic matter and eventual dissolution thereof has the potential to negatively affect the soil carbon sequestration process exacerbating the climate change phenomena (Jastrow, Amonette, & Bailey, 2007)

Heavy metals uptake by plants has also been reported leading to heavy metals finding their way into the food chain (Mozafar et al., 2002; Robinson et al., 2008). Elevated heavy metal concentrations of 25 mg/kg and 50 mg/kg for Cu and Ni respectively, have been found in plant species, which are a source of food for livestock (Robinson et al., 2008). Nine plant species in two decommissioned shooting ranges have been evaluated for their Cu, Zn, Sb and Cd concentrations and the potential health risk to animals made to graze in such shooting ranges (Evangelou et al., 2012). At a clay pigeon shooting range in Tuscany (central Italy) high concentrations of Sb, Ni, Zn, Mn and Cu were found in arthropod communities inhabiting shooting range soils (Migliorinia, Pigino, Bianchi, Bernini, & Leonzio, 2004). A reduction in soil enzymatic activity has also been reported in soils contaminated with heavy metals such as cadmium can have a negative effect in the soil chemistry due to its harmful effect to the microorganisms found in the soil which are important to soil fertility (Vig, Megharaj, Sethunathan, & Naidu, 2003). The mobility, bioavailability and bioaccessibility of heavy metals in shooting range soils is dependent on the physicochemical properties of the soil such as soil pH, cation exchange capacity (CEC), soil organic matter and moisture (Kelebemang et al., 2017). Under favourable conditions of soil properties the chemical weathering, solubility, mobility and bioavailability of heavy metals in shooting range soils can be enhanced (Ma, Hardison, Harris, Cao, & Zhou, 2007).

To the best of our knowledge no research has been carried out on heavy metal pollution of shooting range soils in Botswana except Pb, a study that was carried out by our research group. Botswana has shooting ranges all over the country that are used for shooting practices by the military, police services, prisons services and department of wild life national parks. This study follows our pioneering study that looked into pollution of shooting range soils from Pb which is the main constituent of ammunition (Sehube et al., 2017). Our pioneering study found elevated Pb pollution in shooting range soils with highs of up to 38 406.87 mg/kg far exceeding the set USEPA critical limit of 400 mg/kg Pb in soil (Sehube et al., 2017). Evaluation of the extent of shooting range soil pollution from heavy metals can help strengthen best management practices of shooting ranges. Soil remedial processes can also be employed for soil cleaning and reclamation. Based on the extent of heavy metal pollution, soil remediation methods and techniques such as chemical immobilization, physical soil excavation, phytoremediation and chemical washing techniques have been utilized in fighting shooting range soil heavy metal pollution (Sorvari, Antikainen, & Pyy, 2006). The objectives of this study include (1) quantitative evaluation of heavy metal pollution of shooting range soils found in Botswana (2) environmental pollution risk assessment of such pollution based on such methods as potential ecological risk index (*PERI*), heavy metal pollution load index (*RI*) and contamination factor (*CF*). The findings of this study will provide a baseline data for heavy metal pollution of shooting range soils in Botswana and will also provide a basic measure on the extent of heavy metal pollution emanating from similar ammunition used elsewhere around in world. Appropriate remedial strategies and methods may also be suggested based on the extent of pollution encountered in the studied shooting ranges and shooting ranges using same type of ammunition around the world. In addition best shooting range management practices for future shooting range users will be suggested.

2. Materials and methods

2.1. Location and description of shooting ranges

Five military shooting ranges scattered across Botswana were used in this study. These shooting ranges are mostly used for small arms shooting practices using rifles of calibres 0.50", 5.56 mm, 7.62 mm, 7.65 mm and 9 mm. It is important to note that small arms ammunition contain predominantly the heavy metals under study. The five shooting ranges were Thebephatshwa (TAB) located in the southern part of the country (GPS coordinates: 24° 14′ 42.79" South, 25° 19′ 55.84" East), Shoshong (SHO) in central Botswana (GPS coordinates: 23° 2′ 8.28" South, 26° 30′ 7.12" East), Selibe Phikwe (S/P) in eastern Botswana (GPS Coordinates:27° 50′ 14.39" East, 21° 57′ 58.55" South), Matsiloje, (MAT) with GPS coordinates: 21° 22′ 12.00" South, 27° 52′ 12.00" East in north eastern Botswana and lastly Pandamatenga (PANDA) located in northern Botswana (GPS coordinates 18° 31′ 57.01" South, 25° 39′ 14.22" East). The five shooting ranges and their year of establishment are shown in Figure 1(a) below. It is important to note that all these shooting ranges have been in operation for the past 16–32 years and that there has never been a single study carried out on possible pollution risk from heavy metal accumulation in these shooting ranges. Two of the shooting ranges studied are located in fenced military camps while the other three occur in open areas acting as grazing fields for livestock and small stock.

2.2. Soil sampling procedure and techniques

Berm soil samples were collected to a depth of 20 cm using a soil recovery probe in all the five the shooting ranges. This soil represents the surface layer (0–5 cm), the subsurface layer (5–10 cm) and the deeper soil layer (10–20 cm). The berm accumulates the highest proportion of bullets and shots since it acts as a barrier for bullets and shots after they have hit the target. The berm section of the shooting range was divided into three segments being; berm-upper, berm-middle and berm-bottom (Figure 2). The berm was divided into three sections because some bullets and shots can lend at the bottom, or middle or upper section of the berm. Three circular sampling areas of radius 150 cm were made at the upper, middle and bottom of the berm along a central transect. In addition, four samples were collected from the three sampling points within the sampling area to make three composite samples representative of the upper, middle and bottom berm soils (Figure 2). This sampling technique was preferable because it covers a large area making the sample a true representation of the study site. Control soil samples which served as background samples were collected at a distance of 200 m away from the five shooting ranges. The collected samples were

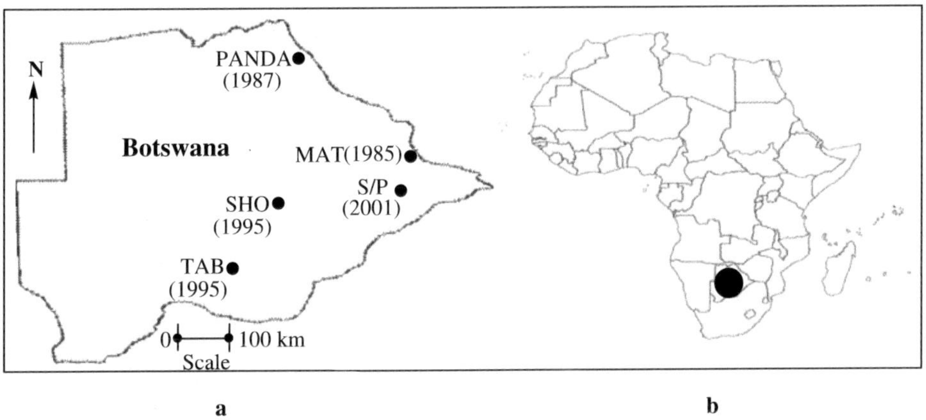

Figure 1. (a) Location of shoot-ing ranges sampled on the map of Botswana. (b) Map of Africa showing location of Botswana.

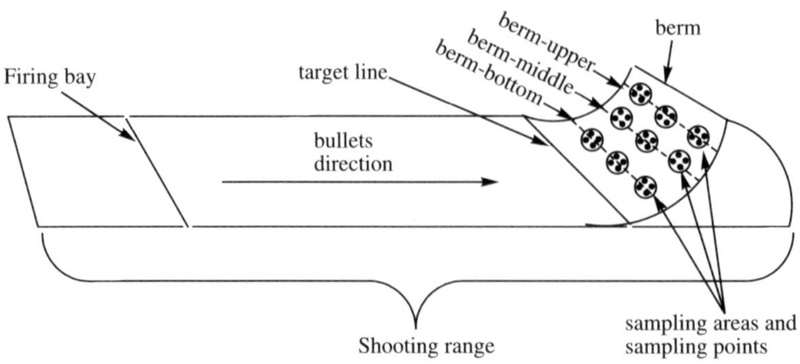

Figure 2. Schematic of sampling points in the berm of the shooting range.

then stored in butyrated zip-lock plastic bags for transportation to Botswana University of Agriculture and Natural Resources (BUAN) for further sample treatment and analysis.

2.3. Soil sample preparation and chemical analysis

The three quadruplicate samples from sampling points in three sampling areas of berm-upper, berm-middle and berm-bottom were mixed together to obtain three composite samples corresponding to each berm section. A total of nine berm soil samples were obtained from each range making a grand total of forty-five soil samples. The soil samples were air dried for a day after which a 2 mm stainless steel sieve was used to remove spent bullets, stones and rock pebbles, dry tree leaves and small tree barks. Soil sample digestion was achieved using the dry heating-block digestion procedure (USEPA Method 3050) by taking a 0.500 g of soil samples in 100 mL of 1:3 HNO_3/HCl mixture for two hours. After sample digestion total Cd, Cu, Ni, As, Sb and Zn concentrations were analysed using inductively coupled plasma optical emission spectrometry (ICP-OES; Perkin Elmer, Optima 2100 DV). Quality check and control was done using certified reference material (CRM) of soil, NCS DC 73,320, obtained from the Botswana Geoscience Institute (BGI).

2.4. Physicochemical properties of shooting range soils

The physicochemical properties of the soil such as pH, electrical conductivity (EC), organic matter (OM) and cation exchange capacity (CEC) were determined using standard methods. A mixture of 0.01 M $CaCl_2$ solution and 40 mL of deionized water were added to 20 g of soil sample and pH measurements were taken using a calibrated crimson basic 20 pH meter at 25°C. The Walkley-Black procedure was used for the determination of the organic matter (OM) content as reported in the literature (Nelson & Sommers, 1982). The soil electrical conductivity (EC) was carried out in the ratio of 1:2 soil to water using the HACH SenseIon EC7 electrical conductivity meter (Tchoumou, Kami-Ouako, Mbilou, & Ossebi, 2017).

2.5. Environmental risk assessment of pollution from heavy metals

2.5.1. Potential ecological risk index

Potential ecological risk index (*PERI*) method was first used by Hakanson in 1980 (Hakanson, 1980). This method provides a powerful tool for estimating environmental pollution by combining the ecological and environmental impacts with toxicology (Ke et al., 2017). The potential ecological risk index of a single heavy metal (E^i_R) is given by Equation (1) below while Equation (2) evaluates the ecological risk (*RI*) emanating from contributions by all the toxic heavy metals present in the soil. The T^i_R signifies the biological toxic response factor of an individual metal while C_i and C_0 are concentrations of individual heavy metal and its background concentration respectively. The T^i_R values of the heavy metals under study are Cd = 30, Cu = Ni = 5 and Mn = Zn = 1 (Ke et al., 2017; Soliman, Nasr, & Okbah, 2015). Table 1 below categorises the potential ecological risk indices with corresponding pollution risk level.

$$E^Q_R = T^i_R \times C_i/C_o \tag{1}$$

Table 1. Categories of *PERI* and *RI* indicating the pollution risk level of heavy metals to the soil (Ke et al., 2017)

Pollution indicator	Pollution risk level				
	Low	**Moderate**	**Considerable**	**High**	**Very high**
E^{i}_{R}	<40	$40 < E_r \leq 80$	$80 < E_r \leq 160$	$160 < E_r \leq 320$	$E_r \geq 320$
RI	<150	$150 < RI \leq 300$	$300 < RI \leq 600$	$RI \geq 600$	

Table 2. Classification of contamination factor and corresponding level of soil contaminate on from heavy metal accumulation (Ke et al., 2017)

Contamination Factor *(CF)*	Contamination level
$CF < 1$	Low contamination
$1 \leq CF < 3$	Moderate contamination
$3 \leq CF \leq 6$	Considerable contamination
$CF > 6$	Very high contamination

$$RI = \sum_{i-1}^{m} E_R^Q \tag{2}$$

2.5.2. Contamination factor

Assessment of pollution level can also be carried out using other pollution indicators such as the contamination factor (*CF*), given in Equation (3) below, where C_i and C_b are the mean concentrations of the heavy metal under investigation and its background concentration respectively. This represents the contamination level of an individual target heavy metal. Contamination factor, *CF*, correlates the background concentration of heavy metals in the soil with contributions from anthropogenic activities. The classification for contamination factor values and the corresponding degree of pollution are shown on Table 2 below.

$$CF = C_i/C_b \tag{3}$$

The resultant overall integrated contamination given by the different target metals in the soil is evaluated using the pollution load index (PLI), given by Equation (4) below. It is determined by calculating the *n*th root of the *n* contamination factors (CF_n) multiplied together (Cao et al., 2015)

$$\text{PLI} = (CF_1 \times CF_2 \times CF_3 \times \ldots CF_n)^{\frac{1}{n}} \tag{4}$$

The PLI > 1 denotes polluted soil while PLI < 1 indicates no heavy metal pollution (Cao et al., 2015)

2.6. Statistical analysis of data

A statistical analysis technique such as the Pearson correlation coefficient (*r*) was used to evaluate the concomitant contributions towards heavy metal pollution from the soil physical and chemical properties as shown on Table 3 below. Pearson correlation coefficient can also help ascertain if the heavy metal pollutants under investigation have common anthropogenic source.

3. Results and discussion

3.1. Physicochemical properties of shooting range soils

The soils in the all the five shooting ranges studied experienced neutral to slightly alkaline pH. Shooting ranges such as S/P and SHO experienced neutral pH values of 6.7 and 6.9 respectively

Table 3. Pearson correlation coefficient range indicating the possible correlation between the properties of the soil and degree of pollution (Pejman, Bidhendi, Ardestani, Saeedi, & Baghvand, 2017)

Pearson correlation coefficient range	Correlation effect
0.5 to 1.0 or −0.5 to −1.0*	High correlation
0.3 to 0.5 or −0.3 to −0.5	Medium correlation
0.1 to 0.3 or −0.1 to −0.3	Low correlation

*A high negative Pearson correlation coefficient signifies direct negative correlation between two variables.

(Table 4). On the other hand, PANDA, MAT and TAB soils were slightly alkaline recording pH values of 8.3, 8.5 and 8.6 respectively. The soil pH values found in the five shooting ranges are consistent with the soil pH for the respective areas in which the shooting ranges are found (Ekosse & Anyangwe, 2012; Joshua, 1991). For example, in PANDA, the soil type is mainly dark clay vertisols with a pH range of 6.5 to 9.0 (Almendros et al. 2003; Moganane, Van Waveren, & Remmelzwaal, 1991). The alkaline soil pH of 8.3 found at the MAT shooting range is consistent with the alkaline soil pH of the soils found in the Matsiloje area which can reach pH values of 9.0 (Mitchell, 1976). The organic matter (OM) found in the five shooting ranges was in the range 0.54–1.88% with S/P soils experiencing the highest OM relative to the other four shooting ranges. The soil organic matter found at the five shooting ranges studied is consistent with average organic matter content of (0.5–1.9%) for the soils found in the study sites (Almendros et al. 2003; Joshua, 1991; Pardo, Ristori, D'acqui, & Almendros, 2003). For example, at the PANDA shooting range, organic matter of 0.54% was measured and previous studies found similar organic matter content in the Pandamatenga area in which the PANDA shooting range is located (Moganane et al., 1991). The soils electrical conductivity (EC) were in the range 33.9 ± 3.3 to 208.9 ± 8.8 μS/cm and the highest electrical conductivity was recorded at S/P shooting range soil (Table 4). The organic matter and pH measurements at the five shooting ranges are within range of our preliminary studies on Pb pollution in the same shooting ranges (Kelebemang et al., 2017; Sehube et al., 2017).

3.2. Total heavy metal concentration in shooting range berm soils

The berm soils of all the five shooting ranges accumulated highest concentrations of Cu in the range 67.4 ± 0.05 mg/kg to 1569 ± 13 mg/kg followed by Mn (25.9 ± 0.1-953.8 ± 2.8 mg/kg) as shown on Figure 3(a–c) below. On average, the berm soil of TAB shooting range collected the highest concentration of Cu (1190.7 mg/kg) compared to the other shooting ranges. On the other hand, all the shooting ranges investigated experienced lower concentrations of Cd of the range 0.2 ± 0.08–5.0 ± 0.06.4 mg/kg. The lowest concentration of Cd (0.2 ± 0.08 mg/kg) was recorded at MAT shooting range. The degree of pollution at a shooting range correlates to a larger extent with the frequency of shooting range use than the duration period the shooting range has been in use. TAB shooting range, even though started operation in 1995, accumulated the highest concentra-

Table 4. Physical and chemical properties of shooting range soils

Shooting Range	$pH(H_2O) \pm \delta^a$	$OM^b \pm \delta(\%)$	$EC^c \pm \delta$ (μS/cm)
TAB	8.6 ± 0.2^d	1.26 ± 0.13	170.8 ± 8.4
SHO	6.9 ± 0.2	0.73 ± 0.06	92.5 ± 8.0
S/P	6.7 ± 0.03	1.88 ± 0.42	208.9 ± 8.8
MAT	8.5 ± 0.1	0.73 ± 0.05	52.0 ± 23.8
PANDA	8.3 ± 0.1	0.54 ± 0.01	33.9 ± 3.3

[a]δ = Standard deviation.
[b]Organic matter.
[c]Electrical conductivity.
[d]Mean ± standard deviation (*n* = 3).

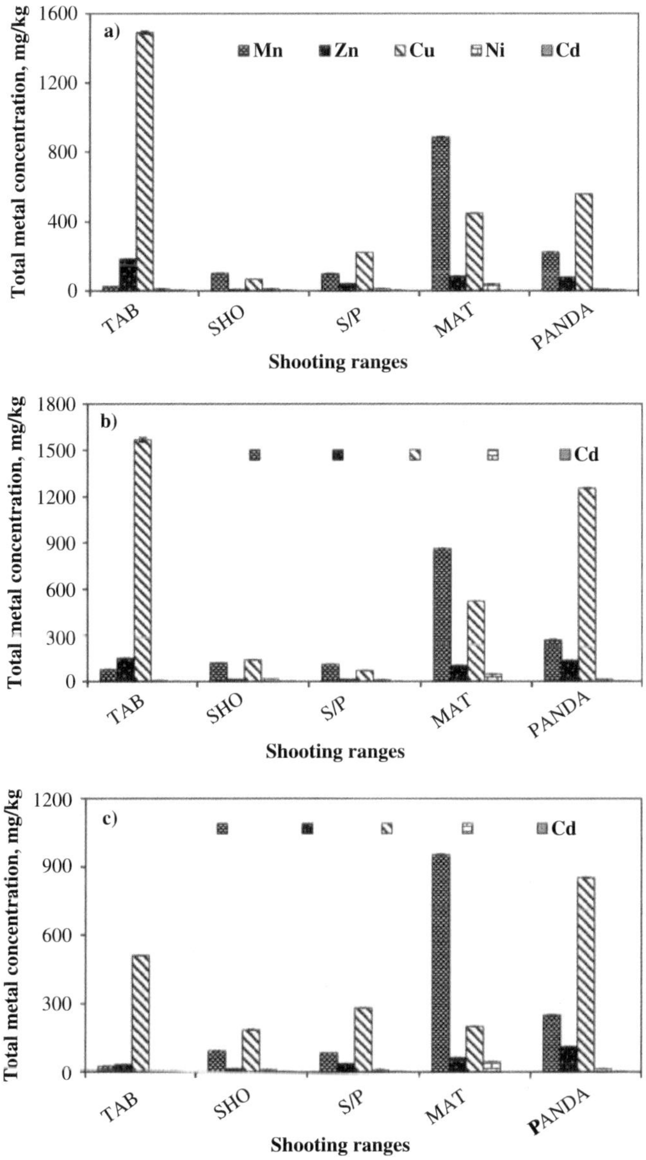

Figure 3. Total concentration of five heavy metals in shooting range berm soils found in Botswana. (a) Upper-berm; (b) Middle-berm and (c) Lower-berm. Mean of $n = 3$; Standard error of the mean, $\delta x = \delta/\sqrt{n}$, where δ = standard deviation.

tion of Cu compared to the other shooting ranges studied that were established some years earlier such as MAT (1985) and PANDA (1987) with measured concentrations of Cu of 390.8 ± 0.6 mg/kg and 888.5 ± 2.3 mg/kg respectively. As shown in our earlier studies (Sehube et al., 2017), TAB shooting range accumulated the highest concentration of Pb (38 386 ± 10 197 mg/kg) due to its frequent use by the Special Forces commando squadron based at the Thebephatshwa air base. The results from our previous studies correlate well with our current study. The middle of the berm accumulated the highest concentration of heavy metals compared to the upper and bottom sections of the impact berm. For an example, TAB shooting range experienced concentrations of 1569 ± 13, 1490.7 ± 9.1 and 512.3 ± 0.5 mg/kg of Cu at the middle, upper and lower part of the impact berm respectively. One of the reasons for this observation is that when shooting at the targets, most of bullets and shots accumulate in the middle of the impact berm making this section of the berm have the highest density of bullets and shots resulting in higher concentrations of heavy metals around the same area. MAT shooting range acquired the highest average concentration of all the five studied heavy metals. Previous studies have discovered elevated concentrations of heavy metals in berm soils of shooting ranges. A study by Islam, Nguyen, Jung, and Park (2016) found elevated concentrations of Cu at four times higher than the allowed Korean maximum contaminant level (MCL) of 150 mg/kg and Cd was found to be two times higher than the Korean warning level (Islam et al., 2016). Similar studies found that concentrations of Cu and Ni in shooting range berm soils far exceeded the Dutch Intervention Values (DIV) of 190 mg/kg and 210 mg/kg respectively (Robinson et al., 2008). It is important to note that this study is the first of its kind in Botswana and that there is no local information in the literature for heavy metals pollution in shooting range soils that could be used for comparisons. As a result, the findings obtained in this study were therefore used to investigate the extent of shooting range soils pollution from heavy metals Cd, Cu, Mn, Ni and Zn. A previous study by Mmolawa, Likuku, and Gaboutloeloe (2011) in the Matsiloje area where the MAT shooting range is found was able to measure soil background concentrations of 0.97 mg/kg Mn; 0.50 mg/kg Ni; 0.11 mg/kg Cu and 0.34 mg/kg Zn. In comparison, the concentrations of Cu, Mn, Ni and Zn found at MAT shooting range soil are much higher than these background concentrations, an indication that heavy metals from ammunition contributed to these high amounts of heavy metals deposit into shooting range soils. For example, the average concentration of Cu in the berm soils of MAT (390.8 ± 0.6 mg/kg) shooting range is 3500 times higher than the background concentration while the Ni concentration (44.7 ± 0.2 mg/kg) at the same shooting range was 90 times higher than the background concentrations. Quality control was performed using a soil certified reference material, NCS DC 73,320. A percentage recovery of over 90% was obtained for all the five heavy metals studied with accuracy of measurements within <3% RSD.

3.3. Effect of pH, organic matter and electrical conductivity on the distribution of heavy metals in shooting range soils

The soil properties such as pH, organic matter (OM) and electrical conductivity (EC) may have had an effect on such elevated levels of heavy metals at MAT shooting range. The alkaline soils (pH 8.6) at TAB shooting range provided for a suitable environment for the corrosion of the elemental heavy metals under study (Table 4 above). Concentrations of heavy metals Cu and Mn of up to 1569 ± 23 mg/kg and 184 ± 0.9 mg/kg respectively were recorded at TAB shooting range. At elevated pH, oxides, hydroxides, carbonates, sulphates and phosphates of the heavy metals under study become insoluble, thereby increasing their retention time in the soil. Their mobility become greatly reduced at elevated pH levels (Sporting Arms and Ammunition Manufacturers' Institute [SAAMI] 1996). Previous studies have found that soils with pH values of ≥ 7.0 experience restricted mobility of heavy metals whereas soils with pH values of ≤ 6.5 undergo enhanced metal mobility (SAAMI 1996). Our findings corroborate our previous studies where it was found that alkaline soils increased the stability of heavy metal oxides such as Pb oxides and carbonates (Sehube et al., 2017). A study by Ma et al. (2007) found out that as soil pH increased the weathering processes of Pb shot was greatly reduced leading to a decrease in mobility of Pb. The amount of organic matter (OM) present in the soil has been found to provide a suitable environment for the weathering of bullets and shots. Organic matter helps in the transformation of heavy metals present in shots and bullets into soluble complexes and thereby increasing the bioavailability and mobility of such heavy metals in the soil (Kelebemang et al., 2017; Sehube et al., 2017). S/P shooting range recorded the highest amount of organic matter (1.88 ± 0.42% OM) while the lowest amount of organic matter (0.54 ± 0.01% OM) was measured at PANDA shooting range.

The electrical conductivity evaluates the concentration of soluble salts present in the soil (Carmo, Lima, & Silva, 2016). Higher electrical conductivity (EC) in the soil implies higher concentrations of heavy metals in the soil are present in soluble form. S/P and TAB shooting range soils recorded the highest electrical conductivity values of 208.9 ± 8.8 μS/cm and 170.8 ± 8.4 μS/cm respectively. Consequently, TAB and S/P shooting ranges accumulated the highest concentrations of heavy metals in the berm soils. Concentrations reaching highs of 1569 ± 23 mg/kg and 282 ± 2 mg/kg Cu were recorded at TAB and S/P shooting range soils respectively. There is a direct relationship between the amount of organic matter in the soil and its electrical conductivity. Organic matter leads to production of organic and inorganic acids that are capable of dissolving heavy metal complexes in the soil registering high values of electrical conductivity giving rise to increased bioavailability and mobility of heavy metals in the soil (Verma, Sharma, & Paramanick, 2015). To the best of our knowledge, no study has been conducted in Botswana shooting ranges that assessed the environmental impact of heavy metals such as Cd, Cu, Mn, Ni and Zn making it difficult to compare our findings with the similar data found elsewhere in Botswana.

3.4. Environmental risk assessment of pollution from heavy metals

3.4.1. Potential ecological risk index and RI

All the five shooting ranges experienced between moderate to very-high ecological pollution risk from Cd. The ecological risk index level due to Cd was in the range 52–8141 as shown on Table 5 below. The ecological risk index (*PERI*) for Cd was highest for S/P (8140.9) followed by TAB (3507.4) and lowest for PANDA (51.9) shooting ranges. It is worth noting that even though all the five shooting ranges accumulated the lowest amounts of Cd compared to the other four heavy metals (Mn, Zn, Cu and Ni), these small amounts of Cd pose the highest ecological risk compared to the other four heavy metals due to its high toxicity effect. Cd has lower USEPA critical limit (7.1×10^1 mg/kg) implying accumulation of smaller amounts of Cd into the soil pose a pollution risk. On the other hand the other heavy metals studied have much higher critical limits in the soil in which Cu = 3.1×10^3 mg/kg, Ni = 8.2×10^2 mg/kg, Zn = 2.3×10^4 mg/kg and Mn = 1.8×10^3 mg/kg. Most of the shooting ranges such as MAT and PANDA showed very low ecological risk from Mn with risk indices of 0.07 and 0.04 respectively even though the two shooting ranges accumulated higher concentrations of Mn (Table 5). This means that even though the shooting ranges may have experienced elevated levels of Mn, such high concentrations do not pose an ecological risk to the environment due in part to the high critical limit of Mn in the soil. The sum of the ecological risk indices (*RI*) of all the heavy metals studied indicates very high ecological risk with *RI* values greater than 600 for the three ranges TAB, SHO and S/P as shown in Table 5 below. The high *PERI* and *RI* indices quantify the potential risk posed by heavy metals to the living organisms inhabiting the shooting range soils and immediate environment. High potential ecological risk index of 1158 for Cd has been reported before in literature implying ecological hazard to biota (Shen et al., 2017). The high *PERI* and *RI* indices indicate that drastic remedial measures need to be taken to arrest the ecological contamination risk posed by these heavy metals to shooting range soils.

3.4.2. Contamination factor

The contamination factors (CF) corroborate the *PERI* and *RI* results. Cadmium (Cd) displayed the highest contamination factor than any other heavy metal studied in the five shooting range soils. *CF* for Cd was determined in the range 2–271 in all the five shooting ranges, indicating moderate to very high contamination of the soil from Cd (Table 6). As seen from the *PERI* results, the *CF* for S/P indicate that this shooting range experienced the highest contamination from Cd than any other shooting range under study. Contamination factor has been used as a good estimate of pollution risk of heavy metals in the soil (Matong, Nyaba, & Nomngongo, 2016; Mmolawa et al., 2011).

Table 5. *PERI* and *RI* indicating the pollution risk level of heavy metals to five shooting range soils under study

	PERI					***RI***
Shooting Range	**Mn**	**Zn**	**Cu**	**Ni**	**Cd**	
TAB	1.4	7.5	18.0	396.2	3507.4	3930.5
SHO	1.4	3.8	11.3	49.3	1484.6	1550.4
S/P	9.0	6.3	46.5	197.3	8140.9	8399.9
MAT	0.007	0.2	1.2	3.2	54.8	59.4
PANDA	0.04	0.4	1.0	0.7	51.9	54.0

Table 6. Contamination factor of the five shooting ranges showing level of contamination of shooting range soils from the five heavy metals investigated

	Contamination Factor *(CF)*				
Shooting Range	**Mn**	**Zn**	**Cu**	**Ni**	**Cd**
TAB	1.4	7.5	3.6	79.2	116.9
SHO	1.4	3.8	2.3	9.9	49.5
S/P	9.0	6.3	9.3	39.5	271.4
MAT	0.007	0.2	0.2	0.6	1.8
PANDA	0.04	0.4	0.2	0.1	1.7

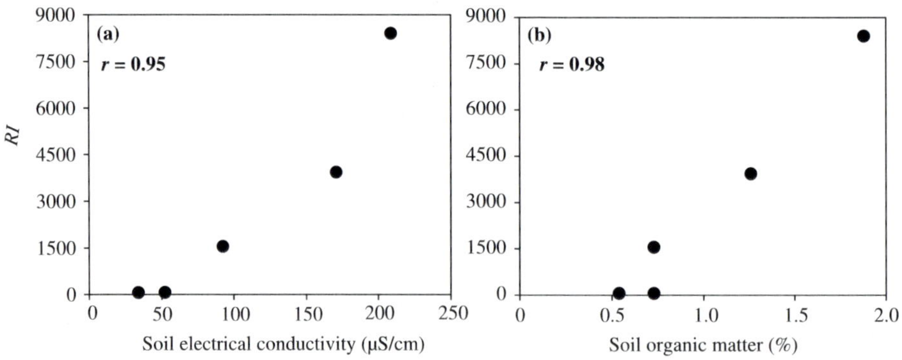

Figure 4. Pearson correlation between; (a) soil EC and (b) soil organic matter to total ecolo-gical risk index (*RI*) of five heavy metals found in five shooting ranges found in Botswana. Mean of *n* = 3 ; Standard error of the mean, δx = δ/√*n*, where δ = standard deviation.

3.5. Statistical assessment of pollution using the Pearson correlation coefficient

The effects of physicochemical properties of the soils such as organic matter (OM) and electrical conductivity (EC) on the pollution risk of heavy metals in shooting range soils were determined using the person correlation coefficient (*r*). There is direct correlation between the ecological pollution risk from heavy metals and the soil organic matter and electrical conductivity as indicated by the high Pearson correlation coefficients. Pearson correlation coefficient (*r*) of 0.95 depicted high ecological risk from heavy metals at high electrical conductivities of the soil (Figure 4). Similarly, correlation coefficient (*r*) of 0.98 suggested positive relationship between the ecolo-gical pollution risk of heavy metals and the soil organic matter (Figure 4). Our previous have shown that high organic matter stabilizes heavy metal complexes and salts in the soil enhancing the accumulation of such heavy metals in the soil (Kelebemang et al., 2017; Sehube et al., 2017)

4. Conclusions

This study is the first in Botswana to assess the environmental pollution risk associated with heavy metal deposition in shooting range soils arising from ammunition. The findings from this study will provide a baseline data in Botswana for subsequent studies on heavy metal contamination of shooting range soils. High concentrations of Cu (1569 ± 13 mg/kg) and Mn (953.8 ± 2.8 mg/kg) were found in shooting range soils. However, the high concentration of heavy metal in the soil does not give a true picture of the environmental impact and risk from such heavy metal. Potential ecological risk (*PERI*) and contamination factor (*CF*) were able to establish that Cd, even though had lower concentrations measured in shooting range soils, posed greater environmental pollution risk than the other four heavy metals studied. The ecological risk (*RI*) emanating from contributions by all the five heavy metals present in the soil indicated very high pollution in three shooting ranges TAB, SHO and S/P. The environmental pollution risk posed by these heavy metals call for remedial actions and best shooting range management practices to be taken to arrest the situation before it spirals out of control. Remedial control measures such as chemical and phytoremediation processes can be employed to control soil pollution of shooting range soils. Chemical remediation processes involving use of chemicals such phosphate salts and lime have been found to immobilize heavy metals in the soil, reducing their mobility and bioavailability. Future studies into the best soil cleaning and soil reclamation strategies and methods should be conducted in order to come up with remedies and measures to control and manage pollution in shooting range soils.

Acknowledgements

The authors are grateful to the Botswana Defence Force (BDF) for providing their shooting ranges to be used in the study. We would like to acknowledge the Botswana University of Agriculture and Natural Resources (BUAN) for their assistance with ICP-OES instrument for heavy metal analysis.

Funding

Financial support for this research work was provided by the Botswana International University of Science and Technology (BIUST), Research Initiation Grant Number BIUST/ds/R&I/26/2016; Botswana International University of Science and Technology [Research Initiation [grant number BIUST/ds/R&I/26/2016]].

Competing Interests

The authors declares no competing interests.

Author details

Pogisego Dinake[1]
E-mail: dinakep@biust.ac.bw
ORCID ID: http://orcid.org/0000-0003-2456-2043
Onneetse Maphane[1]
E-mail: onneetse.maphane@studentmail.biust.ac.bw
Karabo Sebogisi[2]
E-mail: bmadzonga@buan.ac.bw
Obed Kamwi[3]
E-mail: kamwi42@gmail.com

[1] Department of Chemical and Forensic Sciences, Botswana International University of Science and Technology, Biust, Palapye, Botswana.

[2] Department of Crop Science and Production, Botswana University of Agriculture and Natural Resources, Private Bag 0027, Gaborone, Botswana.

[3] Safety Health and Environment, Botswana Defence Force, Bdf, Sir Seretse Khama Barracks, Private Bag x06, Mogoditshane, Botswana.

References

Alexander, T. C., Gulledge, E., & Han, F. (2016). Arsenic occurrence, ecotoxicity and its potential remediation. *Journal of Bioremediation and Biodegradation*, *7*, e174. doi:10.4172/2155-6199.1000e174

Almendros, G., Kgathi, D., Sekhwela, M., Zancada, M. C., Tinoco, P., & Pardo, M. T. (2003). Biogeochemical assessment of resilient humus formations from virgin and cultivated Northern Botswana soils. *Journal of Agricultural and Food Chemistry*, *51*, 4321–4330. doi:10.1021/jf034006u

Broadley, M. R., White, P. J., Hammond, J. P., Zelko, I., & Lux, A. (2007). Zinc in plants. *New Phytologist*, *173*, 677–702. doi:10.1111/nph.2007.173.issue-4

Brocato, J., & Costa, M. (2017). The biological chemistry of nickel: nickel toxicity and carcinogenesis. In D. Zamble, M. Rowinska-Zyrek, & H. Kozlowski (Eds.), *Royal society of chemistry* (pp. 27–42). Croydon, CR0 4YY, UK: CPI Group (UK) Ltd.

Cao, L., Tian, H., Yang, J., Shi, P., Lou, Q., Waxi, L., … Peng, X. (2015). Multivariate analyses and evaluation of heavy metals by chemometric BCR sequential extraction method in surface sediments from Lingdingyang Bay, South China. *Sustainability*, *7*, 4938–4951. doi:10.3390/su7054938

Carmo, D. L., Lima, L. B., & Silva, C. A. (2016). Soil fertility and electrical conductivity affected by organic waste rates and nutrient inputs. *Revista Brasileira de Ciência do Solo*, *40*, 150–152. doi:10.1590/18069657rbcs20150152

de Queiroz Barcelo, J. P., de Souza Osóri, C. R. V., Leal, A. J. F., Alves, C. Z., Santos, E. F., Reis, H. P. G., & Dos Reis, A. R. (2017). Effects of foliar nickel (Ni) application on mineral nutrition status, urease activity and physiological quality of soybean seeds. *Australian Journal of Crop Science*, *11*, 184–192. doi:10.21475/ajcs.17.11.02.p240

Department of Water Affairs and Forestry. (1996). South African water quality guidelines. *Aquatic Ecosystems*, *7*, 45–49.

Ekosse, G. I., & Anyangwe, S. (2012). Mineralogical and particulate morphological characterization of Geophagic clayey soils from Botswana. *Bulletin of the Chemical Society of Ethiopia*, *26*, 373–382. doi:10.4314/bcse.v26i3.6

Evangelou, M. W. H., Hockmann, K., Pokharel, R., Jakob, A., & Schulin, R. (2012). Accumulation of Sb, Pb, Cu, Zn and Cd by various plants species on two different relocated military shooting range soils. *Journal of Environmental Management*, *108*, 102–107. doi:10.1016/j.jenvman.2012.04.044

Fayiga, A. O., & Saha, U. K. (2016). Review: Soil pollution at outdoor shooting ranges: Health effects, bioavailability and best management practices. *Environmental Pollution*, *216*, 135–145. doi:10.1016/j.envpol.2016.05.062

Hakanson, L. (1980). An ecological risk index for aquatic pollution control. A sedimentological approach. *Water Research*, *14*, 975–1001. doi:10.1016/0043-1354(80)90143-8

Hockmann, K., Tandy, S., Lenz, M., Reiser, R., Conesa, H. M., Keller, M., … Schulin, R. (2015). Antimony retention and release from drained and waterlogged shooting range soil under field conditions. *Chemosphere*, *134*, 536–543. doi:10.1016/j.chemosphere.2014.12.020

Islam, M. N., Nguyen, X. P., Jung, H. Y., & Park, J. H. (2016). Chemical speciation and quantitative evaluation of heavy metal pollution hazards in two army shooting range backstop soils. *Bulletin of Environmental Contamination and Toxicology*, *96*, 179–185. doi:10.1007/s00128-015-1689-z

Jastrow, J. D., Amonette, J. E., & Bailey, V. L. (2007). Mechanisms controlling soil carbon turnover and their potential application for enhancing carbon sequestration. *Climatic Change*, *80*, 5–23. doi:10.1007/s10584-006-9178-3

Johnson, C. A., Moench, H., Wersin, P., Kugler, P., & Wenger, C. (2005). Solubility of antimony and other elements in samples taken from shooting ranges. *Journal of Environmental Quality*, *34*, 248–254. doi:10.2134/jeq2004.0398

Joshua, W. D. (1991). *Physical properties of the soils of Botswana*. Gaborone: Soil Mapping and Advisory Services Botswana, Food and Agricultural

Organization (FAO) of the United Nations, United Nations Development Programme (UNDP), Government of Botswana.

Ke, X., Gui, S., Huang, H., Zhang, H., Wang, C., & Guo, W. (2017). Ecological risk assessment and source identification for heavy metals in surface sediment from the Liaohe river protected area, China. *Chemosphere, 175*, 473–481. doi:10.1016/j.chemosphere.2017.02.029

Kelebemang, R., Dinake, P., Sehube, N., Daniel, B., Totolo, O., & Laetsang, M. (2017). Speciation and mobility of lead in shooting range soils. *Chemical Speciation and Bioavailability, 29*, 143–152. doi:10.1080/09542299.2017.1349552

Lafond, S., Blais, J. F., Martel, R., & Mercier, G. (2013). Chemical leaching of antimony and other metals from small arms shooting range soil. *Water Air Soil and Pollution, 224*, 1371–1385. doi:10.1007/s11270-012-1371-6

Laghlimi, M., Baghdad, B., El Hadi, H., & Bouabdli, A. (2015). Phytoremediation mechanisms of heavy metal contaminated soils: A review. *Open Journal of Ecology, 5*, 375–388. doi:10.4236/oje.2015.58031

Lange, B., Van der Ent, A., Baker, A. J. M., Echevarria, G., Mahy, G., Malaisse, F., … Faucon, M. P. (2017). Research review: Copper and Cobalt accumulation in plants: A critical assessment of the current state of knowledge. *New Phytologist, 213*, 537–551. doi:10.1111/nph.14223

Lee, I. S., Kim, O. K., Chang, Y. Y., Bae, B., Kim, H. H., & Baek, K. H. (2002). Heavy metal concentrations and enzyme activities in soil from a contaminated korean shooting range. *Journal of Bioscience and Bioengineering, 94*, 406–411. doi:10.1016/S1389-1723(02)80217-1

Ma, L. Q., Hardison, D., Jr, Harris, W. G., Cao, X., & Zhou, Q. (2007). Effects of soil property and soil amendment on weathering of abraded metallic Pb in shooting ranges. *Water Air and Soil Pollution, 178*, 297–307. doi:10.1007/s11270-006-9198-7

Mariussen, E., Johnsen, I. V., & Strømseng, A. E. (2017). Distribution and mobility of Lead (Pb), Copper (Cu), Zinc (Zn), and Antimony (Sb) from ammunition residues on shooting ranges for small arms located on mires. *Environmental Science and Pollution Research International, 24*, 10182–10196. doi:10.1007/s11356-017-8647-8

Matong, J. M., Nyaba, L., & Nomngongo, P. N. (2016). Fractionation of trace elements in agricultural soils using ultrasound assisted sequential extraction prior to inductively coupled plasma mass spectrometric determination. *Chemosphere, 154*, 249–257. doi:10.1016/j.chemosphere.2016.03.123

Migliorinia, M., Pigino, G., Bianchi, N., Bernini, F., & Leonzio, C. (2004). The effects of heavy metal contamination on the soil arthropod community of a shooting range. *Environmental Pollution, 129*, 331–340. doi:10.1016/j.envpol.2003.09.025

Mitchell, A. J. B. (1976). *The irrigation potential of soils along the main rivers of Eastern Botswana - A reconnaissance assessment.* Surbiton, Surrey, England: Land Resource Study 7, Land Resources, Division, Ministry of Overseas Development, Tolworth Tower.

Mmolawa, K. B., Likuku, A. S., & Gaboutloeloe, G. K. (2011). Assessment of heavy metal pollution in soils along major roadside areas in Botswana. *African Journal of Environmental Science and Technology, 5*, 186–196.

Moganane, B. G., Van Waveren, E., & Remmelzwaal, A. (1991). *The soils of the pandamatenga plains.* Botswana: Soil Mapping and Advisory Services. Field Document: 12, Gaborone. AG:BOT/85/011.

Mozafar, A., Ruh, R., Klingel, P., Gamper, H., Egli, S., & Frossard, E. (2002). Effect of heavy metal contaminated shooting range soils on mycorrhizal colonization of roots and metal uptake by leek. *Environmental Monitoring and Assessment, 79*, 177–191. doi:10.1023/A:1020202801163

Nagajyoti, P. C., Lee, K. D., & Sreekanth, T. V. M. (2010). Heavy metals, occurrence and toxicity for plants: A review. *Environmental Chemistry Letters, 8*, 199–216. doi:10.1007/s10311-010-0297-8

Nelson, D. W., & Sommers, L. E. (1982). Total carbon, organic carbon, and organic matter. In A. L. Page, R. H. Miller, & D. R. Keeney (editors), *Methods of soil analysis, part 2 chemical and microbiological properties* (Vol. 9, pp. 539–577). Madison (WI: ASA.

Okkenhaug, G., Amstatter, K., Bue, H. L., Cornelissen, G., Breedveld, G. D., Henriksen, T., & Mulder, J. (2013). Antimony (Sb) contaminated shooting range soil: Sb mobility and immobilization by soil amendments. *Environmental Science and Technology, 47*, 6431–6439. doi:10.1021/es302448k

Pardo, M. T., Ristori, G., D'acqui, P., & Almendros, G. (2003). An assessment of soil fertility and agronomic constraints in Southern African Savannas: A case study of the Pandamatenga area, Botswana. *South African Geographical Journal, 85*, 35–41. doi:10.1080/03736245.2003.9713782

Pejman, A., Bidhendi, G. N., Ardestani, M., Saeedi, M., & Baghvand, A. (2017). Fractionation of heavy metals in sediments and assessment of their availability risk: A case study in the Northwestern of Persian Gulf. *Marine Pollution Bulletin, 114*, 881–887. doi:10.1016/j.marpolbul.2016.11.021

Robinson, B. H., Bischofberger, S., Stoll, A., Schroer, D., Furrer, G., Roulier, S., … Schulin, R. (2008). Plant uptake of trace elements on a Swiss military shooting range: Uptake pathways and land management implications. *Environmental Pollution, 153*, 668–676. doi:10.1016/j.envpol.2007.08.034

Rodríguez-Seijo, A., Alfaya, M. C., Andrade, M. L., & Vega, F. A. (2016). Copper, chromium, nickel, lead and zinc levels and pollution degree in firing range soils. *Land Degradation and Development, 27*, 172–1730. doi:10.1002/ldr.2497

Rooney, C. P., McLaren, R. G., & Cresswell, R. J. (1999). Distribution and phytoavailability of lead in a soil contaminated with lead shot. *Water Air and Soil Pollution, 116*, 535–548. doi:10.1023/A:1005181303843

Sehube, N., Kelebemang, R., Totolo, O., Laetsang, M., Kamwi, O., & Dinake, P. (2017). Lead pollution of shooting range soils. *South African Journal of Chemistry, 70*, 21–28. doi:10.17159/0379-4350/2017/v70a4

Seshadri, B., Bolan, N. S., Choppala, G., Kunhikrishnan, A., Sanderson, P., Wang, H., … Kim, K. (2017). Potential value of phosphate compounds in enhancing immobilization and reducing bioavailability of mixed heavy metal contaminants in shooting range soil. *Chemosphere, 184*, 197–206. doi:10.1016/j.chemosphere.2017.05.172

Sharma, A., & Sachdeva, S. (2015). Cadmium toxicity and its phytoremediation: A review. *International Journal of Scientific and Engineering Research, 9*, 395–405.

Shen, F., Liao, R., Ali, A., Mahar, A., Guo, D., Li, R., … Zhang, Z. (2017). Spatial distribution and risk assessment of heavy metals in soil near a Pb/Zn smelter in Feng County, China. *Ecotoxicology and Environmental Safety, 139*, 254–262. doi:10.1016/j.ecoenv.2017.01.044

Soliman, N. F., Nasr, S. M., & Okbah, M. A. (2015). Potential ecological risk of heavy metals in sediments from the Mediterranean Coast, Egypt. *Journal of Environmental*

Health Science and Engineering, *13*, 70–81. doi:10.1186/s40201-015-0223-x

Sorvari, J., Antikainen, R., & Pyy, O. (2006). Environmental contamination at finnish shooting ranges—The scope of the problem and management options. *Science of the Total Environment*, *366*, 21–31. doi:10.1016/j.scitotenv.2005.12.019

Sporting Arms and Ammunition Manufacturers' Institute, SAAMI. (1996). Lead mobility at shooting ranges. (Project No. 13081.01), Flintlock Ridge Office Center, 11 Mile Hill Road, Newtown, CT 06470, 22–24.

Sundar, S., & Chakravarty, J. (2010). Review: Antimony toxicity. *International Journal of Environmental Research and Public Health*, *7*, 4267–4277. doi:10.3390/ijerph7124267

Tchoumou, M., Kami-Ouako, M., Mbilou, G. U., & Ossebi, J. (2017). Comparative study of physicochemical parameters of water of four waterways in the Town of Brazzaville, Republic of Congo. *LARHYSS Journal*, *29*, 61–74.

Urrutia-Goyes, R., Argyraki, A., & Ornelas-Soto, N. (2017). Assessing lead, nickel, and zinc pollution in topsoil from a historic shooting range rehabilitated into a Public Urban Park. *International Journal of Environmental Research and Public Health*, *14*, 698–712. doi:10.3390/ijerph14070698

Verma, J. K., Sharma, A., & Paramanick, K. K. (2015). To evaluate the values of electrical conductivity and growth parameters of apple saplings in nursery fields. *International Journal of Applied Sciences and Engineering Research*, *4*, 321–332.

Vig, K., Megharaj, M., Sethunathan, N., & Naidu, R. (2003). Bioavailability and toxicity of cadmium to microorganisms and their activities in soil: A review. *Advances in Environmental Research*, *8*, 121–135. doi:10.1016/S1093-0191(02)00135-1

Wersin, P., Johnson, A. A., & Furrer, G. (2002). Antimony contamination in soil and ground water by shooting range activities. *Geochimica et Cosmochimica Acta*, *66*, A829.

6

Coral nursery and transplantation of the staghorn coral, *Acropora downingi* in Sabah Al-Ahmad Sea City, Kuwait, Arabian Gulf

Manickam Nithyanandan[1,2]*, Lewis Le Vay[3], Dinesh Kumar Raja[1,4], Ratheesh Kesavan[1] and Don Pereira[1]

*Corresponding author: Manickam Nithyanandan, Environmental Department, Sabah Al Ahmad Sea City, La Ala Al Kuwait Real Estate Co. K.S.C., Souk Al Kuwait 6th Floor, Office No. 613, P.O.Box 29964, Safat, Kuwait 13090, Kuwait
E-mail: nandan.ocean@gmail.com
Reviewing editor: Agnes Muthumbi, University of Nairobi, Kenya

Abstract: The present work describes the first attempt at cultivating corals in the northern Arabian Gulf where seawater temperature ranges from 13 to 33°C and salinity exceeds 40 psu, exceeding normal lethal limits for coral survival. Despite the environmental extremes, natural recruitment of corals occurred in Sabah Al-Ahmad Sea City water ways during 2009, this encouraged an attempt to create an in situ coral nursery and establish a coral garden. A mid-water suspended coral PVC nursery (3 m^2) was installed at 3.5 m depth for rearing staghorn coral, *Acropora downingi* for transplantation into artificial lagoons in Sabah Al-Ahmad Sea City. Fragments of *Acropora downingi* colonies damaged by boat anchoring were collected (n = 240) from patch reefs at Min Al Zour and Qit At Binaya, Kuwait. Coral nubbins obtained from fragments were glued individually to discs made of powdered Electric Arc Furnace Slag (EAF Slag) and marine grade cement. Nubbins showed up to 56% survival and a mean skeletal extension of 10.6–13.4 mm (SD ± 0.8) in 10 months. The nursery served as an artificial reef ecosystem. Transplantation of 116 *Acropora downingi* colonies was carried out (in June 2014 and June 2015) and monitored colonies (n = 6) attained an average

ABOUT THE AUTHORS

M. Nithyanandan, a marine ecologist with master's degree in Oceanography from Alagappa University, India. Interested in various aspects of marine ecology, marine biodiversity and habitat restoration in Arabian gulf. Currently working in Kuwait Institute for Scientific Research under the Ecosystem based management of marine sciences program. Recent research interest is to study the benthic ecology of Kuwait Bay and has published several articles in international peer reviewed journals.

Professor Lewis Le Vay, Director, Centre for Applied Marine Sciences in Bangor University, UK. An expert in marine biology and aquaculture.

Dinesh Kumar Raja, a marine biologist and PADI divemaster experienced in studying marine ecology of Andaman islands and Kuwait, Arabian Gulf.

Ratheesh Kesavan, a marine biologist studying marine ecology of Kuwait, Arabian Gulf.

Don Pereira, a Research Diver and PADI dive master with a good experience in coral and seagrass restoration techniques in Arabian Gulf.

PUBLIC INTEREST STATEMENT

Coral reefs in the Arabian/Persian gulf are very unique as they thrive in extreme environmental conditions (low–high temperature and high salinity). Pollution, coastal development, recreational activities, etc. threaten the existence of reefs, which can impact fisheries. In this study, for the first time in northern Arabian Gulf, we conducted a "coral nursery" and transplantation experiment for staghorn coral (*Acropora downingi*) in the largest coastal township (>70 Km^2 area) project in Kuwait known as "Sabah Al-Ahmad Sea City". Coral colonies damaged by boat anchoring were collected from nearby reefs and glued to discs made of steel slag and cement, reared in the nursery for a year and transplanted onto a rock island. Transplanted staghorn corals successfully grew in spite of extreme temperature and salinity in Kuwait waters. The results of this experiment will serve as a baseline to design future reef restoration efforts in the region.

geometric mean diameter (GMD) of 73.6 mm (SD ± 2.91) in one year. The first batch of *A. downingi* transplants (June 2014) showed a survival rate of 43% but the second batch (June 2015) showed 89.5% survival. Temperature induced mortality, detachment rate reduced after using marine grade cement and additive mixture instead of epoxy.

Subjects: Bioscience; Earth Sciences; Environmental Studies and Management

Keywords: coral nursery; coral transplantation; *Acropora downingi*; Kuwait; reef restoration

1. Introduction

Growing multi-species of coral in a nursery is the first phase in coral gardening, which adopts "silviculture" methods (Rinkevich, 2008). Several species of corals worldwide have been successfully grown from nubbins (10–30 mm length) to a transplantable size in floating nurseries prior to transplantation onto denuded reefs (Mbije, Spanier, & Rinkevich, 2010; Putchim, Thongtham, Hewett, & Chansang, 2008; Shafir, Rijin, & Rinkevich, 2006a, 2006b; Shaish, Levy, Gomez, & Rinkevich, 2008). In the Arabian Gulf (hereafter Gulf), coral nursery experiments are so far limited to southern tropical waters (Sen & Yousif, 2016; Wilson & Marimuthu, 2012) with a few preliminary transplantation experiments in Iran (Vajed Samei, Koosha, & Behrooz, 2012) and Abu Dhabi, United Arab Emirates (Katakura, Yuriko, Sumihiro, Hiroshima, & Iwata, 2013; Sen & Yousif, 2016). The present work describes the first establishment of an in situ coral nursery in the extreme environmental conditions of the northern Gulf (Kuwait), for habitat enhancement. Corals in Kuwait survive in extreme environmental conditions with winter minimum (13°C) and summer maximum sea surface temperatures of (33°C) and salinity (>40 psu) exceeding lethal limits for corals elsewhere and challenging reef building (Carpenter, Harrison, Hodgson, Al-Saffar, & Al-Hazeem, 1997).

Sabah Al Ahmad Sea City (hereafter Sea City) is the largest coastal township development in Kuwait with over >70 km^2 of artificial lagoons, beaches, islands and will house 100,000 people. An array of man-made marine habitats now exists in Sea City including intertidal sand beaches, rock revetments, groynes, mud flats, salt marsh, seagrass and mangrove habitats and subtidal sand, mixed rock and rock habitats. Over 1,000 species of benthic fauna and 100 species of fish (Jones & Nithyanandan, 2013; Jones, Nithyanandan, & Williams, 2012) have been recorded to date. Since 2009 natural recruitment of nine species of scleractinian corals—"*Dipsastrea speciosa* (=*Favia speciosa*), *D. pallida* (=*F. pallida*), *Siderastrea savignyana*, *Porites* sp, *Cyphastrea serialia*, *C. microphthalma and Platygyra daedalea*"—have been observed colonizing on the rock revetments of Sea City with a mean colony density of 10 (SD ± 5) colonies m^{-2} (Jones & Nithyanandan, 2015) covering an area of <200 m^{-2}. However, no recruitment of the branching corals such as *Acropora* spp and *Stylophora pistillata* has occurred, which are essential to the creation of three dimensional habitat increasing complexity and niches for various marine organisms (Sheppard, 2015).

This study aims at establishing an in situ coral nursery for the mariculture of the branching coral, *Acropora downingi* and subsequent transplantation into Sea City waterways to enhance the environmental value. Further this study also reports the impact of sea surface temperature-SST (°C) on the survival and growth rate of *A. downingi* in northern gulf.

2. Materials and methods

2.1. Site selection

Two donor coral sites: on offshore patch reef at Min Al Zour (28°45′N; 48°24°E) and Qit at Binaya (28° 38°N; 48° 25°E) off Al Khiran (Figure 1) were selected as they are dominated by branching and non-branching coral genera. The nursery site was selected in a shallow channel of 5 m depth with a rocky bottom, adjacent to the second rock island in phase A3 of Sea City (Figure 1). The second

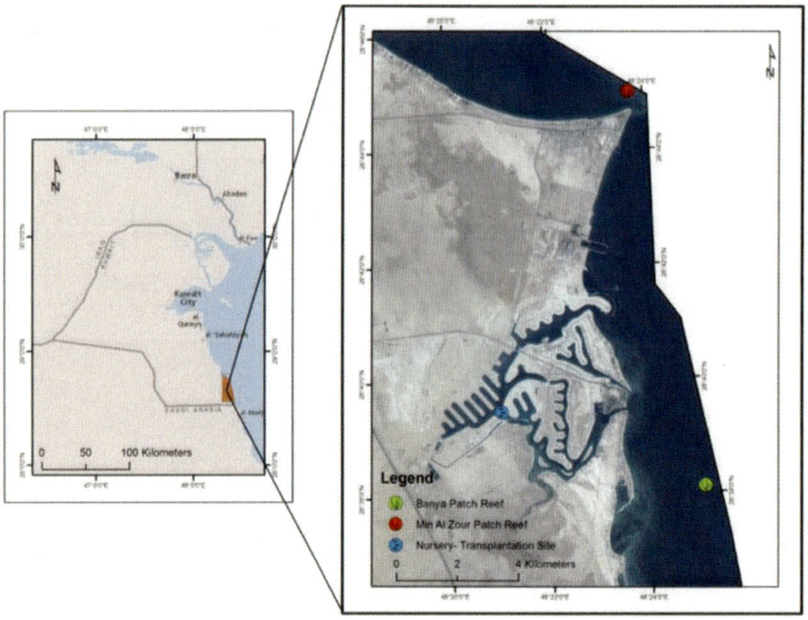

Figure 1. *Acropora downingi* offshore donor sites (Al Zour and Qit at Binaya patch reefs) and nursery-transplantation sites at Sabah Al Ahmad Sea City.

rock island 50 m away from the nursery site with an inclined open rock surface was selected as a transplantation site (Figure 1). Both nursery and transplantation sites were 8 km away from the donor sites.

2.1.1. Physicochemical parameters

Prior to installation of the nursery, currents (ms^{-1}) were measured during spring tides using a Valeport ® 106 self reading current meter in April 2013. Daily sea temperature (°C) (SST), salinity (psu), dissolved oxygen (mgl^{-1}) and pH were recorded at 1 m depth using YSI 556 multi-probe system and bi-weekly measurements of total suspended solids (TSS), nutrients—Nitrates (NO3), nitrites (NO_2), phosphates (PO_4), ammonia (NH_3) and silicates (SiO_3) following American Public Health Association (APHA, 1995) methods were carried out during May 2012–July 2015.

2.1.2. Nursery design and substrate preparation

A cost-effective nursery design was adopted based on Edwards (2010) and Shaish et al. (2008) and modified as a "Suspended table nursery" (Figure 2a) with a 3 m^2 frame made out of 6 × 25 mm diameter PVC pipes, six legs (1 m) with end cups and a plastic mesh (3 m × 3 m × 30 mm mesh size)

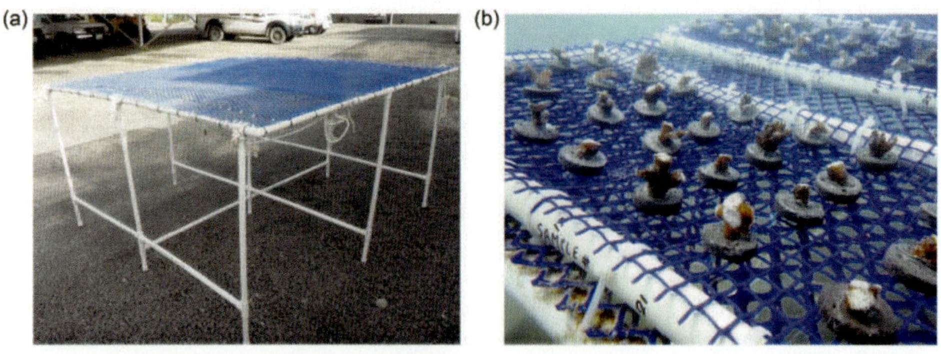

Figure 2. (a) Table nursery used in the experiment, (b) nubbins immediately after being placed in the nursery showing stress symptoms by secreting intense mucous threads from severed portions. (Photos M. Nithyanandan.)

stretched over the frame and secured with cable ties. Seven rearing trays were constructed using PVC (600 × 300 mm) pipes and plastic mesh (600 mm × 300 mm × 30 mm mesh size). The nursery was suspended at 3.5 m depth with marine grade ropes attached to concrete sinkers. Buoys were (4 × 1) attached to nursery to maintain buoyancy and to suspend it at 1.5 m above the lagoon bed. Although table nurseries are of a fixed type (Shaish et al., 2008), we suspended the table nursery in the water column and its legs offered framework support and stability. Seven rearing trays were used in the experiment and rest of the space was kept empty for further addition.

Slag, a by-product obtained during manufacturing of steel has a proven potential as a cost-effective substrate/substrate ingredient for recent coral rehabilitation experiments worldwide (Mohammed, Hamed, Habib, Ezz El-Arab, & El-Moshley, 2012; Okamoto et al., 2010; Oyamada, Okamoto, & Iwata, 2014). Powdered EAF slag was mixed with marine grade cement (3:1) and the mixture was poured into PVC rings (30 mm diameter, 10 mm thick) into which four small holes were made by inserting a thick metal wire and cured overnight. Slag discs were attached using a fish line to the rearing trays in arrays of 4 × 8 rows with 20–30 mm interval between discs (Figure 2b), and this method is based on Kerby (2014). Each rearing tray was marked with an individual number and row numbers for identification.

2.1.3. Coral nubbin preparation and transplantation

Acropora downingi ramets of 50–100 mm size broken off by boat anchoring were collected during June–August 2013 from various areas of the patch reefs (depth 5–6 m) in Min Al Zour and Qit at Binaya, Kuwait to ensure diversity in genotypes. Collecting anchor damaged colonies to serve as donors eliminated pressure on existing natural population of *A. downingi*. Fragments were placed in a 60 L cooler box filled with fresh seawater and aerated vigorously to minimize stress levels which were indicated by extensive mucous secretion after fragmentation. Four people were involved in cutting the fragments into nubbins (10–30 mm size) using electrical pliers and gluing to slag discs using Super Glue® (b, Shafir et al., 2006a). Further, two SCUBA divers placed trays with discs in the nursery and attached them using cable ties. Intensive mucous secretion by newly attached nubbins was observed as a stress reaction upon exposure to air for 5–10 min while attaching to EAF slag discs (Figure 2b).

After a year in the nursery, during May 2014 the first batch of 30 colonies of different genotypes were randomly removed from the nursery, cleaned of foulers and transplanted onto rock surfaces at 3 m depth using marine grade epoxy resin (Gomez, Cabaitan, Yap, & Dizon, 2013) in two alternate rows with 200–300 mm interval between rows and nubbins. At least a 50–60 mm thick layer of epoxy had to be used to hold the nubbins in position against gravitational and current forces on the inclined rock substrate. In June 2015, a second batch of 86 nubbins were transplanted using marine grade cement (Yap, 2009) mixed with a water proofing additive (3:1), Water Proof.150 BLC® from Sodamco® which cures in less than 5 min. Manufacturer specification indicates that the curing process is directly proportional to temperature for additive use, hence the transplantation was carried out in early summer (May–June). Significant difference in sample size was due to availability of extra manpower for transplantation during the second phase (June 2015) of transplantation.

2.1.4. Maintenance and monitoring

Two divers conducted weekly maintenance dives to check nursery's structural stability, buoyancy and remove algae and foulers on the substrate and amongst the coral nubbins using a brush. Divers regularly waved their hands to remove any debris settling between nubbins (Shafir et al., 2006b). A similar cleaning method was also adopted for transplanted colonies.

In the nursery, 20 nubbins were randomly tagged with a code number to record growth and number of branches produced monthly from September 2013 to July 2014. A twist tie was attached to the base of each tagged nubbin and growth was recorded by measuring skeletal extension beyond the twist tie to the nearest mm using a plastic Vernier caliper. Monthly digital photographs were taken with a ruler placed beside each colony for calibration and growth analysis, using the software Adobe Acrobat Pro®. Species of colonizing benthic community and fishes observed in the nursery were recorded as abundant (>10 individuals), common (sighted during every monitoring/maintenance dive) or rare (sighted only once during the entire monitoring period). In the case of transplanted colonies (first batch, n = 30), six individuals were tagged randomly and growth was monitored for 12 months (August 2014–June 2015) by recording maximum horizontal and perpendicular length (mm) to calculate geometric mean diameter (GMD) in mm following Clark and Edwards (1995), for a three dimensional growth pattern. The number of branches was counted monthly to derive mean number of branches produced. No growth monitoring was carried out for the second batch transplants. Growth rate differences between genotypes were not studied as it was beyond the scope of the present study. Nursery and transplanted colonies were also monitored for monthly survival, detachment, bleaching and mortality.

2.1.5. Statistical analysis

A two tailed *t*-test was performed to understand the influence of SST (°C) on nubbin mortality and detachment in the nursery and similarly salinity (psu) on mortality, detachment, skeletal extension (mm) of nubbins and average GMD (mm) of transplants. Linear regression analysis was performed for monthly mean SST (°C) against average GMD (mm) and number of branches against average GMD (mm) in MS Office Excel 2007.

3. Results

A peak flow velocity of 0.45 ms^{-1} was recorded during the spring tides at the nursery and transplantation sites. The suspended table nursery was moved in all directions by currents. Minimum and maximum values of physicochemical parameters (Table 1) indicate that the nursery and transplantation sites are exposed to a wide temperature, salinity range, optimum nutrient levels and low total suspended solids (TSS).

After a year in the nursery *A. downingi* nubbins (n = 240) showed a survival rate of 56%, detachment 37.9% and mortality 6.3% (Table 2). Mortality occurred in summer (September 2013) and winter (December 2013). No significant correlation was observed between SST (°C) and nubbin mortality and detachment rates (*t*-test, $p > 0.05$). Nubbins produced axial corallites at the cut point and produced branches with radial corallites in 2 months. The first batch of transplanted colonies (n = 30) showed 43.3% survival (Figure 3a), 46.6% detachment and 10% mortality (Figure 3b), whereas the second batch (n = 86) showed 89.5% survival (Figure 3a), 4.7% detachment and 5.8% mortality during the post-transplantation period (Figure 3b). Similarly, mortality in the first batch of transplants occurred during cold temperatures in January–February 2015 and in the second batch due to a partial bleaching event in August 2015 (Figure 3b), however the rates of mortality in nubbins and first batch of transplants were much lower compared to detachment rate influenced by the mechanical disturbance during maintenance (in nubbins) and use of poor quality epoxy (in transplants). However the use of 3:1 marine grade cement and water proof BLC® (additive) reduced the detachment rate in the second batch transplants. In the nursery, nubbins self-attached in 30 days with tissue and skeleton encrusting the EAF slag disc (Figure 5a), whereas in the case of transplants self-attachment to the substrate occurred in 60 days.

Sea surface temperatures below 18°C and above 31°C showed an impact on growth in the nursery (Figure 3a). Moreover SST values recorded for the sea city nursery and transplantation sites have a mean annual difference of 3.0°C (Figure 4a), lower compared to southern Gulf waters reported by Bauman et al. (2013). Similarly higher salinity (>42 psu) recorded in the nursery-transplantation site is 2.3 PSU higher than the southern gulf waters (Bauman et al., 2013). However, there is no significant impact of salinity on the survival, detachment and growth of nubbins (*t*-test, $p > 0.05$) and transplants (Figure 4b, *t*-test, $p > 0.05$). In the case

Table 1. Minimum and maximum values of physicochemical parameters recorded during May 2012–July 2015. Values are expressed in ±SD

Year	Sea Surface Temperature °C	TSS mgl^{-1}	NO_2 μgl^{-1}	NO_3 μgl^{-1}	NH_3 μgl^{-1}	PO_4 μgl^{-1}	SiO_3 μgl^{-1}
2012	17.7–34.8 ± 4.5	8.7–19.3 ± 2.46	2.5–7.5 ± 1.2	10–15 ± 1.8	21.7–54.5 ± 10.5	6.1–7.6 ± 0.3	160.4–504.7 ± 95.8
2013	12.9–32.8 ± 5.2	5.5–18.2 ± 2.4	2.6–4.5 ± 0.4	10–18.3 ± 1.9	25–58.2 ± 9.3	8.7–10.2 ± 0.5	199–541.1 ± 108.7
2014	14.1–34.7 ± 5.7	12.2–18.4 ± 1.7	2.6–9.9 ± 1.4	10–15 ± 1.4	18.3–51.3 ± 8.7	9.3–11.2 ± 13.2	131.6–492.3 ± 105.6
2015	13.3–33.3 ± 5.5	11.0–15.9 ± 1.3	2.2–5.2 ± 0.7	10–13.3 ± 0.93	26.7–58.2 ± 10.2	11.0–13.2 ± 0.5	223.3–673.7 ± 150.4

Table 2. Monthly survival, mortality and detachment (%) of *A. downingi* nubbins in the nursery

Month	Mean Temperature (°C)	Survival %	Mortality (%)	Detachment (%)
September-13	29.44	94	4.2	1.25
October-13	24.39	94	0	0
November-13	22.67	94	0	0
December-13	16.4	76	0.83	16.67
January-14	16.33	76	0	0
February-14	16.5	76	0	0
March-14	21.09	56	1.25	19.17
April-14	24.21	56	0	0
May-14	27.98	56	0	0
June-14	29.39	56	0	0

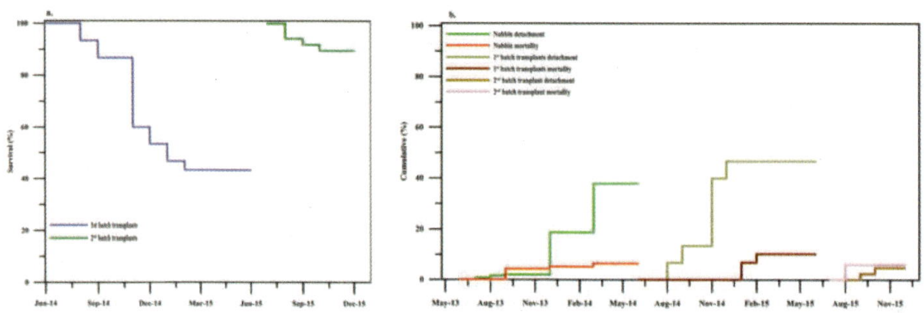

Figure 3. Survival, mortality and detachment (%), (a) survi-val of transplants and (b) detachment and mortality in nursery reared nubbins and transplants.

of transplants negative growth was witnessed during hottest summer, August–September 2014 and coldest winter months, January–February 2015 (Figure 4a).

Monthly mean skeletal extension of nursery reared *A. downingi* ranged from 0.03 (SD ± 4.0 mm) to 0.98 mm (SD ± 3.9 mm) and the annual skeletal extension was 2.8 mm yr^{-1} (Figure 5a, b). Whereas post transplantation, the monitored *A. downingi* colonies (n = 6) average GMD (mm) increased from 24.1 (SD ± 9.85 mm) to 73.6 (SD ± 29.09 mm) with an annual growth rate (GMD) of 50 mm yr^{-1} and the highest average GMD attained was from 23.5 mm to 102 mm (SD ± 20.8 mm) (by nubbin no. 8, Figure 5c, d) in 1 year. Linear regression of mean SST (°C) against average GMD (mm) of transplants showed negative correlation (Figure 6a; r^2 = 0.079, $p > 0.05$). Significant positive correlation is seen between average number of branches and average GMD (mm) of transplants (Figure 6b; r^2 = 0.928, $p < 0.05$). Partial bleaching occurred in transplanted colonies during August 2014 as the mean maximum sea surface temperature (°C) reached 34.7°C (Table 1); however, colonies completely recovered by November 2014.

The suspended nursery attracted colonization of benthic invertebrates and vertebrates which eventually established a diverse community (Table 3). Barnacles and bryozoans dominated the community and intensively competed for space with the nubbins. Barnacle infestation was more intense due to the shallow nursery site. Many fish species visited the nursery for food, and a few species such as *Apogon taeniatus, Pseudochromis persicus* and cryptic species such as Gobies and Blennies (Laith et al., 2015) sought a permanent refuge in the nursery. Intensive algae cover in the nursery was grazed by herbivore top shells such as *Priotrochus obscura*, but as they were low in abundance several algae species continued to dominate over several periods although regular cleaning procedures were adopted. During winter (November–February) drift, algae fouled the nursery.

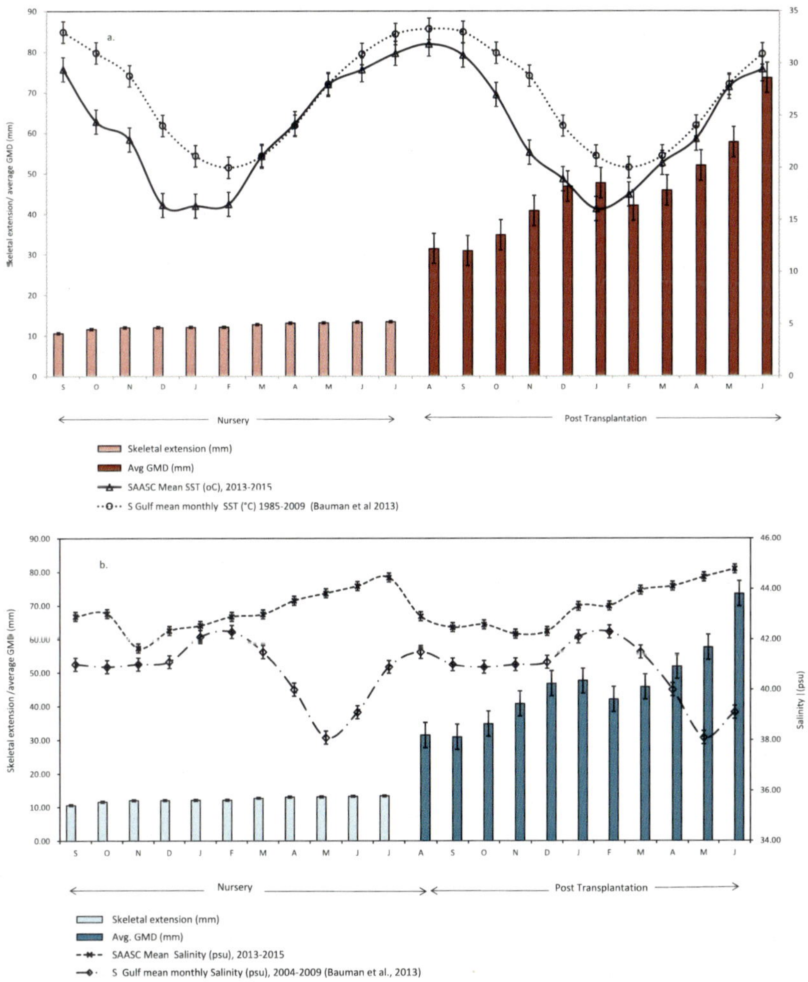

Figure 4. (a) SST (°C) impact on growth-mean skeletal extension (mm)/average GMD (mm) *in A. downingi* (b) Salinity (psu) impact on growth-mean skeletal extension (mm)/average GMD (mm) in *A. downingi.*

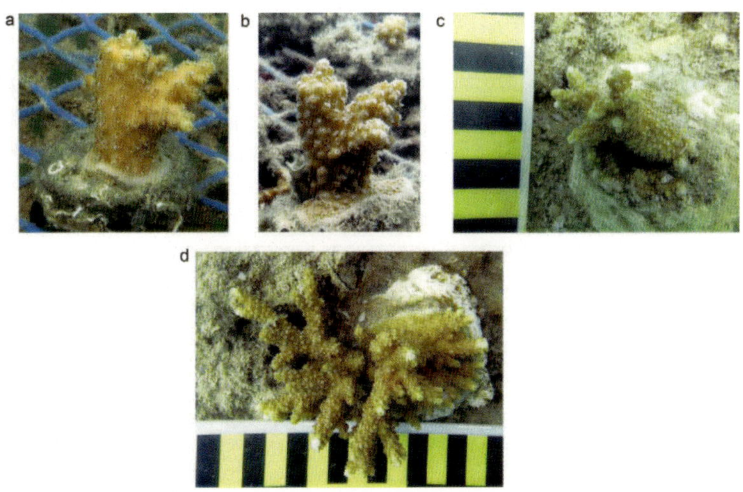

Figure 5. (a) *A. downingi* nubbin in the nursery self-attached to EAF slag substrate in a month;(b) one year old nubbin with impacted growth; (c, d) nubbin (average GMD 23.5 mm) trans-planted onto a open inclined rock surface using epoxy (June 2014) and in a year it attained an average GMD of 102 mm (SD ± 20.08 mm). Alternate yellow and black bars in the scale is 10 mm each. (Photos R. Dinesh Kumar.)

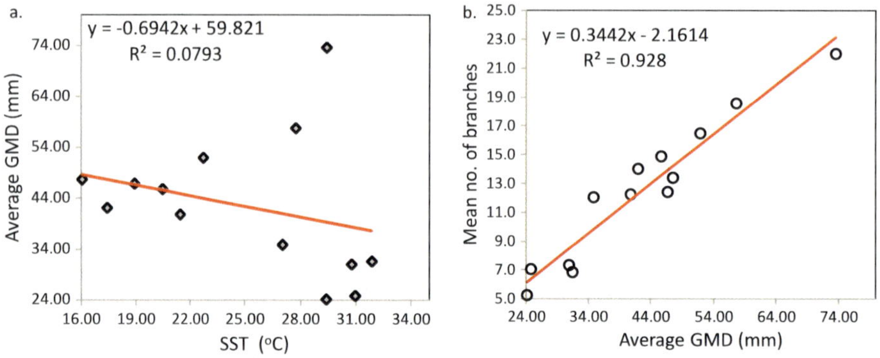

Figure 6. Linear regression analysis of transplanted *A. downingi*; (a) average GMD (mm) vs. SST (°C), (b) mean number of branches vs. average GMD (mm).

4. Discussion

The Kuwait Environmental Protection Society (KEPS), a voluntary organization in Kuwait, has carried out preliminary experiments adopting biorock® technique to grow coral fragments in the offshore islands of Kuwait for reef restoration (KEPS, personal communication).

The movement of nursery in all direction aided by currents helped in removal of debris and prevented accumulation of fine sediments. This increased water circulation was also beneficial to the nubbins as reported by Shafir et al. (2006a). Ambient nutrient levels recorded show nutrient levels similar in those of the open sea southern waters of Kuwait (Al-Yamani, Bishop, Ramadhan, Al-Husaini, & Al-Ghadban, 2004).

The survival and detachment rates (%) of *A. downingi* in the nursery and post transplantation are comparable with results obtained for *Acropora* spp from Gulf of Eilat, Red Sea (Shafir et al., 2006a, 2006b), Tanzania (Mbije, Spanier, & Rinkevich, 2013; Mbije et al., 2010), Egypt (Mohammed et al., 2012) and Thailand (Sujirachato, Thamrongnawasawat, Thongtham, Jantrarotai, & Worachananant, 2013). The higher detachment rate was not due to SST variations as *Acropora* spp thrives in extreme temperature conditions in both the northern and southern Gulf (Al-Yamani et al., 2004; Bauman et al., 2013) and is mostly due to poor adhesion and mechanical disturbances caused by fishes and divers during maintenance (Shaish et al., 2006b). However, nursery survival was lower compared to *A. formosa* reared in a shallow site using a suspended nursery in the Philippines (Shaish et al., 2008). Mortality of nursery reared nubbins in summer is probably due to temperature related stress, but in post spring (March 2014) additional mortality due to drift algae covering the nubbins demanded regular cleaning. In this case, increased maintenance also lead to increased detachment rate (Table 2). Mortality rate in the present study was related to stress caused by higher summer temperature (Table 2) and a partial bleaching episode that occurred in summer of 2014 and 2015. Temperature extremes challenge corals physiologically, lower seawater temperature recorded during winter in the northern Gulf (Saudi Arabia) waters showed lethal effects on *Acropora pharaonis* and *Platygyra daedalea* (Coles & Fadlallah, 1991). In their study in Philippines, Yap, Alino, and Gomez (1992) demonstrated that warm temperature can also induce mortality and bleaching in transplanted corals.

The SST (°C) values in the present study showed impact on growth of nubbins in the nursery at temperatures below 18°C and above 31°C (Figure 3a), which are lower than those recorded in southern gulf waters where coral community structure is much influenced by SST and recurrent bleaching events (Bauman et al., 2013). Corals in northern Gulf continue to survive in temperatures above the normal threshold for survival (Carpenter et al., 1997) reported for the region (Bauman et al., 2013; Riegl et al., 2012). Similarly mean salinity (psu) variations recorded in this study are higher compared to southern Gulf waters (Bauman et al., 2013) with no impact on growth (Figure 3b), perhaps indicating adaptability of Kuwait corals to a higher salinity regime (D'Angelo et al., 2015). The negative growth recorded in this study (Figure 4) was probably induced by mechanical disturbances during maintenance, fishes and partial bleaching rather than SST extremes.

Table 3. List of benthic community established in the suspended nursery (A = Abundant, C = Common and R = Rare)

Taxa	Occurrence
Phylum: Porifera	A,C
Phylum: Cnidaria	
Hydrozoa	A,C
Calliactis polypus	R
Phylum: Annelida	
Protohydroides sp	A, C
Phylum: Bryozoa	
Schizoporella errata	A, C
Phylum: Crustacea	
Amphibalanus amphitrite	A, C
Hyastenus hilgendorfi	R
Penaeus semisulcatus	R
Portunus segnis	R
Phlyum: Mollusca	
Pinctada cf radiata	C
Azumapecten ruschenbergerii	C
Priotrochus obscurus	R
Clypeomorus bifasciata	C
Sepia pharaonis	R
Phylum: Echinodermata	
Ophiothrix spp	C
Aquilonastra burtoni	C
Family: Sabellidae	C
Phylum: Vertebrata	
Class: Ascidacea	
Didemnum sp	A,C
Phallusia nigra	A,C
Class: Pisces	
Herklotsichthys lossei	R
Austrobatrachus dussumieri	R
Epinephelus coioides	R
Pseudochromis persicus	R
Cheilodipterus persicus	C
Apogon taeniatus	c
Ganthodon speciosus	R
Plectorhinchus sordidus	C
Diplodus sargus kotschyi	A,C
Acanthopargus bifasciatus	C
Scolopsis taeniatus	C
Acanthopagrus latus	C
Upeneus tragula	C
Parablennius opercularis	C
Petroscrites anclyodon	C
Gobiidae	C

In the present study, use of marine grade cement mixed with water proof additive (3:1), is a new approach to shorten the curing period. Although marine epoxy is an ideal adhesive for coral transplantation experiments (Dizon, Edwards, & Gomez, 2008), in this study the slow curing property of 10–15 min increased detachment rate due to strong currents. Hence, adhesives with a rapid curing property can be very effective, if transplantation experiments are to be carried out in high energy environments. The curing process is directly proportional to temperature, hence the transplantation experiments were conducted during early summer (May–June).

Self-attachment time of 60 days in *A. downingi* transplants is longer than results obtained for *Acroproa* spp in Philippines (Guest, Dizon, Edwards, Franco, & Gomez, 2011). The colonies in this study were transplanted on an inclined rock surface and the amount of epoxy/cement layer was very thick (>5 cm), hence extra energy and time is spent to encrust and reach the rock. In coral transplants self-attachment time is influenced by a combination of factors such as growth form, growth rate and environmental parameters, etc. (Guest et al., 2011).

During the post-transplantation period Average Geometric Mean Diameter (mm) was much higher compared to results obtained for *A. tenuis* transplanted in Philippines (Guest, Baria, Gomez, Heyward, & Edwards, 2014). Growth rate of inversely and normally transplanted *A. downingi* at Hengam Island, Iran (Southern Gulf) was measured as an increment of weight (Vajed Samei et al., 2012), so that methods of growth measurements adopted in the present study do not allow comparison. Annual growth rate of *Acropora downingi* in this study indicates that growth is up to 50 mm yr^{-1} which is comparable to the growth rate of 50–100 mm yr^{-1} for *Acropora* spp recorded by Riegl (2002) in Dubai, UAE. Survival and growth rates of *A. downingi* presented in this study are the first available for a coral species in the Northern Gulf where a high variation in SST in the range of 11.1–35°C is reported (Al-Yamani et al., 2004; Devlin et al., 2015). This is an even higher variation (23.9°C) than that was recorded (17°C) in Abu Dhabi during the recent nursery and transplantation experiment (Sen & Yousif, 2016).

Although growth impact on nubbins and transplants during hot summer and cold winter months were less prominent (Figure 4a), a similar trend was observed in Philippines wherein growth rates were not affected significantly by seasonal fluctuations in temperature (Yap et al., 1992). On the contrary a recent study on calcification rates in *A. downingi* in Hengam Island, Iran predicts that temperature fluctuations and irradiance levels can impact calcification rates seasonally, with winter months having higher calcifications rates than fall, spring and summer (Vajed Samei et al., 2016). However, environmental conditions in Kuwait waters are different from Hengam Island, Iran hence adaptability of corals to these conditions should vary considerably which needs further investigation. Other potential factors which may impact growth include biofouling, shallow nursery site and maintenance induced damage (Shafir et al., 2006b). Also the small size of fragments (10–30 mm) used in this study for nursery rearing could account for reduced growth rate as witnessed for *Acropora crevicornis* in Florida, US (Herlan & Lirman, 2008).

Only partial bleaching of *A. downingi* nubbins and transplants was observed in the summer of 2014 and 2015 despite the mean maximum SST (°C) value (Table 1) being equivalent to the bleaching threshold reported for Gulf corals (Riegl et al., 2012). However, post bleaching recovery in November 2014 was a month earlier than that was observed for *A. arabensis* nubbins reared in Abu Dhabi (Wilson & Marimuthu, 2012). Worldwide branching corals are highly susceptible to large-scale bleaching phenomenon (Edwards, 2010) due to SST anomalies and the same is true for the Gulf (Riegl et al., 2012). However, resilience and adaptability to environmental extremes appears to be a robust trait for northern Gulf coral populations which will be helpful for targeting future restoration experiments.

In situ nursery rearing and transplantation of *A. downingi* in the artificial lagoons of Sea City is successful, as is the use of EAF slag as a potential artificial substrate on which to rear corals as found by Mohammed et al. (2012) in Egypt and Japan (Okamoto et al., 2010; Oyamada et al., 2014). Intensive fouling by algae, barnacles, ascidians *Phallusia nigra, Didemnum* spp, the bryozoan

Schizoporella spp and polychaete *Hydroides* spp was probably influenced by EAF slag that deposits a layer of $CaCo_3$ over the upper surface upon reaction with seawater which attracts invertebrates (Mohammed et al., 2012), thus intensifying the maintenance regime. Hence, deeper sites with low biofouling impact are crucial for successful coral nursery operation as they yield better results (Edwards, 2010; Shafir et al., 2006a), however biofouling impacts can also be curtailed by restricted use of anti-fouling paints (Shafir, Abady, & Rinkevich, 2009) to reduce maintenance costs. There is scope for experimentation on nursery designs appropriate to shallow Gulf environments and selection of appropriate species for nursery rearing and transplantation.

The diverse benthic community colonization in the suspended nursery attracted several species of fish for food and shelter. Shafir and Rinkevich (2010) observed that floating mid-water nurseries in Red Sea attracted colonization of several species of benthic organisms and fishes thus functioning as an artificial reef. Thus, apart from functionality in rearing corals, suspended nurseries can serve to enhance biodiversity by acting as an artificial reef.

Although the Gulf coastline has been subjected to drastic modification due to intensive coastal development (Burt, 2014; Sheppard et al., 2010; Van Leveiren et al., 2011), some of these man-made habitats such as Sea City get to enhance marine biodiversity and create additional marine habitats when designed carefully. The present study demonstrates for the first time in Kuwait that it is possible to enhance the environmental value of artificially created marine habitats by adopting the "coral gardening concept" (Rinkevich, 2008) and rear corals in situ despite the extreme environmental conditions of the northern gulf to restore reefs. Further experiments on non-branching species, nursery designs to suit extreme environmental conditions in the gulf are essential for initiating future reef restoration efforts.

Acknowledgements
We would like to thank Mr. Fawaz Al Marzouq, Chairman of La Ala Al Kuwait Real Estate Co. K.S.C. for permission to publish the data and facilities provided, M/s. Ian Williams, Project Director, Sabah Al Ahmad Sea City for his constant support and encouragement, Chris Rose for his useful suggestions, Drs. David Jones, Spain, Baruch Rinkevich Israel Oceanographic Institute, Israel, Jerald Wilson, King Abdul Aziz University, Jeddah, Saudi Arabia, R.P. Kumarran, Marine Mammal Consultant, India and N. Marimuthu, Zoological Survey of India for their valuable advice in the various stages of this study. We thoughtfully acknowledge Dr. Shaker Al-Hazeem, Kuwait Institute for Scientifc Research for his valuable comments on the manuscript. We also extend our thanks to Dr. Saif Udin, Kuwait Institute for Scientific Research for the preparation of map, M/s. Bapu Rao Sontakke, Hassan Ali Hassan and Jag Jeevan Singh for their support during fieldwork.

Funding
The authors received no direct funding for this research.

Competing interests
The authors declare no competing interests.

Author details
Manickam Nithyanandan[1,2]
E-mail: nandan.ocean@gmail.com
ORCID ID: http://orcid.org/0000-0003-0582-8344
Lewis Le Vay[3]
E-mail: l.levay@bangor.ac.uk
Dinesh Kumar Raja[1,4]
E-mail: dineshantha@hotmail.com
Ratheesh Kesavan[1]
E-mail: foratish@gmail.com
Don Pereira[1]
E-mail: donpereira@hotmail.com
[1] Environmental Department, Sabah Al Ahmad Sea City, La Ala Al Kuwait Real Estate Co. K.S.C., Souk Al Kuwait 6th Floor, Office No. 613, P.O.Box 29964, Safat, Kuwait 13090, Kuwait.
[2] Ecosystem Based Management of Marine Living Resources Program, Environment and Life Sciences Research Center, Kuwait Institute for Scientific Research, P.O. Box 1638, Salmiya 22017, Kuwait.
[3] Centre for Applied Marine Sciences, College of Natural Sciences, Bangor University, Menai Bridge, Anglesey LL59 5AB, Wales, UK.
[4] Environment World Company, Suite 26, 3rd Floor, Al-Mubarak Commercial Complex, Street 50, Block 1, Jleeb Al-Shuyoukh, Kuwait.

References
Al-Yamani, F. Y., Bishop, J., Ramadhan, E., Al-Husaini, M., & Al-Ghadban, A. (2004). *Oceanographic atlas of Kuwait waters* (203 p). Kuwait: Kuwait Institute for Scientific Research.

American Public Health Association. (1995). *Standard methods for the examination of water and waste water* (19th ed.). Washington, DC: Author.

Bauman, A. G., Pratchett, M. S., Baird, A. H., Riegl, B., Heron, S. F., & Feary, D. A. (2013). Variation in the size structure of corals is related to environmental extremes in the Persian Gulf. *Marine Environmental Research, 84*, 43–50. doi:10.1016/j.marenvres.2012.11.007

Burt, J. A. (2014). The environmental costs of coastal urbanization in the Arabian Gulf. *City, 18*, 760–770. doi:10.1080/13604813.2014.962889

Carpenter, K. E., Harrison, P. L., Hodgson, G., Al-Saffar, A. H., & Al-Hazeem, S. H. (1997). *The corals and coral reef fishes of Kuwait*. Kuwait: Kuwait Institute for Scientific Research.

Clark, S., & Edwards, A. J. (1995). Coral transplantation as an aid to reef rehabilitation: Evaluation of a case study in the Maldive Islands. *Coral Reefs, 14,* 201–213. doi:10.1007/BF00334342

Coles, S. L., & Fadlallah, Y. H. (1991). Reef coral survival and mortality at low temperatures in the Arabian Gulf: New species-specific lower temperature limits. *Coral Reefs, 9,* 231–237. doi:10.1007/BF00290427

D'Angelo, C., Hume, B. C. C., Burt, J., Smith, E. G., Achterberg, E. P., & Wiedenmann, J. (2015). Local adaptation constrains the distribution potential of heat-tolerant Symbiodinium from the Persian/Arabian Gulf. *The ISME Journal, 9,* 2551–2560. doi:10.1038/ismej.2015.80

Devlin, M. J., Massoud, M. S., Hamid, S. A., Al-Zaidan, A., Al-Sarawi, H., Al-Enezi, M., ... Lyons, B. P. (2015). Changes in the water quality conditions of Kuwait's marine waters: Long term impacts of nutrient enrichment. *Marine Pollution Bulletin, 100,* 607–620. doi:10.1016/j.marpolbul.2015.10.022

Dizon, R. M., Edwards, A. J., & Gomez, E. D. (2008). Comparison of three types of adhesives in attaching coral transplants to clam shell substrates. *Aquatic Conservation: Marine and Freshwater Ecosystems, 18,* 1140–1148. doi:10.1002/aqc.v18:7

Edwards, A. J. (2010). *Reef rehabilitation manual* (ii + 166 pp). St. Lucia: Coral Reef Targeted Research & Capacity Building for Management Program.

Gomez, E. D., Cabaitan, P. C., Yap, H. T., & Dizon, R. M. (2013). Can coral cover be restored in the absence of natural recruitment and reef recovery. *Restoration Ecology, 22,* 142–150. doi:10.1111/rec.12041

Guest, J. R., Baria, M. V., Gomez, E. D., Heyward, A. J., & Edwards, A. J. (2014). Closing the circle: Is it feasible to rehabilitate reefs with sexually propagated corals? *Coral Reefs, 33,* 45–55. doi:10.1007/s00338-013-1114-1

Guest, J. R., Dizon, R. M., Edwards, A. J., Franco, C., & Gomez, E. D. (2011). How quickly do fragments of coral "self attach" after transplantation? *Restoration Ecology, 19,* 234–242. doi:10.1111/rec.2011.19.issue-2

Herlan, J., & Lirman, D. (2008). Development of a coral nursery program for the threatened coral *Acropora crevicornis* in Florida. In *Proceedings of the 11th international coral reef symposium* (pp. 1244–1249). Ft. Lauderdale, FL.

Jones, D. A., & Nithyanandan, M. (2013). Recruitment of marine biota onto hard and soft artificially created subtidal habitats in Sabah Al-Ahmad Sea City, Kuwait. *Marine Pollution Bulletin, 72,* 351–356. doi:10.1016/j.marpolbul.2012.11.001

Jones, D. A., & Nithyanandan, M. (2015). If it is possible to create and manage marine ecosystem in a desert, why is the whole gulf ecosystem under threat? In *Qatar University life science symposium 2015* (Abstract). doi:10.5339/qproc.2015.qulss2015.14

Jones, D. A., Nithyanandan, M., & Williams, I. (2012). Sabah Al-Ahmad Sea City Kuwait: Development of a sustainable man-made coastal ecosystem in a saline desert. *Aquatic Ecosystem Health & Management, 15* (S1), 84–92. doi:10.1080/14634988.2012.663706

Katakura, N., Yuriko, T., Sumihiro, T., Hiroshima, H., & Iwata, I. (2013). Development and practical applications of transplantation technique for coral and seagrass in sub-tropical island. *Journal of Japan Society for Civil Engineers Series B3 (Ocean Engineering), 69,* 1078–1083 (In Japaense).

Kerby, A. B. (2014). *Best practices manual for Caribbean Acropora restoration* (p. 39). Puntacana: Punctacana Foundation.

Laith, A. J., Nithyanandan, M., Raja, D. K., Kesavan, R., & Pereira, D. (2015). First record of the pixie triplefin Enneapterygius pusillus (Pisces: Tripterygiidae) and confirmation of the presence of cheekspot blenny Parablennius opercularis (Blenniidae) in the north Arabian Gulf. *Marine Biodiversity Records, 8,* e8. doi:10.1017/S1755267214001419

Mbije, N. E., Spanier, E., & Rinkevich, B. (2013). A first endeavour in restoring denuded, post bleached reefs in Tanzania. *Estuarine and Coastal Shelf Science, 128,* 41–51. doi:10.1016/j.ecss.2013.04.021

Mbije, N. E. J., Spanier, E., & Rinkevich, B. (2010). Testing the first phase of the 'gardening concept' as an applicable tool in restoring denuded reefs in Tanzania. *Ecological Engineering, 36,* 713–721. doi:10.1016/j.ecoleng.2009.12.018

Mohammed, T. A., Hamed, A. A., Habib, N. F., Ezz El-Arab, M. A., & El-Moshley, K. H. M. (2012). Coral rehabilitation using steel slag as a substrate. *International Journal of Environment Protection, 2,* 1–5.

Okamoto, M., Yap, M., Roeroe, A. K., Nojima, S., Oyamada, K., Fujiwara, S., & Iwata, I. (2010). In situ growth and mortality of juvenile Acropora over 2 years following mass spawning in Sekisei Lagoon, Okinawa (24°N). *Fisheries Science, 76,* 343–353. doi:10.1007/s12562-010-0222-x

Oyamada, K., Okamoto, M., & Iwata, I. (2014). Development of restoration technology for coral reefs using "Marine Block™". *JFE Technical Report, 19,* 118–125.

Putchim, L., Thongtham, N., Hewett, A., & Chansang, H. (2008). Survival and growth of *Acropora spp.* In mid-water nursery and after transplantation at Phi Phi islands, Andaman Sea, Thailand. In *Proceedings of the 11th international coral reef symposium 2* (pp. 1263–1266). Ft. Lauderdale, FL.

Riegl, B. M. (2002). Effects of the 1996 and 1998 positive sea-surface temperature anomalies on corals, coral diseases and fish in the Arabian Gulf (Dubai, UAE). *Marine Biology, 140,* 29–40. doi:10.1007/s002270100676

Riegl, B. M., Purkis, S. J., Al-Cibahy, A. S., Al-Harthy, S., Grandcourt, E., Al-Sulaiti, K., ... Abdel-Moati, A. M. (2012). Coral bleaching and mortality thresholds in the SE Gulf: Highest in the world. In B. M. Riegl & S. J. Purkis (Eds.), *Coral reefs of the Gulf: Adaptations to climatic extremes, coral reefs of the world 3* (pp. 95–105). Netherlands: Springer Science + Business Media BV.

Rinkevich, B. (2008). Management of coral reefs: We have gone wrong when neglecting active reef restoration. *Marine Pollution Bulletin, 56,* 1821–1824. doi:10.1016/j.marpolbul.2008.08.014

Sen, S., & Yousif, O. M. (2016). Development of a coral nursery as a sustainable resource for reef restoration in Abu Al Abyad Island, Abu Dhabi, United Arab Emirates, Arabian Gulf. *Galaxea, 18,* 3–8. doi:10.3755/galaxea.18.1_3

Shafir, S., Abady, S., & Rinkevich, B. (2009). Improved sustainable maintenance for mid-water coral nursery by the application of an anti-fouling agent. *Journal of Experimental Marine Biology and Ecology, 368,* 124–128. doi:10.1016/j.jembe.2008.08.017

Shafir, S., Rijin, J. V., & Rinkevich, B. (2006a). A mid-water coral nursery. In *Proceedings of the 10th intenational coral reef symposium* (pp. 1674–1679). Okinawa.

Shafir, S., Rijin, J. V., & Rinkevich, B. (2006b). Steps in the construction of underwater coral nursery, an essential component in reef restoration acts. *Marine Biology, 149,* 679–687. doi:10.1007/s00227-005-0236-6

Shafir, S., & Rinkevich, B. (2010). Integrated long-term mid-water coral nurseries: A management instrument evolving into a floating ecosystem. *University of Mauritius Research Journal, 16,* 365–385.

Shaish, L., Levy, G., Gomez, E., & Rinkevich, B. (2008). Fixed and suspended coral nurseries in the Philippines: Establishing the first step in the "gardening concept" of reef restoration. *Journal of Experimental Marine Biology Ecology*, *358*, 86–97. doi:10.1016/j.jembe.2008.01.024

Sheppard, C. (2015). Coral reefs in the Gulf are mostly dead now, but can we do anything about it? *Marine Pollution Bulletin*. doi:10.1016/j.marpolbul.2015.09.031

Sheppard, C., Al-Husiani, M., Al-Jamali, F., Al-Yamani, F., Baldwin, R., Bishop, J., ... Zainal, K. (2010). The Gulf: A young sea in decline. *Marine Pollution Bulletin*, *60*, 13–38. doi:10.1016/j.marpolbul.2009.10.017

Sujirachato, P., Thamrongnawasawat, T., Thongtham, N., Jantrarotai, P., & Worachananant, S. (2013). Survival rate of coral fragments transplanted by different methods. *Galaxea*, *15*, 351–358. doi:10.3755/galaxea.15.351

Vajed Samiei, J., Koosha, D., & Behrooz, A. (2012). Short term growth rate of *Acropora downingi* in the coral reef of Hengam island, the Persian Gulf. *Journal of Persian Gulf (Marine Science)*, *3*, 49–54.

Vajed Samiei, J., Saleh, A., Shirvani, A., Fumani, N. S., Hashtroudi, M., & Pratchett, M. S. (2016). Variation in calcification rate of *Acropora downingi* relative to seasonal changes in environmental conditions in the northeastern Persian Gulf. *Coral Reefs*, *35*, 1371–1382. doi:10.1007/s00338-016-1464-6

Van Leveiren, H., Burt, J. D., Feary, D. A., Cavalcante, G., Marquis, E., Benedettii, L., ... Sale, P. F. (2011). *Managing the growing impacts of coastal development on fragile costal and marine ecosystems: Lessons from the Gulf. A policy report*. Hamilton, ON: UNU-INWEH.

Wilson, J., & Marimuthu, N. (2012). Post bleaching recovery of transplanted coral nubbins in Abu Al Abyad island of Abu Dhabi, United Arab Emirates. *Galaxea*, *14*, 1–2. doi:10.3755/galaxea.14.1_1

Yap, H. T. (2009). Local changes in community diversity after coral transplantation. *Marine Ecology Progress Series*, *374*, 33–41. doi:10.3354/meps07650

Yap, H. T., Alino, P. M., & Gomez, E. D. (1992). Trends in growth and mortality of three coral species (Anthozoa: Scleractinia), including effects of transplantation. *Marine Ecology Progress Series*, *83*, 91–101. doi:10.3354/meps083091

7

Environmental framing on Twitter: Impact of Trump's Paris Agreement withdrawal on climate change and ocean acidification dialogue

Sojung Claire Kim[1*] and Sandra L. Cooke[2]

*Corresponding author: Sojung Claire Kim, Department of Communication, George Mason University, Fairfax, VA 22030, USA
E-mail: skim205@gmu.edu

Reviewing editor: Ioannis N Daliakopoulos, Environmental Engineering, Technical University of Crete, Greece

Abstract: The role of social media in communicating emerging environmental issues has received little attention. One such issue is ocean acidification (OA), the process by which carbon dioxide (CO_2) acidifies oceans. Although scientists consider OA to be as dangerous as climate change (CC) and both problems are caused by excess CO_2 emissions, public awareness of OA is low. We investigated public discussions about CC and OA on Twitter, identifying frames and tweeter characteristics. Tweeting patterns before and after President Trump's 1 June 2017 announcement of the U.S.'s withdrawal from the international Paris Climate Agreement were compared because of the potential for diverse framing of this globally communicated event. For CC tweets, Political/Ideological Struggle (PIS) and Disaster (DS) frames were prevalent, with PIS frames increasing threefold after Trump's announcement. DS, Settled Science (SS), and Promotional frames were prevalent among OA tweets, with SS decreasing and PIS increasing after the announcement. Our findings suggest that Trump's decision sparked discourse on CC and facilitated expressions of politicized opinions on Twitter. We conclude that with a careful understanding of issue familiarity among its publics, social media can be effective for disseminating information and opinion of established and emerging environmental issues, complementing traditional media outlets.

ABOUT THE AUTHORS

Sojung Claire Kim (Ph.D., University of Wisconsin-Madison) is an Assistant Professor in the Department of Communication at George Mason University. Her research interests broadly lie on intersections of interactive media, health and risk communication, and strategic messaging. Research contexts primarily concern challenging health and environmental issues such as cancer, addiction, and climate change, as well as verbal and nonverbal interplay of doctor-patient communication.

Sandra L. Cooke (Ph.D., Lehigh University) is an Assistant Professor of Environmental Science in the Biology Department at High Point University. Primarily trained as an aquatic ecologist, her research interests include the effects of ultraviolet radiation and other stressors in freshwater systems, the ecology of aquatic invasive species, and the communication and public perception of critical issues in environmental science.

PUBLIC INTEREST STATEMENT

Ocean acidification (OA) is the process by which carbon dioxide (CO_2) dissolves into and acidifies the world's oceans. Scientists consider OA to be just as dangerous a problem as climate change (CC) and both problems share the same main cause (excess CO_2 emissions), but most non-scientists are unaware of OA. We investigated how people discuss the well-known problem of CC and the lesser-known problem of OA on Twitter. We also compared CC and OA tweeting patterns before and after President Trump's announcement of the U.S.'s withdrawal from the Paris Climate Agreement. The results suggest that Trump's withdrawal decision sparked debate of CC on Twitter and that tweets about CC usually have a political bent or incorporate disaster themes. OA tweets also invoke disaster themes, but also tend to be "promotional" in nature (e.g. announcing a conference on OA) and are less political.

Subjects: Environment & Society; Environmental Politics; Communication Research Methods; Communication Theory; EnvironmentalCommunication

Keywords: environmental framing; climate change; Ocean acidification; Trump Paris agreement withdrawal; social media; content analysis

There are over 3 billion social media users worldwide (Kemp, 2017). Social media applications include collaborative projects (e.g. Wikipedia), blogs, content communities (e.g. YouTube), social networking sites (e.g. Facebook), virtual social worlds (e.g. Second Life), and virtual gaming (Kaplan & Haenlein, 2010). While Facebook is the most-used platform, use of other social media platforms such as the microblogging service Twitter has steadily increased worldwide: as of the second quarter of 2017, there are 328 million monthly active Twitter users (Statista, 2017). This rising popularity is one of the reasons that Twitter is of particular interest to scholars who study online communication.

Social media has played an important role in climate change (CC) communication (Schäfer, 2012). Social network analysis of CC discussions on Twitter show that users can be classified into climate "activist" and "skeptic" groups and that these groups more often interact among themselves in "echo chambers," which tend to increase polarization on the topic (Williams, McMurray, Kurz, & Lambert, 2015). However, some between-group interactions (e.g. open forums) do occur, which can stimulate debate on CC (Williams et al., 2015). These discussion dynamics on social media have practical implications for offline CC communication because social media can influence mainstream media on scientific issues and because social media may have a greater influence on individuals' perceptions of issues due to its public, participatory nature (Williams et al., 2015).

Despite active discussions of CC on social media, its role in communicating lesser-known environmental issues has not been well-studied, including issues that are closely related to CC. One example is ocean acidification (OA), which is the process by which carbon dioxide (CO_2)—primarily produced from fossil fuel burning—acidifies the world's oceans. OA is similar to CC in that both problems are caused by excessive CO_2 emissions from human activities and thus the solution to both problems centers on emissions reduction strategies. The passage of the Federal Ocean Acidification Research and Monitoring Act by U.S. Congress in 2009 and the introduction of additional OA legislation in 2017 demonstrates that politicians are aware of the problem. Although scientists consider OA to be just as "dangerous" a problem as CC (Monaco Declaration, 2008) with negative economic impacts already evident in some regions such as the U.S. Northwest Pacific Coast, public awareness of OA has been low (Capstick, Pidgeon, Corner, Spence, & Pearson, 2016; The Ocean Project, 2012). This may result in less conversation about the issue and little knowledge to take necessary actions among the public. OA may therefore be described as an "emerging" environmental issue in the public mindset (Yates et al., 2015).

The purpose of this study is, therefore, to compare how Twitter users communicate about CC and OA and to compare the content framing and external URL sources of tweets of these two interrelated issues before and after a major global event—U.S. President Donald Trump's announcement on 1 June 2017 that he is withdrawing the U.S. from the 2015 Paris Agreement. The Paris Agreement is a multi-national agreement concerning carbon emissions reductions within the United Nations Framework Convention on Climate Change. This study also broadens our understanding of social media and environmental issues by shedding light on how different actors (e.g. scientists, politicians, activists, the general public) tweet about an established contested issue (CC) in comparison to an emerging issue (OA). A comparison of how the two issues are discussed on social media and who is engaging in the discourse could be useful for scientists and other advocates aiming to increase public awareness of OA. Furthermore, given that OA may be an effective frame for increasing support for carbon emissions reduction policies (Mossler, Bostrom,

Kelly, Crosman, & Moy, 2017), it may be valuable to examine if and how Twitter users connect the two issues when thinking about anthropogenic CO_2 emissions.

1. Social media, framing, and environmental issues

Social media has emerged as an important tool for advocacy, information dissemination, and protest of critical issues such as risk, health, and environmental communication. Although environmental issues are less personal than these other issues and more "psychologically distant" (Schuldt, McComas, & Byrne, 2016), people nevertheless use Twitter and other social media platforms to communicate about environmental topics. When the topics are contentious, discourse on social media may become segmented among users with varying opinions on the issue, as was shown in a Twitter analysis of hydraulic fracturing, a generally well-known and controversial method of extracting natural gas (Hopke & Simis, 2017). Most of the social media discourse on the publicly contentious issue of CC tends to occur within like-minded user groups, although some discussion between groups with differing perspectives does occur (Pearce, Holmberg, Hellsten, & Nerlich, 2014; Williams et al., 2015). Twitter can also be used as an organizing mechanism for collective environmentalism, as demonstrated by its use in two major CC protests related to the 2009 United Nations climate talks in Copenhagen (Segerberg & Bennett, 2011). Social media also has a potential use in dialogic communication of environmental issues, enabling scientists to communicate directly with the public rather than relying on journalists (Lee, VanDyke, & Cummins, 2017).

Although most social media platforms tend to be less textually developed and content-rich compared to traditional forms of communication, it is possible to identify frames in social media postings, including in tweets (Twitter postings well-known for their 140 character limit). The ability to hyperlink to other online sources and utilize hashtags (searchable keywords that are hyperlinked to other tweets with the same hashtag) enhances how a Twitter user can frame their message. Framing refers to making selected aspects of a perceived reality more salient in order to "promote a particular problem definition, causal interpretation, moral evaluation, and/or treatment recommendation" (Entman, 1993, p. 52). Although a frame may not perform all of these functions (defining a problem, diagnosing causes, making moral judgments, and suggesting solutions), a text that performs just one of these actions is increasing the salience of a selected piece of information (Entman, 1993). Framing, therefore, is a central concept to the field of communication because it helps researchers understand the influence of a communicating text (Entman, 1993). The effect of a frame can vary because it depends on how a receiver interprets a communicator's text. But for many frames a "dominant meaning" can be discerned—that with the highest probability of being noticed by most people (Entman, 1993). Framing analysis can take a number of methodological approaches including, for example, developing a framing schema and then systematically and quantitatively applying (i.e. coding) that schema to some texts of interest (Kitzinger, 2007). Other approaches, however, are more of a qualitative rhetorical or discourse analysis (Kuypers, 2010).

Several studies have examined the framing of CC related messages on Twitter. An analysis of tweets in response to the Intergovernmental Panel on Climate Change's (IPCC) Fifth Assessment Report[1] found that the most common frame used was "Settled Science," which acknowledges the scientific consensus on CC and focuses on the validity of scientific findings (O'Neill, Williams, Kurz, Wiersma, & Boykoff, 2015). This contrasts with print media coverage, for which the dominant frame in response to the IPCC report was "Political or Ideological Struggle," which focuses on policies and other strategies to address CC and the conflicts surrounding those strategies (O'Neill et al., 2015). A study of geographic differences in CC frames on Twitter found that "hoax" frames (asserting that CC is not real) are more prevalent in the U.S. compared to the U.K., Australia, and Canada and in "red" states versus "blue" states (Jang & Hart, 2015). Other frames identified were "real" (asserting that CC is really happening), "cause" (discussing the causes of CC, including human activities), "impact" (emphasizing effects of CC), and "action" (messages about policies and other actions to combat CC), which was the most common of the five frames (Jang & Hart,

2015). Collectively, these framing analyses indicate that there is a high degree of frame contestation in CC Twitter dialogue—that is, one or a few frames dominating the discourse among multiple competing frames—as opposed to frame parity, where all frames are equally prevalent (Entman, 2003). Broadly speaking, frame contestation tends to be the norm, at least in the news media (Entman, 2003).

Studies of online communication of CC have also examined how users engage with traditional versus new media in their messages. An analysis of the "media ecology" of 60,000 CC tweets found that mainstream media comprised the majority of URLs shared and that there was a lack of source diversity (Veltri & Atanasova, 2015). Similarly, tweets in response to the IPCC Fifth Assessment Report primarily cited mainstream media sources, although new media sources such as blogs and science news websites comprised over 40% of the URLs (Newman, 2016). On the other hand, tweets with hashtags related to two CC protests mostly contained links to "middle media" sources (e.g. alternative news media, NGOs, climate information sites) rather than mainstream media (Segerberg & Bennett, 2011).

2. Established vs. emerging issues: Climate change and ocean acidification

The way that individuals engage with, process information about, and ultimately make sense of a particular issue is affected by their level of familiarity with the topic (Marcu, Uzzell, & Barnett, 2011; Moscovici, 2000) We define a particular topic as "established" if there is high public awareness of the issue and if individuals have—to borrow from terminology used in the Risk Information Seeking and Processing theory (RISP)—a high perceived level of information sufficiency regarding the topic (Griffin, Dunwoody, & Neuwirth, 1999). In contrast, an "emerging" issue has low awareness and high unfamiliarity among the public, which corresponds to lower levels of information sufficiency when individuals are confronted with the issue. According to the RISP model, information sufficiency is the amount of information people believe they need to deal with a given risk, and it is a key factor that influences information seeking intentions and behaviors (Griffin et al., 1999). The RISP theory could be broadened beyond information seeking intentions about risks to posit that the way individuals seek information about, engage with, and communicate about a specific issue will depend on one's awareness and information sufficiency regarding the issue. Therefore, it could be posited that people may process information and communicate about established and emerging issues in different ways.

When comparing communication of established and emerging issues, it is important to note that low awareness and low information sufficiency do not necessarily mean that people do not have opinions about emerging issues. For example, although polling data indicate that Americans know little to nothing about nanotechnology, they nevertheless form opinions on the topic that appear to be shaped by their value systems (such as religious beliefs) and heuristic cues (such as media representations) (Runge et al., 2013). Experimental work suggests that the ideological context in which individuals encounter an unfamiliar topic influences their willingness to seek information about the topic from countervailing sources (Yeo, Xenos, Brossard, & Scheufele, 2015). Specifically, people exposed to information without clear ideological cues on the unfamiliar topic of nanotechnology were more likely to seek further information only from sources consistent with their ideological value predispositions (Yeo et al., 2015). For established issues, on the other hand, people already have ideological frameworks in place that help them process new information about the issue (Yeo et al., 2015), and thus the context of the new information is not as influential.

Regarding our study topics, CC is clearly an established issue, at least in the developed world where CC awareness is over 90% (Lee, Markowitz, Howe, Ko, & Leiserowitz, 2015). In contrast, public awareness of OA in the U.S. and Europe is very low (Capstick et al., 2016; Gelcich et al., 2014; The Ocean Project, 2012) despite scientists' views that this problem is just as critical as CC and despite the fact that, like CC, OA is caused by excessive anthropogenic CO_2 emissions. Consistent with the topic's low public awareness, OA receives much less coverage in the mainstream and social media compared to CC. To illustrate, a search of the ProQuest database on 18 July 2018 of the terms "climate change" or "global warming" yielded over 2.6 million results within non-scholarly media sources (e.g. newspapers, magazines, wire feeds, blogs, podcasts, websites), whereas a search of "ocean acidification" within the same sources yielded just over 15,000 results. One consequence of CC's prevalence in the popular media is that media use plays an important role on people's information sufficiency on CC (Ho, Detenber, Rosenthal, & Lee, 2014).

3. U.S. Paris Agreement withdrawal announcement and political polarization in Twitter

It has been argued that science communication is equivalent to political communication, especially on polarizing topics such as CC, and thus it is important to understand the political context in which communication of scientific topics takes place (Scheufele, 2014). This is particularly true for communication on Twitter because it is a platform where politically segregated communication networks may form (Conover et al., 2011; Gruzd & Roy, 2014). Furthermore, Twitter is a powerful medium for real-time discourse on global events (Becker, Naaman, & Gravano, 2011) and is a useful tool for gatewatching: monitoring newsworthy organizations for items of interest to a user who selects and disseminates (and may comment on) the chosen information (Bruns, Highfield, & Lind, 2012).

Since the election of U.S. President Donald Trump, his administration's approach toward science and environmental issues has proved to be a critical turning point for many of these issues. This may be best exemplified by Trump's decision concerning the 2015 Paris Agreement, which is a non-binding agreement signed by 195 nations. The agreement centers on carbon emissions reductions within the United Nations Framework Convention on Climate Change (United Nations Framework Convention on Climate Change, 2017). It is also known as COP21 because it was developed at the 21st Conference of the Parties to the UN Framework Convention on Climate Change, which—with 179 nations represented—was one of the largest gatherings of world leaders in recent history (World Economic Forum, 2016). Each country that signed and later ratified the agreement has pledged to reduce their carbon emissions by amounts that the country specified. The U.S., which is the global leader in per capita carbon emissions, agreed (under the Obama administration) to reduce their emissions 26–28% below 2005 levels by 2025 (United Nations Framework Convention on Climate Change, 2016).

During his candidacy in 2016, Trump declared his intentions to withdraw from the Paris Agreement if elected (McGrath, 2016), but there was disagreement on this issue among some of his advisors in the early months of his presidency (Davenport, 2017). Furthermore, even some energy industry leaders argued that remaining in the agreement would be best for U.S. energy interests (Volcovici, 2017). However, on 1 June 2017, Trump formally announced his decision to withdraw the U.S. from the agreement. The announcement sparked responses worldwide in traditional and social media outlets, including Twitter (Jain, 2017). This global political event is a driving force to discuss CC support and could potentially translate into thinking more about closely related but lesser known issues such as OA, for which the Paris Agreement has critical implications (Magnan et al., 2016). More importantly, this political event could further politicize the discourse of CC, especially on Twitter where CC discussions are already segmented (Williams et al., 2015). In other words, regarding frames on Twitter, the Paris Agreement announcement could result in frame changing, which is the tendency of frames used in news coverage of a momentous issue to change over time, resulting in different aspects of the issue being emphasized at different points in time (Houston, Pfefferbaum, & Rosenholtz, 2012). The present study, therefore, examines Twitter discussions on CC and OA before and after the Paris Agreement withdrawal announcement in an effort to identify effects of this political event.

4. Hypotheses and research questions

Given that CC is more established and politically contested than OA, it is reasonable to expect that CC tweets receive more attention from the public and their contents and argument tones tend to be more diverse and abundant compared to OA tweets. Also, a wide range of different media sources often discusses CC, while we predict that more narrowly focused science and environmental news outlets discuss OA. We assume that this applies to Twitter, hypothesizing that CC tweets will include links from more diverse media sources, including mainstream media, than OA tweets. Moreover, we want to examine if CC Twitter users would differ from users who tweet about OA regarding their professional background and the level of involvement in the issue. Because OA is a lesser-known issue relative to CC, we assume that more involved issue publics and/or experts such as scientists or environmental journalists will tweet about OA more than the general publics (Lotan, Graeff, Ananny, Gaffney, & Pearce, 2011; Vis, Faulkner, Parry, Manyukhina, & Evans, 2013). *The main content characteristics* (e.g. content frames, additional external links) of CC and OA tweets are, therefore, examined under H1, and the *main user characteristics* (e.g. topical influencers, user types) of CC and OA tweets are explored under H2.

H1: CC tweets will encompass (a) a greater diversity of content frames and (b) more mainstream external media sources, compared to OA tweets.

H2: CC Twitter (a) users as well as (b) topical influencers will exhibit a greater diversity of user characteristics, compared to OA users, which will likely be more experts.

Further, as the core investigative focus of this study, we explore *potential changes* in different characteristics of CC and OA tweets before and after Trump's announcement of leaving the Paris Agreement. We also examine if CC and OA Twitter users would mention the Paris Agreement withdrawal announcement and if so, their positions on it. Therefore, the following research questions are proposed:

RQ1: What CC and OA *content characteristics* are prevalent before and after President Trump's withdrawal announcement?

RQ2: What CC and OA *user characteristics* are prevalent before and after President Trump's withdrawal announcement?

RQ3: Will CC and OA Twitter users form *different attitudes* (e.g. supportive, neutral, oppose, no mention) concerning support of the Paris Agreement before and after President Trump's withdrawal announcement?

5. Methods

5.1. Data collection

Relevant tweets were collected over six weeks in May and June of 2017 (5/7/2017–6/15/2017) using DiscoverText, which is online text analytical software that retrieves Twitter data through Application Programming Interface (API)-based search (Shulman, 2011). The time window was chosen to capture communication patterns of the public on CC and OA and their reactions to Trump's decision on the Paris Agreement withdrawal on 1 June 2017. Hashtags #climatechange and/or #globalwarming were used for identifying CC tweets and a hashtag #oceanacidification was used to search for OA tweets.

A total of 221,668 CC tweets and 1,104 OA tweets were collected. To prepare the data for further analysis, all of the retweets (tweets with RT in the beginning of text) and exact duplicates were deleted. When deleting duplicates, we kept one tweet with the highest influence score (follower/following ratio) among all the duplicates. After deletion, 71,098 CC tweets (20,441 tweets before Paris Agreement withdrawal and 50,657 after the withdrawal) and 359 OA tweets (251 before Paris Agreement withdrawal and 108 after the withdrawal) remained.

Because there was huge discrepancy in tweet volume for CC and OA topics, we chose a random sampling approach. To have an equal number of tweets in each coding category (100 each for before and after Trump's announcement for CC and OA), 400 tweets were randomly selected for further coding and analysis. Previous literature has used random selection of subsets (e.g. Segerberg & Bennett, 2011; Veltri & Atanasova, 2015) or used purposive sampling such as influence rating (O'Neill et al., 2015) or retweet frequency (e.g. So et al., 2016) as a systematic way to choose sub-tweets. It was essential for this study to employ the random selection method since we were particularly interested in how the general public—not influencers or opinion leaders on the issues—would communicate and discuss CC and OA in Twitter. After excluding irrelevant and deleted accounts at the time of coding, a final sample of 396 tweets (n = 198 each) was included in the analyses. Although the final sample size seems relatively small, random sampling ensures a fair comparison of content and user characteristics of CC and OA tweets. This analysis, therefore, offers an exploratory "snapshot" of OA and CC Twitter dialogue and may be considered a case study that provides good insights into how a key event can shape the framing of established and emerging issues.

5.2. Coding procedures, categories, and reliability

The unit of analysis was each tweet. Metadata such as the date of creation; username; user description; number of followers, following (friends), and likes; and hashtag(s) were collected along with the number of retweets and tweet contents.

In terms of coding categories, content frames were adopted and updated from a study that examined framing in media responses to the IPCC Fifth Assessment Report on CC (O'Neill et al., 2015). Table 1 describes the frames in greater detail with examples. If included, external media sources (none, science news, mainstream media, new media, social media, government reports/ academic articles, advocacy, images like cartoons, videos, or others)[2] of CC and OA tweets were also coded. User characteristics were adapted from Hopke and Simis (2017) and were coded into either the public, for profit business/industry, government agency/official, non-profit/foundation/ advocacy, politician, celebrity, academic/scientist, journalist/media, individual activist, or unclear/ don't know. As for topical influencers for both CC and OA, an influence score of each user's follower/following ratio was used to examine their characteristics. This score calculation method was adopted from previous studies (e.g. Anger & Kittl, 2011), as detailed in the Data Analysis section below. Finally, Paris Agreement support (pro-Paris, neutral, anti-Paris) was coded.

Two coders discussed the meaning of each coding category and coded 10% of all chosen tweets. The initial intercoder reliability for content frame and user characteristics was somewhat low, so another round of in-depth discussions was performed. Approximately 8 h of in-person training, with detailed instructions, were devoted to reduce any discrepancies between the coders. The final intercoder reliability was calculated using Cohen's Kappa. For CC tweets, content frame score was 0.97, user characteristics score was 0.90, and external media sources score was 1. For OA tweets, content frame score was 0.89, user characteristics score was 1, and external media sources score was 0.82. Any Kappa scores higher than 0.6 were considered good while scores above 0.75 were considered excellent (Bakeman, 2000).

5.3. Data analysis

This study used a content analysis of social media messages to investigate communication patterns of one established but contentious issue, CC, and one novel issue, OA. Comparisons were made before and after Trump's announcement of leaving the Paris Agreement to examine potential changes in different content and user characteristics of CC and OA tweets.

SPSS version 24.0 was used for data analysis. A series of crosstab analyses were run, along with frequencies and descriptive statistics to test the proposed hypotheses and research questions. As for the topical influence score calculation, the follower/following ratio was chosen among different user performance indicators. The ratio enables a comparison of how many

Table 1. Content frame descriptions and example tweets

Frame	Description	Sample tweets
Settled Science (SS)	Emphasis on CC or OA science, the expert consensus, and the need for action (although action may not be emphasized as much as the science). Uncertainty or skepticism quashed or ignored.	"#Climatechange real & significant threat requires local, global #action" #Cville thanks @CvilleCityHallâ€¦ https://t.co/LrghzlplYM
Uncertain Science (US)	Focus on uncertainty—in CC/OA science, impacts or solutions. May question anthropogenic nature of CC and OA, or discuss natural variability. We cannot act, should not act, or will struggle to act.	Ha! I'm not a scientist, but this tells me all I need to know about #climatechange #ClimateChangeFactsâ€¦ https://t.co/ZPjqujB77E
Political/Ideological Struggles/ Activism (PIS)	A conflict over the way the world should work; over solutions or strategy to address climate change (above disagreements over science). A battle for power (for example, between nations or personalities or activist groups).	Proves @realDonaldTrump IS an IDIOT; The #ParisAgreement is NOT about Jobs, it's about #ClimateChange.
Disaster/Security (DS)	Predicted (or observed) impacts are dire. Impacts are numerous, discussed in detail, and threaten all aspects of life. Includes depictions or descriptions of extreme weather. (If focus is on economic cost, then Economic frame not DS).	Completely heartbreaking images of the Quebec flood. Human resiliency as people continue to fill sandbags. #ClimateChange Impact on lives.
Opportunity (O)	CC/OA poses beneficial opportunities. May come from action or inaction.	Infiniti Research Identifies Market Opportunities in the Solar Powerâ€¦ https://t.co/X2ZG95hQBt #Photovoltaicâ€¦ https://t.co/Ns783rWVZD
Economic (E)	Mentions an effect of CC/OA that has economic costs. Costs may results from action or inaction against CC/OA. May discuss growth, prosperity, investments, and/or markets. Likely to mention uncertainty.	West Coast Waters on Acid Trip Fishing Industry in Peril: Hotspots ofâ€¦ https://t.co/HDZWCWYIDV #OceanAcidificationâ€¦ https://t.co/bTVGqAyciZ
Role of Science (ROS)	Explores the role science plays in society. May debate transparency, funding or public awareness; especially in relation to things like Paris Agreement.	Concern grows as more #climatechange information disappears from @EPA website. https://t.co/gQNd5nhr0k https://t.co/CAFx68NGi8 Hard to get primary information (info that hasnt been interpreted w/bias) on climate change. Media hysteria driving opinion #climatechange
Health (H)	CC/OA poses danger to human health, for example, malnutrition, air quality. Urgent action required.	By 2030 #climatechange is expected to cause + 1/4 million deaths per year due to malaria, malnutrition, dehydration,â€¦ https://t.co/pyyCLgDA6E
Promotion/ Piggybacking (PP)	CC/OA issues not addressed directly, emphasis is on self-promotion or promotion of a tangential issue/event/product.	Pumped to find out about @aahearn's show! "New Podcast 'Terrestrial' Makes #ClimateChange Personal" #enviro #susty https://t.co/PTu3XWnlIl

(Continued)

Frame	Description	Sample tweets
Engagement/Empowerment (EE)	Emphasizes the influence of personal or small-scale actions (but not necessarily activism/PIS). Or, may convey encouragement, positive messages about addressing CC/OA.	Proud to live in a state that values a healthy environment. We must keep working to be even greener. #ClimateChange https://t.co/LypdkUxWrQ
None/Unknown (NU)	If there is not enough information in the tweet to identify a frame, or tweet has no frame.	BytorNSnowdog Beautiful photo! Thanks for sharing! #ClimateChange

Note: These frames were derived from O'Neill et al. (2015). Also, due to its relevance, the Promotion/piggybacking (PP) frame was adapted from Fellenor et al. (2017) and added to the content frame category. Last, the Engagement/Empowerment (EE) frame was added inductively after reviewing a few relevant tweets.

individuals have signed up to receive updates from User A with how many others User A has been following. The higher the ratio, the more individuals are attentive to User A's Twitter updates, without having to subscribe to their updates first (Anger & Kittl, 2011). Although it is ideal if this ratio is considered with interaction ratios (e.g. the amount of retweets, mentions), the follower/following ratio still offers insights into the total amount of followers as well as the level of interest from others in a user.

Last, to offer more quantitative support, the diversity of frames and user characteristics of CC and OA tweets overall (H1a and H2a) and before and after the Paris announcement (RQ1 and RQ2) were measured by calculating the Shannon Diversity Index (H), a common metric of species diversity used in ecology. As a metric of frame parity, the Shannon Equitability Index (E_H) was also calculated, which ranges from 0 (complete dominance of a single frame) to 1 (all frames used equally). E_H is the ratio of H (the observed diversity in a sample of n species or content frames) to H_{max} (the maximum possible diversity, which occurs when all n species or content frames are equally abundant).

6. Results

6.1. H1: Content characteristics

Results for H1a are presented in Table 2, which shows numbers and percentages of tweets using the different content frames. For CC tweets, Political/Ideological Struggles/Activism (PIS) was the most dominant frame, followed by Disaster/Security (DS) frame. For OA tweets, interestingly, different patterns emerged: DS was the most dominant frame, followed by Promotional/Piggybacking (PP) and Settled Science (SS). Overall, differences between CC and OA frames were statistically significant (χ^2 (10, 396) = 93.23, $p < 0.001$). The given percentages indicate that CC tweets were more narrowly concentrated on just two frames (PIS and DS) than the OA ones. The diversity indices (H) of 1.78 for CC and 1.84 for OA frames support this interpretation.

The analyses also revealed that around 83% of CC and 90% OA tweets had external media links. Among them, CC tweets contained 21.9% social media links, followed by 20.9% new media, and 15.3% mainstream media links. OA tweets showed different patterns, such that 31.8% had new media external links, which included online-only media, 19.2% linked to social media, and 16.2% linked to external images or photos. Only around 5% of the OA tweets had mainstream media links. The differences between CC and OA tweet external media sources were statistically significant (χ^2 (9, 394) = 26.37, $p < 0.01$). See Table 3 for further details on H1b results.

Table 2. Frequencies and percentages of CC and OA tweets by content frame

Content Frame	CC (%)	OA (%)
Settled Science	15 (7.6)	32 (16.2)*
Uncertain Science	10 (5.1)	2 (1.0)*
Political/Ideological Struggle/ Activism	96 (48.5)	23 (11.6)*
Disaster/Security	24 (12.1)	60 (30.3)*
Opportunity	7 (3.5)	9 (4.5)
Economic	6 (3.0)	8 (4.0)
Role of Science	2 (1.0)	0 (0)
Health	4 (2.0)	0 (0)*
Promotion/Piggybacking	15 (7.6)	45 (22.7)*
Engagement/Empowerment	14 (7.1)	14 (7.1)
Unknown/Unclear	5 (2.5)	5 (2.5)
Total	198 (100)	198 (100)

Note: * indicates that a subset of frame categories whose column proportions differ significantly from each other at the 0.05 level.

Table 3. Frequencies and percentages of CC and OA tweets by external URL media sources

External URL Media Source	CC (%)	OA (%)
None	33 (16.8)	19 (9.6)*
Science News	5 (2.6)	10 (5.1)
Mainstream Media	30 (15.3)	10 (5.1)*
New Media	41 (20.9)	63 (31.8)*
Social Media	43 (21.9)	38 (19.2)
Govt. Academic	7 (3.6)	14 (7.1)
Advocacy	7 (3.6)	6 (3.0)
Image	21 (10.7)	32 (16.2)
Video	8 (4.1)	6 (3.0)
Other	1 (0.5)	0 (0)
Total	198 (100)	198 (100)

Note: * indicates that a subset of media source categories whose column proportions differ significantly from each other at the 0.05 level.

6.2. H2: User characteristics

For H2a, crosstab analyses revealed that 60.1% of CC Twitter users were individual users (the public), followed by around 11% non-profit and 8.6% academics/scientists. For OA users, around 33% were academics/scientists, 29.8% were non-profit/advocacy/foundation, and 13.6% were individual users (the public). OA users were more diverse than CC users in terms of user type (H = 1.69 for OA and H = 1.36 for CC) and the differences between CC and OA users were statistically significant (χ^2 (8, 396) = 110.29, $p < 0.001$). See Table 4 for details.

For H2b, top 10 topical influencers for both CC and OA topics were explored by calculating an influence score of each user using the follower/following ratio (Anger & Kittl, 2011). For CC tweets, three of the top influencers were academics/scientists and two of them were individual journalists/ the media, with an H of 1.70. For OA tweets, five of them were non-profit/advocacy/foundations and two were government officials/agencies, with an H of 1.36. Although the CC users were more

Table 4. Frequencies and percentages of CC and OA Twitter users by user characteristic

User Characteristic	CC (%)	OA (%)
Individual (The Public)	119 (60.1)	27 (13.6)*
For Profit	11 (5.6)	7 (3.5)
Govt. Agency/Official	3 (1.5)	10 (5.1)*
Non-Profit/Advocacy/Foundation	22 (11.1)	59 (29.8)*
Politician	1 (0.5)	2 (1.0)
Academic/Scientist	17 (8.6)	66 (33.3)*
Journalist/Media	12 (6.1)	10 (5.1)
Activist	13 (6.6)	16 (8.1)
Unknown/Unclear	0 (0)	1 (0.5)
Total	198 (100)	198 (100)

Note: * indicates that a subset of user characteristic categories whose column proportions differ significantly from each other at the 0.05 level.

diverse than the OA users, there were more academic/scientist influencers among the CC users. Therefore, H2b was partially supported. See Table 5 for details.

6.3. RQ1—RQ3: Trump's Paris agreement withdrawal impact on tweeting patterns

For RQ1, the most prevalent content frame of CC tweets for both before (24 out of 98; 24.5%) and after (72 out of 100; 72%) the Paris Agreement announcement was PIS. After the announcement, however, the frequency of the PIS frame almost tripled, which was reflected in a decline in H from 2.13 to 1.14 and a decline in E_H from 0.89 to 0.52. In contrast, the diversity and parity of frames used in OA tweets changed very little after the announcement (H = 1.77 and E_H = 0.81 before; H = 1.80 and E_H = 0.82 after). The DS frame was the most prevalent among OA tweets both before (31 out of 102; 30.4%) and after (29 out of 96; 30.2%) the announcement. Also, SS-framed OA tweets decreased from 23.5% (24 out of 102) before the announcement to 8.3% (8 out of 96) after, and PIS tweets increased from 5.9% (6 out of 102) to 16.7% (17 out of 96). Overall, the content frame differences between before and after the announcement for CC tweets were statistically significant (χ^2 (10, 198) = 53.77, $p < 0.001$), as well as for OA tweets (χ^2 (8, 198) = 20.47, $p < 0.01$). For external media sources, there were statistically significant differences between time for CC (χ^2 (9, 196) = 20.55, $p < 0.05$) and OA tweets (χ^2 (8, 198) = 32.79, $p < 0.001$). For instance, CC tweets with no external links increased after the announcement (23 out of 100; 23%) compared to before (10 out of 96; 10.4%), and there were no advocacy-related external links after the announcement, compared to 7.3% before. For OA tweets, there were more mainstream media links after the announcement (8 out of 96; 8.3%), compared to before (2 out of 102; 2%), and there were more links to other social media postings after the decision (24 out of 96; 25%) than before (14 out of 102; 13.7%).

For RQ2, there were no statistically significant differences between before and after the announcement for CC Twitter users, which were mostly the public. For OA users, however, there were marginally significant differences ((χ^2 (8, 198) = 15.57, $p < 0.05$), such that before the announcement, non-profit/advocacy/foundations (37 out of 102; 36.3%) comprised the plurality of tweeters, whereas after the announcement, academics/scientists (31 out of 96; 32.3%) comprised the plurality of tweeters. The diversity of OA user types also increased from H = 1.51 before the announcement to H = 1.80 after.

For RQ3, there were significantly more supportive tweets (43 out of 100; 43%) and also opposing tweets (4 out of 100, 4%) of the Paris Agreement among CC users after the announcement, compared to supportive (9 out of 97; 9.3%) and opposing (0 out of 97) tweets before the announcement ((χ^2 (3, 197) = 35.14, $p < 0.001$). For OA users, there were significantly more

Table 5. Top 10 topical influencers on CC and OA by follower/following ratio

Climate Change				
Top Rank	Twitter User ID	Followers	Following	User Characteristics
1	AL77_infosec	1246	1	Academic/Scientist
2	rahmstorf	11,642	33	Academic/Scientist
3	emmie_madam	684	2	The Public
4	IDSAIndia	18,580	119	Govt. Agency/Official
5	yclimatechanges	8015	53	Academic/Scientist
6	GdnVoluntary	51,554	1522	Journalist/Media
7	OxfamAmerica	362,959	11,231	Non-Profit/Advocacy/ Foundation
8	KQED	71,136	3032	Journalist/Media
9	ViennaSecret	987	43	The Public
10	NatachaQS	16,169	907	For Profit
Ocean Acidification				
Top Rank	Twitter User ID	Followers	Following	User Characteristics
1	oaicc_project	1116	6	Non-Profit/Advocacy/ Foundation
2	franceonu	87,120	1512	Govt. Agency/Official
3	TerraMarProject	50,027	898	Non-Profit/Advocacy/ Foundation
4	UNMediaLiaison	38,273	909	Non-Profit/Advocacy/ Foundation
5	SenatorCantwell	61,387	1498	Politician
6	WHOI	7931	194	Non-Profit/Advocacy/ Foundation
7	KCTS9	83,683	2629	Journalist/Media
8	oceanfdn	37,669	1312	Non-Profit/Advocacy/ Foundation
9	PlymouthMarine	7814	706	For Profit
10	franceintexas	2382	235	Govt. Agency/Official

Note: If a same user appears more than once, the user with the latest follower number is chosen for inclusion.

supportive (18 out of 96; 18.8%) and less no-mention (77 out of 96, 80.2%) tweets after the announcement compared to supportive (1 out of 102; 1%) and no-mention (101 out of 102; 99%) tweets before the announcement ((χ^2 (2, 198) = 19.28, p < 0.001). Please see Figures 1–3 for visualization of these findings.

7. Discussion

This study investigated how the public discusses significant and timely global environmental issues, CC and OA, on Twitter, specifically with what frames and by whom. In particular, we explored if CC tweets differed from OA tweets, the two being closely related topics with the same cause, in terms of content frame and inclusion of external media links. Specific Twitter user characteristics were also compared. Further, all these characteristics of CC and OA tweets were compared for potential differences before and after President Trump's announcement on 1 June 2017 of the Paris Climate Agreement withdrawal, a decision that has implications for both CC and OA.

First, regarding content characteristics, PIS was the most dominant frame for CC tweets, whereas DS frame was the most dominant one for OA tweets. This pattern was consistent pre- and post- Trump's withdrawal announcement, in that CC tweets used PIS frames regardless of

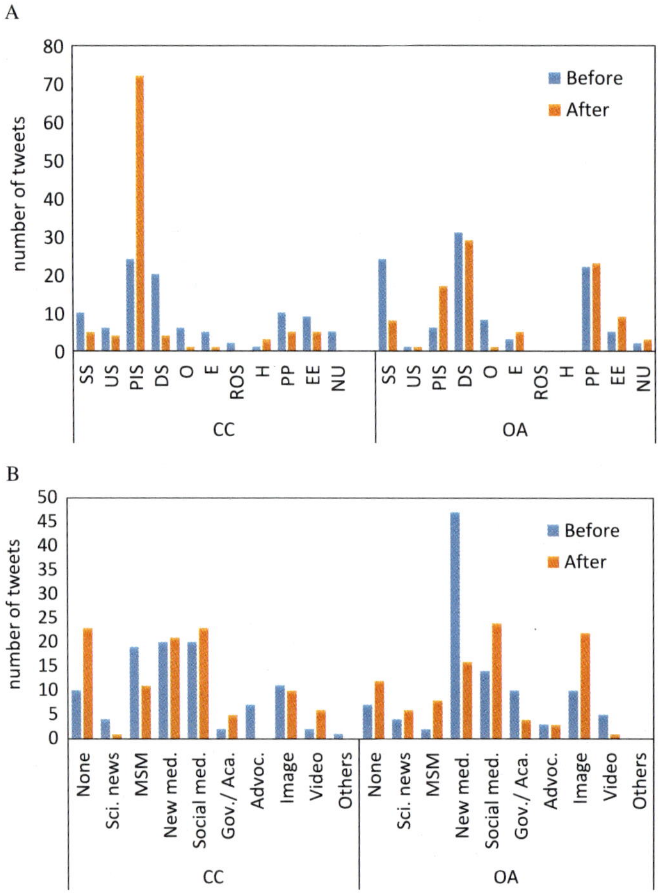

Figure 1. Frames (A) and exter-nal URL media source type (B) of CC and OA tweets posted before and after President Trump's Paris agreement with-drawal announcement. Frame codes are identified in Table 1; External URL source codes are: Sci. news = Science news, MSM = Mainstream media, Social med. = Social media, Gov./Aca. = Government or academic website, Advoc. = Advocacy media.

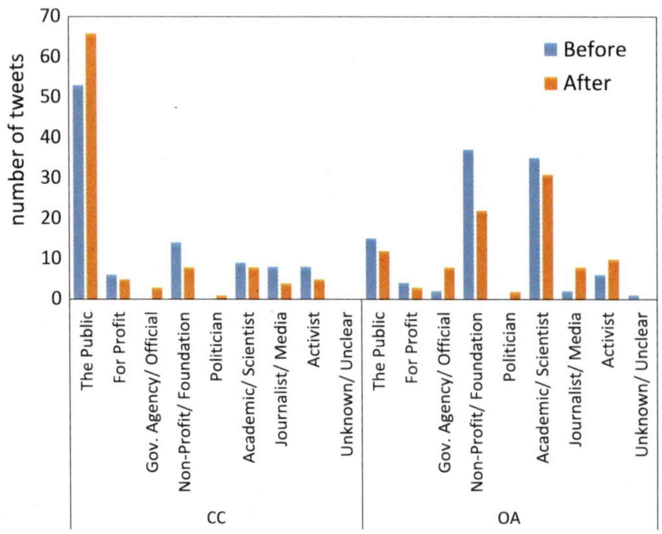

Figure 2. Characteristics of Twitter users tweeting about CC and OA before and after President Trump's Paris agree-ment withdrawal announcement.

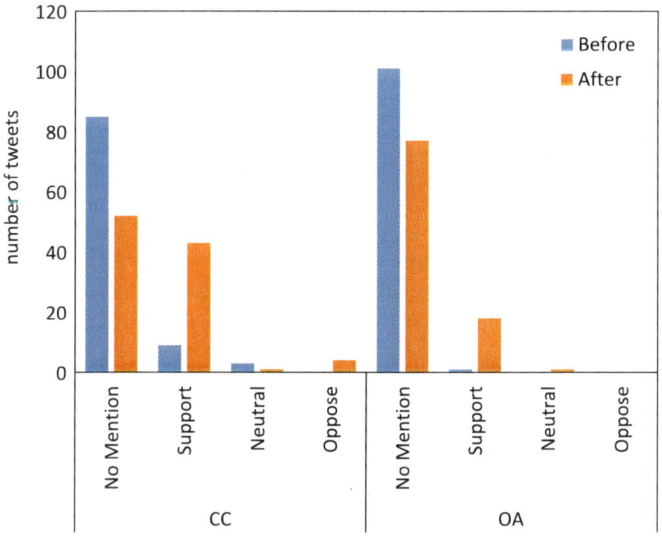

Figure 3. Opinions of the Paris Agreement contained in tweets about CC and OA posted before and after President Trump's withdrawal announcement.

timing. However, it was interesting to note that the number of CC tweets using the PIS frame almost tripled after the withdrawal announcement, indicating increased discourse on CC from political and ideological perspectives that likely reflected the controversial nature of President Trump's political decision. OA tweets using the PIS frame also almost tripled following the announcement, which could suggest that this emerging issue is becoming politicized, but overall, PIS framed tweets only comprised 11.6% of all OA tweets. Instead, the DS frame was the most used among OA tweets, followed by the PP and SS frames. The PP frame occurs when a Twitter user discusses a topic as a means to promote his or her related agenda or events (e.g. professional conference, community workshop, demonstration or protest for a cause) (Fellenor et al., 2017). This frame can be used in Twitter as a convenient tactic to facilitate self-promotion and this was the case for OA tweets. DS, which was the most used frame for OA tweets (both pre- and post-Trump's announcement) typically uses apocalyptic rhetoric, which can lead to labeling the communicators as alarmists and be counterproductive to efforts to promote awareness, concern, and action for the issue (Foust & O'Shannon Murphy, 2009). Although this type of frame may be effective to draw individuals' attention to the issue initially, scientists and other advocates who aim to increase public awareness of OA in an honest and accurate manner should be mindful of using potentially counterproductive frames on Twitter and other types of media due to the possibility of evoking unnecessary fear or worry instead of increasing knowledge on the issue.

For external links, CC tweets had more links containing mainstream media sources than OA tweets, as expected. For both topics, however, social and new media sources comprised the majority of external links. Overall, CC tweets contain slightly more diverse media links than OA tweets, possibly due to its well-known, contested nature. After the Trump announcement, OA tweets' external media sources also became more diverse and inclusive of social, new, and mainstream media, compared to tweets mainly with new media external links before the announcement.

Regarding user characteristics, CC tweeters were mostly individuals both pre- and post- Trump decision. However, OA tweeters were mostly non-profit, advocacy, and foundations before the announcement, but were mostly academics and scientists after the announcement, with an increase in the diversity of user types after the announcement. Further, academics such as university professors or research scientists, as well as the media were topical influencers for CC, whereas 70% of the top 10 topical influencers for OA were non-profits, advocacy groups, foundations, and government agencies/officials. This was reflected in the relatively low diversity score for

OA topical influencers compared to CC influencers. As hypothesized in H2, we did indeed find that relatively few public users tweet about OA, especially compared to CC, but specialized advocacy groups, rather than scientists, were the main OA influencers. Such issue publics or engaged groups likely have a mission to promote outreach and awareness on OA and may draw followers who are interested in broader ocean or environmental issues, not just OA, by disseminating information and promoting relevant activities.

Last, the Trump withdrawal announcement on 1 June 2017 appeared to make public opinion of the Paris Agreement more polarized: there was more support of as well as opposition to the Paris Agreement among CC tweeters after the announcement than before. For OA users, there was more support and more mention of the Paris Agreement after the withdrawal announcement compared to before. The response to this international political event on social media supports the notion of environmental communication as political communication (Scheufele, 2014), and these findings may suggest that as OA becomes more well-known, it could become a politically contentious issue, like CC.

Broadly, our findings suggest that individuals differ in the ways they communicate about established versus emerging issues on social media. The overall higher frame parity among OA tweets compared to CC tweets suggests that emerging issues may not be well known or salient enough for a particular frame (or few frames) to dominate the discourse. Additionally, the prevalence of the PIS frame in tweeting about CC suggests that ideology and its connection to activism are influential in shaping how people perceive and communicate about established contested issues, and Twitter, as one of the popular social media platforms, may be a good place to engage in or promote such conversations on established yet contested issues. On the other hand, the relatively lower frequency of the PIS frame among OA tweets suggests that people may be less likely or not yet equipped with knowledge to view emerging issues through an ideological lens. This is consistent with research showing the importance of ideological cues in how the public seeks further information on the emerging issue of nanotechnology (Yeo et al., 2015). The increase in polarized discourse of CC following Trump's withdrawal announcement lends further support to the role of ideology in shaping discourse on established and emerging issues.

Our findings also provide insights into the effect of momentous news events on frame changing and frame parity on Twitter. Both topics experienced frame changing after the Paris withdrawal announcement that can largely be attributed to the increase in PIS frame dominance. This decrease in frame parity was much greater for CC tweets than for OA tweets, which is perhaps unsurprising given that CC is the core issue of the Paris Agreement. But the agreement also has important implications for OA (Magnan et al., 2016), a fact that appeared to be recognized by some OA tweeters, as there was an increase in OA tweets mentioning (and supporting) the agreement after the withdrawal announcement. But the finding of greater frame changing among the established issue of CC suggests that momentous news events may not be as influential on the framing of emerging issues on Twitter, even when the news event is highly relevant to the issue.

Although this study produced informative data on CC and OA social media interactions and framing, the study has some limitations. First, we examined tweets from 7 May to 15 June 2017, a six-week data collection period that enabled us to capture opinions closely preceding and following the Paris Agreement announcement. Although six weeks is not a short time frame, a longer time period could have allowed us to study if (and how) the tweeting patterns of CC and OA change both in the short term and in the long term. This event will be covered in the media over the next few years, as the U.S. cannot formally withdraw from the Paris Agreement until 2020 (United Nations Framework Convention on Climate Change, 2017). This will give scholars ample opportunities to build upon our findings and further explore frame changing, especially if the Paris Agreement decision experiences Downs' classic "issue-attention cycle," in which an emerging issue initially receives high public attention that then wanes in salience over time (Downs, 1972). Second, although the follower/following ratio used in this study can capture the popularity and connectivity of a Twitter user, previous research (Anger & Kittl, 2011; Cha, Haddadi, Benevenuto, & Gummadi, 2010) suggests that this influence metric needs to be examined along with retweets and/or mentions—other metrics that are more content- and name-value driven—to accurately reflect other aspects of influence on Twitter such as active audience engagement. A third limitation is that our sample size was somewhat small, especially given the noisy and variable nature of Twitter data. Related to this is the fact that because CC tweets greatly outnumbered OA tweets, the randomly selected sample for OA tweets comprised a larger proportion of the total population of tweets than for CC, an issue that can be addressed in future studies using more robust statistical tools. Also, relevant tweets may have been missed because some Twitter users could have discussed CC and OA but with other hashtags not used in this study such as #climatedenial, #ParisAgreement, or #savetheocean to express their concerns on the same issues. Future research could use advanced computational methods such as machine learning technique to analyze a greater number of relevant tweets with a range of different hashtags, so that the complex nature and more nuanced patterns of CC and OA dialogue on Twitter can be understood.

Despite these limitations, our findings provide timely and important exploratory findings as well as theoretical and practical insights into the understanding of these environmental communications in Twitter, one of the most used social media platforms (Statista, 2017). Specifically, by investigating relevant frames of CC and OA tweets based on Entman (1993, 2003)'s and O'Neill et al. (2015)'s framing approach, the study can contribute to confirming and expanding the explanatory power of framing theory to the understanding of social media messages. PIS and DS frames were used the most to either put a political spin on the public understanding of CC or increase fear or worry associated with the issue to facilitate or discourage action. A fair amount of PP frames encouraging attendance for educational workshops or conferences was found for OA tweets. Understanding which frames are popular and prevalent for each issue in social media platforms such as Twitter can be essential for those who want to strategically use interactive media to further improve public awareness and seek support for these important environmental issues. Based on our results, with a careful understanding of different levels of issue familiarity among its publics, social media can be a potentially effective communication vehicle for disseminating information and opinion on both well-established and lesser-known, emerging environmental issues, complementing traditional and mainstream media outlets.

Negative consequences of CC and OA have been ongoing and confirmed by scientists and can be detrimental to the individual, societal, and economic health of the U.S. and the world (Cooley & Doney, 2009; Doney, Balch, Fabry, & Feely, 2009; Monaco Declaration, 2008). Since the political decision of Trump's Paris Agreement withdrawal announcement sparked substantial CC discourse and facilitated social media use to express individuals' opinions, this study can also provide timely recommendations to scientists, educators, and/or outreach professionals that effective environmental communication is essential but can only be achieved when understanding its complex, versatile nature in the context of current societal and political climates.

Funding
This work was supported by High Point University under Research Advancement Grant [number 17-003].

Competing Interests
The authors declares no competing interests.

Author details
Sojung Claire Kim[1]
E-mail: skim205@gmu.edu
ORCID ID: http://orcid.org/0000-0002-9622-3035
Sandra L. Cooke[2]
E-mail: scooke@highpoint.edu
ORCID ID: http://orcid.org/0000-0003-0909-8887
[1] Department of Communication, George Mason University, 4400 University Drive, Fairfax, VA 22030, USA.
[2] Department of Biology, High Point University, One University Parkway, High Point, NC 27268, USA.

Notes

1. The IPCC comprises the work of thousands of scientists from 195 nations. It was established by the United Nations Environmental Program and World Meteorological Organization in 1988 to provide scientific assessments on climate change (IPCC, 2017b). Released in 2013, the Fifth Assessment Report is the IPCC's most recent report that details the scientific, technical, and socio-economic aspects of climate change, including its causes, impacts, and potential solutions (IPCC, 2017a).

2. "Science news" refers to the websites of legacy journals such as Nature, legacy science magazines such as Scientific American, and environmental focused news websites such as Climate Central and Grist. "Mainstream media" refer to media outlets such as CNN or The Washington Post, as well as magazines such as The Atlantic or US News. "New media" refers to individual blogs as well as online-only media platforms, such as The Huffington Post or Buzzfeed. "Government/academic" refers to governmental agencies, such as NASA, as well as universities or research institutes. "Advocacy" refers to advocacy-oriented websites, such as Azaaz.org or Reality Drop, as well as environmental advocacy organizations, including Sierra Club.

References

Anger, I., & Kittl, C. (2011). *Measuring influence on Twitter*. Paper presented at the Proceedings of the 11th International Conference on Knowledge Management and Knowledge Technologies, Graz, Austria.

Bakeman, R. (2000). Behavioral observation and coding. In H. T. Reis & C. M. Judd (Eds.), *Handbook of research methods in social and personality psychology* (pp. 138–159). Cambridge: Cambridge University Press.

Becker, H., Naaman, M., & Gravano, L. (2011). Beyond trending topics: Real-world event identification on Twitter. *Proceedings of the Fifth International AAAI Conference on Weblogs and Social Media, 11*(2011), 438–441.

Bruns, A., Highfield, T., & Lind, R. A. (2012). Blogs, Twitter, and breaking news: The produsage of citizen journalism. *Produsing Theory in a Digital World: the Intersection of Audiences and Production in Contemporary Theory, 80*, 15–32.

Capstick, S. B., Pidgeon, N. F., Corner, A. J., Spence, E. M., & Pearson, P. N. (2016). Public understanding in Great Britain of ocean acidification. *Nature Climate Change, 6*(8), 763–767. doi:10.1038/nclimate3005

Cha, M., Haddadi, H., Benevenuto, F., & Gummadi, P. K. (2010). Measuring user influence in twitter: The million follower fallacy. *Icwsm, 10*, 10–17.

Conover, M., Ratkiewicz, J., Francisco, M. R., Gonçalves, B., Menczer, F., & Flammini, A. (2011). Political polarization on twitter. *Icwsm, 133*, 89–96.

Cooley, S. R., & Doney, S. C. (2009). Anticipating ocean acidification's economic consequences for commercial fisheries. *Environmental Research Letters*, 4(2), 024007. doi:10.1088/1748-9326/4/2/024007

Davenport, C. (2017, March 2). Top Trump advisers are split on Paris Agreement on climate change. *New York Times*, Retrieved from https://www.nytimes.com/2017/03/02/us/politics/climate-change-trump.html

Doney, S. C., Balch, W. M., Fabry, V. J., & Feely, R. A. (2009). Ocean acidification: A critical emerging problem for the ocean sciences. *Oceanography, 22*(4), 16–25. doi:10.5670/oceanog

Downs, A. (1972). Up and down with ecology: The "issue-attention cycle". *Public Interest, 28*, 38–50.

Entman, R. M. (1993). Framing: Toward clarification of a fractured paradigm. *Journal of Communication, 43*(4), 51–58. doi:10.1111/j.1460-2466.1993.tb01304.x

Entman, R. M. (2003). Cascading activation: Contesting the White House's frame after 9/11. *Political Communication, 20*(4), 415–432. doi:10.1080/10584600390244176

Fellenor, J., Barnett, J., Potter, C., Urquhart, J., Mumford, J. D., & Quine, C. P. (2017). The social amplification of risk on Twitter: The case of ash dieback disease in the United Kingdom. *Journal of Risk Research*, 1–21. doi:10.1080/13669877.2017.1281339

Foust, C. R., & O'Shannon Murphy, W. (2009). Revealing and reframing apocalyptic tragedy in global warming discourse. *Environmental Communication, 3*(2), 151–167. doi:10.1080/17524030902916624

Gelcich, S., Buckley, P., Pinnegar, J. K., Chilvers, J., Lorenzoni, I., Terry, G., ... Duarte, C. M. (2014). Public awareness, concerns, and priorities about anthropogenic impacts on marine environments. *Proceedings of the National Academy of Sciences, 111*(42), 15042–15047. doi:10.1073/pnas.1417344111

Griffin, R. J., Dunwoody, S., & Neuwirth, K. (1999). Proposed model of the relationship of risk information seeking and processing to the development of preventive behaviors. *Environmental Research, 80*(2), S230–S245. doi:10.1006/enrs.1998.3940

Gruzd, A., & Roy, J. (2014). Investigating political polarization on Twitter: A Canadian perspective. *Policy & Internet, 6*(1), 28–45. doi:10.1002/1944-2866.POI354

Ho, S. S., Detenber, B. H., Rosenthal, S., & Lee, E. W. J. (2014). Seeking information about climate change. *Science Communication, 36*(3), 270–295. doi:10.1177/1075547013520238

Hopke, J. E., & Simis, M. (2017). Discourse over a contested technology on Twitter: A case study of hydraulic fracturing. *Public Understanding of Science, 26*(1), 105–120. doi:10.1177/0963662515607725

Houston, J. B., Pfefferbaum, B., & Rosenholtz, C. E. (2012). Disaster news: Framing and frame changing in coverage of major US natural disasters, 2000–2010. *Journalism & Mass Communication Quarterly, 89*(4), 606–623. doi:10.1177/1077699012456022

IPCC. (2017a). IPCC: Intergovernmental Panel on Climate Change: Activities. Retrieved from https://www.ipcc.ch/activities/activities.shtml

IPCC. (2017b). IPCC: Intergovernmental Panel on Climate Change: Organization. Retrieved from https://www.ipcc.ch/organization/organization.shtml

Jain, S. (2017, June 2). Twitter heats up after Donald Trump exits Paris climate agreement. Retrieved from https://www.ndtv.com/offbeat/twitter-heats-up-after-donald-trump-exits-paris-climate-agreement-1706926

Jang, S. M., & Hart, P. S. (2015). Polarized frames on "climate change" and "global warming" across countries and states: Evidence from Twitter big data. *Global Environmental Change, 32*, 11–17. doi:10.1016/j.gloenvcha.2015.02.010

Kaplan, A. M., & Haenlein, M. (2010). Users of the world, unite! The challenges and opportunities of social media. *Business Horizons, 53*(1), 59–68. doi:10.1016/j.bushor.2009.09.003

Kemp, S. (2017). Number of social media users passes 3 billion with no signs of slowing. Retrieved from https://thenextweb.com/contributors/2017/08/07/number-social-media-users-passes-3-billion-no-signs-slowing/

Kitzinger, J. (2007). Framing and frame analysis. In E. Devereaux (Ed.), *Media studies: Key issues and debates* (pp. 134–161). Los Angeles: Sage Publications.

Kuypers, J. A. (2010). Framing analysis from a rhetorical perspective. In P. D'Angelo & J. A. Kuypers (Eds.), *Doing news framing analysis: Empirical and theoretical perspectives* (pp. 286–311). New York, NY: Routledge.

Lee, N. M., VanDyke, M. S., & Cummins, R. G. (2017). A missed opportunity?: NOAA's use of social media to communicate climate science. *Environmental Communication*, 1–10. doi:10.1080/17524032.2017.1308407

Lee, T. M., Markowitz, E. M., Howe, P. D., Ko, C.-Y., & Leiserowitz, A. A. (2015). Predictors of public climate change awareness and risk perception around the world. *Nature Climate Change* 5(11), 1014.

doi:10.1038/nclimate2728

Lotan, G., Graeff, E., Ananny, M., Gaffney, D., & Pearce, I. (2011). The Arab spring| The revolutions were tweeted: Information flows during the 2011 Tunisian and Egyptian revolutions. *International Journal of Communication*, *5*, 31.

Magnan, A. K., Colombier, M., Billé, R., Joos, F., Hoegh-Guldberg, O., Pörtner, H.-O., ... Gattuso, J.-P. (2016). Implications of the Paris agreement for the ocean. *Nature Climate Change*, *6*(8), 732–735. doi:10.1038/nclimate3038

Marcu, A., Uzzell, D., & Barnett, J. (2011). Making sense of unfamiliar risks in the countryside: The case of Lyme disease. *Health & Place*, *17*(3), 843–850. doi:10.1016/j.healthplace.2011.03.010

McGrath, M. (2016, May 27). Donald Trump would 'cancel' Paris climate deal. *BBC News*, Retrieved from http://www.bbc.com/news/election-us-2016-36401174

Monaco Declaration. (2008). *Second international symposium on the ocean in a high-CO2 world. Monaco*, Retrieved from https://www.iaea.org/nael/docrel/MonacoDeclaration.pdf

Moscovici, S. (2000). *Social representations: Explorations in social psychology*. Cambridge, UK: Polity Press.

Mossler, M. V., Bostrom, A., Kelly, R. P., Crosman, K. M., & Moy, P. (2017). How does framing affect policy support for emissions mitigation? Testing the effects of ocean acidification and other carbon emissions frames. *Global Environmental Change*, *45*, 63–78. doi:10.1016/j.gloenvcha.2017.04.002

Newman, T. P. (2016). Tracking the release of IPCC AR5 on Twitter: Users, comments, and sources following the release of the working group I summary for policymakers. *Public Understanding of Science*, *26*(7), 815–825. doi:10.1177/0963662516628477

O'Neill, S., Williams, H. T., Kurz, T., Wiersma, B., & Boykoff, M. (2015). Dominant frames in legacy and social media coverage of the IPCC fifth assessment report. *Nature Climate Change*, *5*(4), 380–385. doi:10.1038/nclimate2535

The Ocean Project. (2012). America and the Ocean: Ocean acidification. Retrieved from http://theoceanproject.org/wp-content/uploads/2012/09/Special_Report_Summer_2012_Public_Awareness_of_Ocean_Acidification.pdf

Pearce, W., Holmberg, K., Hellsten, I., & Nerlich, B. (2014). Climate change on Twitter: Topics, communities and conversations about the 2013 IPCC working group 1 report. *PloS one*, *9*(4), e94785. doi:10.1371/journal.pone.0094785

Runge, K. K., Yeo, S. K., Cacciatore, M., Scheufele, D. A., Brossard, D., Xenos, M., ... Li, N. (2013). Tweeting nano: How public discourses about nanotechnology develop in social media environments. *Journal of Nanoparticle Research*, *15*(1), 1381. doi:10.1007/s11051-012-1381-8

Schäfer, M. S. (2012). Online communication on climate change and climate politics: A literature review. *Wiley Interdisciplinary Reviews: Climate Change*, *3*(6), 527–543.

Scheufele, D. A. (2014). Science communication as political communication. *Proceedings of the National Academy of Sciences*, *111*(Supplement 4), 13585–13592. doi:10.1073/pnas.1317516111

Schuldt, J. P., McComas, K. A., & Byrne, S. E. (2016). Communicating about ocean health: Theoretical and practical considerations. *Philosophical Transactions of the Royal Society B*, *371*(1689), 20150214.

Segerberg, A., & Bennett, W. L. (2011). Social media and the organization of collective action: Using Twitter to explore the ecologies of two climate change protests. *The Communication Review*, *14*(3), 197–215. doi:10.1080/10714421.2011.597250

Shulman, S. (2011). *DiscoverText: Software training to unlock the power of text*. Paper presented at the Proceedings of the 12th annual international digital government research conference: Digital government innovation in challenging times., New York, NY, USA.

So, J., Prestin, A., Lee, L., Wang, Y., Yen, J., & Chou, W.-Y. S. (2016). What do people like to "share" about obesity?: A content analysis of frequent retweets about obesity on Twitter. *Health Communication*, *31*(2), 193–206. doi:10.1080/10410236.2014.940675

Statista. (2017). Number of monthly active Twitter users worldwide from 1st quarter 2010 to 3rd quarter 2017 (in millions). Retrieved from https://www.statista.com/statistics/282087/number-of-monthly-active-twitter-users/

United Nations Framework Convention on Climate Change. (2016). NDC registry (Interim): United States of America. Retrieved from http://www4.unfccc.int/ndcregistry/pages/Party.aspx?party=USA

United Nations Framework Convention on Climate Change. (2017). The Paris agreement. Retrieved from http://unfccc.int/paris_agreement/items/9485.php

Veltri, G. A., & Atanasova, D. (2015). Climate change on Twitter: Content, media ecology and information sharing behaviour. *Public Understanding of Science*, *26*(6), 721–737. doi:10.1177/0963662515613702

Vis, F., Faulkner, S., Parry, K., Manyukhina, Y., & Evans, L. (2013). Twitpic-ing the riots: Analysing images shared on Twitter during the 2011 UK riots. In K. Weller, A. Bruns, J. Burgess, M. Mahrt, & C. Puschmann (Eds.), *Twitter and Society* (pp. 385–398). New York: Peter Lang.

Volcovici, V. (2017). U.S. coal companies ask Trump to stick with Paris climate deal. *Reuters*, Retrieved from https://www.reuters.com/article/us-usa-trump-coal/u-s-coal-companies-ask-trump-to-stick-with-paris-climate-deal-idUSKBN1762YY

Williams, H. T., McMurray, J. R., Kurz, T., & Lambert, F. H. (2015). Network analysis reveals open forums and echo chambers in social media discussions of climate change. *Global Environmental Change*, *32*, 126–138. doi:10.1016/j.gloenvcha.2015.03.006

World Economic Forum. (2016). What is the Paris agreement on climate change? Retrieved from https://www.weforum.org/agenda/2016/09/what-is-the-paris-agreement-on-climate-change/

Yates, K. K., Turley, C., Hopkinson, B. M., Todgham, A. E., Cross, J. N., Greening, H., ... Johnson, Z. (2015). Transdisciplinary science: A path to understanding the interactions among ocean acidification, ecosystems, and society. *Oceanography*, *28*(2), 212–225. doi:10.5670/oceanog.2015.43

Yeo, S. K., Xenos, M. A., Brossard, D., & Scheufele, D. A. (2015). Selecting our own science: How communication contexts and individual traits shape information seeking. *The ANNALS of the American Academy of Political and Social Science*, *658*(1), 172–191. doi:10.1177/0002716214557782

8

Early detection of chloroform hot spots in the Montevideo drinking water network

Marcel Achkar Borras[1], Gustavo Seoane[2], Mariana Gomez-Camponovo[3]*, Eleuterio Umpierrez Vazquez[4] and Nicolas Perez[5]

*Corresponding author: Mariana Gomez-Camponovo, CENUR Litoral Norte. Universidad de la Republica, Social Medicine Unit, Florida 1065. Paysandú, UY 60000, Uruguay
E-mail: mgomez@cup.edu.uy

Reviewing editor: Conor Buggy, School of Public Health, Physiotherapy and Sports Science, University College Dublin, Ireland

Abstract: The use of chlorine to disinfect water produces a series of by-products, particularly trihalomethanes. This is important given that there is a recognized association with different types of cancer after prolonged exposure, as well as with probable adverse effects on reproduction, especially in relation to offspring. The concentrations of these organic compounds vary greatly depending on the season and the conditions involved in providing water for consumption. This study was aimed at determining the geographic distribution pattern of chloroform in the city of Montevideo, and correlating that with the characteristics of the water distribution network. A total of 400 samples were taken from a sampling network between 2009 and 2015. Chloroform was identified by analyzing these samples using the head-space gas chromatography/mass spectrometry method. Data regarding piping length, diameter, and type of material were obtained. A geographic information system was constructed and hot spots were analyzed using the Getis–Ord G_i^* statistic. A neighborhood piping density index was also proposed. The analysis found two zones in the city: hot and cold spots. The proposed index showed an increase in the G_i^* statistic as the neighborhoods' piping density increased, with a strong correlation. In addition, the highest G_i^* statistic values corresponded to

ABOUT THE AUTHOR

An interdisciplinary group was formed of professionals with extensive research experience, consisting of chemists, geographers, epidemiologists, and engineers. From the perspective of their disciplines, each has contributed to addressing the complex issue of the quality of water for human consumption. Specifically, the working group has been researching organic byproducts formed by the chlorination process used to disinfect water, and possible effects on health. To understand how THMs behave in a drinking water system, a dense monitoring network was developed in the city of Montevideo, the capital of Uruguay. Advances in the development of geostatistical models have made it possible to delimit zones in the city according to the presence of organic byproducts in water for consumption. In order to address the problem, this working group aims to use this methodology in other cities in the country where the sources of raw water and the management of the supply networks vary.

PUBLIC INTEREST STATEMENT

Providing people with water that is suitable for consumption is increasingly challenging for all societies due to larger populations as well as a gradual deterioration in the sources of the raw water that is extracted for the treatment process. Compounding this are growing urban centers that have aging supply networks alongside networks with new characteristics, creating a complex maze that needs to be characterized. In the case of Uruguay, the disinfection process is performed by chlorinating water.
Trihalomethanes are some of the known by-products from this process, which must be properly monitored in order to be maintained at acceptable levels, thereby reducing the risk of health disorders. This work contributes to understanding the formation of trihalomethanes in the network and to determining the geographic zones that show a capability of reaching higher levels of these compounds over time. Mitigation and control plans can be developed based on this data, which will have an impact on population health.

larger amounts of iron piping in neighborhoods. This work was able to determine that the hot spots were associated with the piping density in the neighborhoods as well as with the type of piping material, particularly iron.

Subjects: Environmental Health; Urban Studies; Cities & Infrastructure; Cities & Infrastructure; Urban Studies; Urban Geography; Environmental Geography; Geographic Information Systems; Environmental Modeling; Environmental health

Keywords: Trihalomethanes; chloroform; disinfection by-products; geostatistic; water distribution network

1. Introduction

A total of 700 by-products are known to be produced as a result of using chlorine to disinfect water for human consumption (as is done in Uruguay). The most abundant of these are trihalomethanes (THMs), whose association with health problems has been recognized (Bove, Rogerson, & Vena, 2007a, 2007b; Wang, Deng, & Lin, 2007). Exposure to these by-products can occur in different ways, including ingestion, inhalation, and skin absorption (Botton et al., 2015)

Epidemiological investigations have studied a likely association between prolonged exposure to disinfection by-products (DBP) in chlorinated water and the risk of cancer (Villanueva et al., 2004; Wang et al., 2007). Studies have specifically reported on colorectal cancer (King, Marrett, & Woolcott, 2000; Chen, Weiping, Xinyan, Yao, & Jiang, 2005; Bove et al., 2007b; Kuo, Tiao, Trong-Neng, & Yang, 2009; Kuo, Chen, Shu-Chen, Wang, & Yang, 2010; Md, Bayzidur, & Tim, ; Villanueva et al., 2017) and bladder cancer (Bove et al., 2007a; Chang, Shu-Chen, Wang, & Chun-Yuh, 2007; Goebell, Villanueva, Rettenmeier, Rübben, & Kogevinas, 2004; Michaud et al., 2007; Villanueva et al., 2004, 2006; Villanueva, Kogevinas, & Grimalt, 2001).

Adverse reproductive effects have also been the subject of attention (Chisholm, Cook, Bower, & Weinstein, 2008; Grellier et al., 2010; Nieuwenhuijsen et al., 2009; Nieuwenhuijsen, Toledano, Eaton, Fawell, & Elliott, 2000).

In Uruguay, the national norm (UNIT Norm 833:2008) is based on 2008 guidelines published by the World Health Organization (WHO (World Health Organization), 2008), although a more recent guide has been published (WHO (World Health Organization), 2017).

THMs are formed by a reaction between the organic matter present in water and chlorine used for disinfection (Guilherme & Rodriguez, 2015). This produces chloroform (CHCl3), chlorodibromo-methane (CHClBr2), bromodichloromethane (CHBrCl2), and bromoform (CHBr3). Chloroform is the most abundant of these compounds (Baytak, Sofuoglu, Inal, & Sofuoglu, 2008; Jang, Choi, Ro, & Ka, 2012; Ristoiu et al., 2009; Uyak, Ozdemir, & Toroz, 2007) (Knight, Watson, Carswell, Comino, & Shaw, 2011; Gomez-Camponovo et al., 2014), representing as much as 80% of THMs, on average (Rodriguez, Vinette, Jean-B., & Bouchard, 2003).

The concentration of these compounds depends on several factors, including the source and type of raw water, the concentration of organic matter, the pH of the water (Cool, Lebel, Sadiq, & Rodriguez, 2015; Guilherme & Rodriguez, 2015), bromide and iodide levels, water temperature (Guilherme & Rodriguez, 2015; Ristoiu et al., 2009), initial chlorine concentration (Brown, Bridgeman, & West, 2011), and contact time with the disinfectant agent (Sadiq & Rodriguez, 2004) (considered to be positively correlated with the amount of time it remains in the piping) (Guilherme & Rodriguez, 2015; Loyola-Sepulveda, Lopez-Leal, Jorge., Bravo-Linares, & Mudge, 2009). THM levels also increase as the distance from the chlorination point increases (Rodriguez et al., 2003). This set of parameters affects the decay of chlorine, which occurs due to reactions with compounds present in the bulk water (known as bulk decay), and which previous works on this subject have recognized as the principal determinant of the formation of THMs (Krasner et al., 2006).

Several researchers are currently considering other variables, including those that affect the reaction of chlorine at the pipe wall (known as wall decay).

This type of decay involves corrosion reactions with the wall material itself, as well as with adhering biofilms and accumulated sediments. It depends on multiple factors, such as pipe material, diameter (and water velocity), service age, inner coating and the presence of attached biofilms, corrosion (Kiene, Lu, & Levi, 1998), and initial chlorine concentrations (Brown et al., 2011). It is important not to underestimate the effect of the material and the pipe service age (which are related to corrosion). These factors can cause chlorine consumption to vary from one distribution system to another (Al Jasser 2007). In addition, material on the pipe wall, such as corrosion products and biofilm slime, is known to exert significant chlorine demand (Rossman, Brown, Singer, & Nuckols, 2001). Similarly, since iron is highly reactive, it is known to have a higher free chlorine demand than PVC piping (Wang et al., 2012). Also, iron tubercles have been shown to contain organic material, including precursors of DBP (Rossman et al., 2001). And the effect of service age on effective wall decay constants is most evident in cast iron pipes. For these pipes, the effect of service age on chlorine demand should also be considered, especially when there is enough wall decay to render the bulk decay negligible (Al Jasser 2007).

In addition, iron rust not only increases pipe surface porosity and roughness but also acts as a nutrient, thereby contributing to the formation of biofilm (Wang, Masters, Edwards, Falkinham, & Pruden, 2014; Wang et al., 2012). In iron pipes, the higher chlorine demand associated with non-DBP reactions with corrosion products (ferrous corrosion) and biofilm decreases the amount of chlorine available to react with the residual natural organic matter in the water, thereby decreasing the formation of DBPs. However, Rossman et al. (2001) reported that organic material attached to the walls of these pipes can actually (moderately) increase THM levels, which would otherwise be produced by bulk solution reactions.

In this context, drinking water systems have been evaluated over the past 20 years based on the structural integrity (Charisiadis et al., 2015) of the systems, the characteristics of the networks, the materials with which they are made (Baird, 2012; Hallam, West, Forster, Powell, & Spencer, 2002; Legay, Rodriguez, Sérodes, & Levallois, 2010; Rossman et al., 2001; Wang et al., 2012), the age of the piping (Al-Jasser 2007; Baird, 2012), its diameter (Bassey & Egbe Jerome, 2016; Hallam et al., 2002), and the size of the network (Legay et al., 2010).

While evidence of medium- and long-term effects on health has been mounting, significant difficulties and questions have arisen as to how to assign and evaluate exposure (Hinckley, Bachand, Nuckols, & Reif, 2005), and how to predict organic pollutant levels in drinking water distributed throughout a territory.

While it is challenging to identify and delimit areas with differences in the variability of THMs, doing this would enable the establishment of a prediction system and the eventual implementation of timely actions. The importance of linking the presence of these compounds with the territory has also been reported (Cool et al., 2015), which needs to take into account the surrounding conditions and the way in which water is provided in a locality (Cool et al., 2015; Guilherme & Rodriguez, 2015).

One solution to this problem is to identify the spatial pattern of the geographic distribution of the compounds and delimit the areas that have a certain uniformity in the contents of THM in drinking water, and then analyze these trends in terms of the structure of the network, type of materials, and diameter of the pipes. Specifically, this work analyzes the spatial distribution of THMs in the drinking water network in the city of Montevideo.

It has been recommended that all the variables mentioned above be integrated into a geographic information system (GIS) (Charisiadis et al., 2015; Legay et al., 2010; Sandoval, Lucio, Jaime, León, & Bruno, 2013) so that the data and their georeferenced attributes can be computationally treated

(Montoya et al., 2009). Based on this, a geostatistical analysis can be performed in order to predict the spatial distribution of a phenomenon, in this case the pollutants of interest (Diggle, 2003; Werneck, 2008).

Geostatistics provides indicators that make it possible to identify whether the levels of a particular phenomenon are higher in certain parts of a territory. These include global indicators, such as spatial autocorrelation measurements (Tsai, Lin, Chu, & Perng, 2009), as well as local indicators.

Sometimes the process needs to be analyzed at the local level (Getis & Ord, 1992), as is the case of the present study.

This particular case used the $G_i(d)$ (Ord & Getis, 1995) indicator of autocorrelation. This indicator is disaggregated at the level at which the measurements are taken, in this case the neighborhood, which represents the geographic unit. By using this indicator, dependence can be determined in different areas based on data from neighboring areas, which can be used to compare the various neighborhoods (Tsai et al., 2009). This technique can detect hot spots as well as cold spots (Buehler et al., 2015).

An extensive sampling network in the city of Montevideo provided data for the present study, which was based on 51 points (Gomez-Camponovo et al., 2014) and a total of 8 sampling campaigns. In addition, new variables provided by the water supplier (Obras Sanitarias del Estado [OSE] [http://www.ose.com.uy/]) were gradually incorporated into the analysis, specifically: the formation of THM, hydraulics, structures, the type of piping material, and pipe diameter.

Based on the results that have been obtained thus far (Gomez-Camponovo et al., 2014), some hypotheses can be proposed about the distribution of chloroform in drinking water. The present study includes new data from recent sampling campaigns and uses geostatistical tools to generate additional knowledge about variations in chloroform and its spatial relationship in a drinking water network. Thus, the present study is aimed at determining the geographic distribution pattern of chloroform in the city of Montevideo based on (1) the analytical data that were available from the years 2009–2015 and (2) the physical characteristics of the drinking water network, that is, the density of the network and piping diameter and materials. It also aims to simplify the analysis and facilitate the transfer of technology by consolidating the data from the network and constructing an index.

2. Materials and methods

Uruguay is a country in South America with an area of 318,413 km^2 and a population of 3,440,157, with 1,377,617 people living in its capital, Montevideo (NIS (Uruguay National Institute of Statistic), 2014). This city is a geographically large city in relation to the total population, with a total urban area of 230 km^2. It is divided into 63 neighborhoods with an average area of 3.64 km^2 each.

The city has only one source of water, which is treated by a plant owned by Obras Sanitarias del Estado (OSE, Spanish acronym). Booster chlorination stations are also located in the city.

2.1. Sampling network

Environmental samples were collected from private homes, schools, and shops near the sampling points belonging to the OSE, where routine controls are performed.

Since the locations of the sampling points were determined so as to encompass the entire distribution network in the urban area, an extensive sampling network was established throughout the city. Fifty-one samples were taken in each campaign, as presented in Figure 1, which shows the Montevideo metropolitan area, the drinking water supply network, and the distribution of the sampling points throughout that space.

2.2. Sampling procedures

Tap water was collected in 250 mL glass amber bottles with tapered PTFE caps. The bottles were previously washed with phosphate-free detergent, rinsed with deionized water and acetone, and

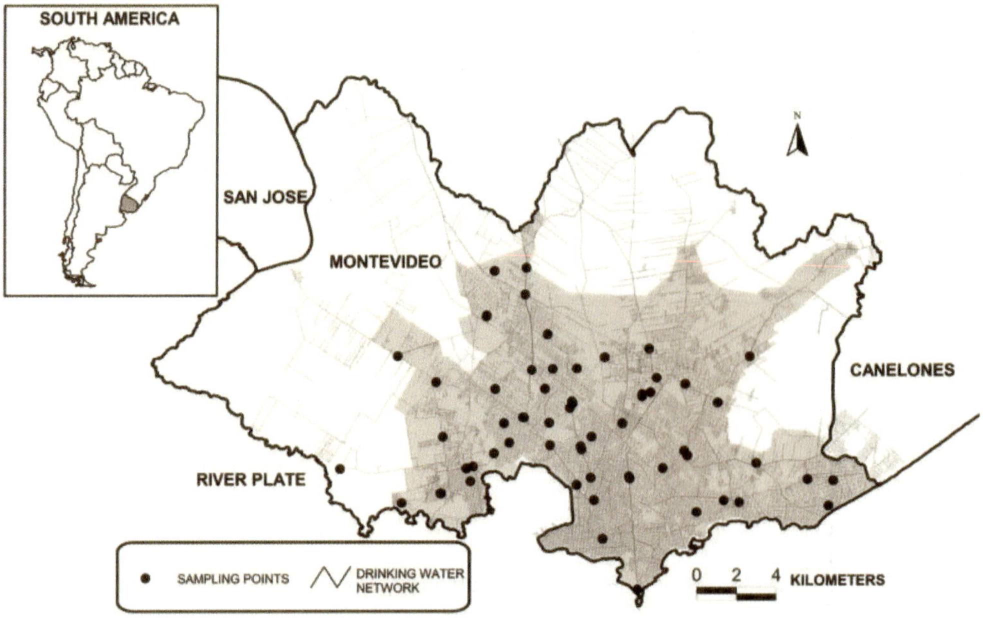

Figure 1. Sampling points in the drinking water network, Montevideo, Uruguay.

placed in an oven overnight at 60°C. When sampling, the bottles were completely filled, leaving no headspace. The samples were stored at 4°C until analysis (Ristoiu et al., 2009).

2.3. Analytical method

Certified reference material ERS-06 (2,000 µg/mL each component, in acetone) from a Cerilliant THM mix and Milli Q-water were used for the THM calibration curve.

The samples were analyzed at certified (ISO 9001:2015) laboratories belonging to the Water Analysis Unit (Unidad de Analisis de Agua) at the University of the Republic's School of Chemistry (Facultad de Química, Universidad de la Republica).

The methodology used was similar to EPA method 524.2 for GC/MS (Cho, Kong, & Seong-Geun, 2003), with slight modifications, which were validated based on ISO 17025:2005.

Headspace-gas chromatography/mass spectrometry (HS-GC/MS) was used to detect TMH (Caro, Serrano, & Gallego, 2007).

The samples were stored at 4°C and were allowed to reach room temperature before conducting the analysis. Then, 10.0 mL of each sample were transferred into 20 mL headspace vials and immediately sealed with Teflon-coated septa and aluminum crimp caps.

The vials were incubated at 60°C for 10 min, with agitation 30 s on and 30 s off, and then 1 mL of the vapor phase was injected in splitless mode.

HS-GC/MS (GC/MS Focus DSQII, Triplus Thermo) equipment was used. The GC was equipped with a capillary column (Agilent HPMS-5: 25 m, i.d. 0.25 mm, and 0.25 mm film).

The temperature program was isothermal at 35°C for 5 min, followed by a ramping up to 150°C at a rate of 10°C/min and holding for 2 min. Helium was used as a carrier gas (grade 5.0, flow rate 1.5 mL/min).

The mass spectrometer conditions were electron impact ionization (70 eV), source temperature at 200°C, and interface temperature at 230°C, in scan mode.

Xcalibur GC-MS software was used to analyze the data. The chloroform was quantified according to their respective area, which was obtained with single-ion extract ion monitoring. Calibration curves were generated with the standard 200 mg/mL Cerilliant ERS-062. Chloroform (*m/z* 83) was quantified using ion, and the limit of quantification was 1µg/L; the linearity was good, with a correlation coefficient (r^2) of 0.9996.

By employing this analytical methodology, the various volatile organic components (THM) found in the water were efficiently separated for their correct identification (Culea, Cozar, & Ristoiu, 2006) using mass spectrometry.

2.4. Geographic information system

ArcGIS software (ESRI (Environmental Systems Research Institute, Inc.), 1996) was used to construct a GIS. The GIS integrated the primary sampling information and the secondary information obtained from the Montevideo Geographic Information Department (Montevideo City Council 2009) (IMM, Spanish acronym). Information about the water distribution network and the piping distribution, diameter, and material was obtained from the drinking water supplier (OSE). The projection of the different layers was corrected and standardized using georeferenced control points (important crossings, notable points) distributed throughout the city. UTM reference system zone 21S was used. The sampling points were incorporated into a point vector layer based on the information collected with the GPS during sampling.

2.5. Spatial analysis

With the chloroform data obtained from the different samples, an analysis of hot spots was performed using ArcGIS® software (https://www.arcgis.com/). Specifically, the G_i* statistic was used, which is a global method that quantifies the degree of spatial autocorrelation in an area. The G_i* statistic measures the local variation in the autocorrelation corresponding to a study area and calculates a value for each geographic entity (neighborhood). The intention is to identify points having a statistically significant local deviation from the global mean of the set. For example, consider points with a Gaussian distribution and values distributed across space. We have three sets of values: points well below the mean, points close to the mean value, and points well above the mean. While a single point with a high or low value may be interesting, for the geographic entity to be statistically significant it must be surrounded by other values that are also high or low. The local mean of a geographic entity and its neighbors is compared proportionally to the mean of the entire set. The local mean is calculated based on a weighting matrix, which is a function of the distance, *d*, between the point of interest and its neighbors. In this matrix, $\omega_{ij}(d)$, points at a distance less than *d* are assigned a value of 1 and those that are further away are assigned a value of 0. The G_i* statistic uses the spatial dependence relationship with distance to calculate the weights $\omega_{ij}(d)$. This boundary parameter *d* thereby defines the distance within which regions are considered to be neighbors and the regions outside of which they are not considered neighbors. An intuitive version of the Getis-Ord G_i* statistic (Getis & Ord, 1992) is calculated according to Equation (1):

$$G_i^*(d) = \frac{\sum_{j=1}^{n} \omega_{ij}(d).x_j}{\sum_{j=1}^{n} x_j} \tag{1}$$

This describes a set of *n* entities x_i, each of which represents a value or attribute associated with spatial position *i*. The G_i*(*d*) is the local statistic, *G*, for the entity located at position *i* within distance *d*. The spatial contiguity matrix *W* contains the elements $\omega_{ij}(d)$, which only select the sum of the neighbors with a distance less than *d*. In order to improve the test statistic, Ord and Getis (1995) developed a transformed version of the G_i* by subtracting the expected value from the G_i* statistic and dividing by the standard deviation and a normalization, which depends on the number of neighbors at distance *d*. This transformation, called the standardized G_i*, is given by Equation (2):

$$G_i^*(d) = \frac{\sum_{j=1}^{n} \omega_{ij}(d).x_j - \bar{X} \sum_{j=1}^{n} \omega_{ij}(d)}{S\sqrt{\frac{n \sum_{j=1}^{n} \omega_{ij}^2(d) - \left(\sum_{j=1}^{n} \omega_{ij}(d)\right)^2}{n-1}}} \quad (2)$$

Here, n is the total number of observations, $\bar{X}$ represents the mean value, and S is the standard deviation calculated for the overall set. The standardized G_i* returns a positive Z-score when the local mean is larger than the global mean and a negative score when it is not. Its value is measured in S units, which is the distance from the local mean in standard deviations.

To detect the degree of local autocorrelation between each geographic entity and its neighbors, a (G_i*)Z score test was performed, which is the standardized G_i* variable. The distribution of the G_i* statistic after its (G_i*)Z standardization follows a standard normal distribution (Cliff & Ord, 1982).

A significant positive (G_i*)Z value indicates that high values tend to be clustered while significant negative values indicate that low values tend to be clustered, resulting in the rejection of the null hypothesis of no spatial autocorrelation.

When a spatial autocorrelation exists, a cluster of high values are called hot spots and a cluster of low values are called cold spots. When the local mean is very different than the global mean, and that difference is too large to be the result of randomness, a statistically significant Z score (G_i* statistic) is obtained.

Thus, the resulting Z scores indicate where the entities with high and low values are spatially clustered (ESRI 2012). This analysis uses the Getis-Ord G_i* statistic (Getis & Ord, 1992) to identify statistically significant hot and cold spots with neighborhoods as the set of entities and chloroform values in the neighborhoods as the field of analysis.

2.6. Piping density index for the neighborhood

The sizes of the neighborhoods and the density of their piping are very different. To take this into account, a piping index was constructed to represent the density of the network in each neighborhood. Since the distribution network may contain pipes with different diameters, the index (Equation 3) represents the sum of all possible diameters:

$$I = \sum_i \frac{\frac{\pi}{4} D_i^2 . L_i}{A.XT} \quad (3)$$

where L_i is the length of piping of diameter D_i, XT is the total length of the piping in the neighborhood, and A is the total area of the neighborhood. Thus, the index represents the total volume of the piping in the neighborhood, standardized by geographic area and by total length of the piping in the neighborhood.

3. Results

A total of 400 samples were measured as a result of all of the sampling campaigns, which were performed in June, September, and December of 2009, June and November of 2010, April and July of 2011, and November of 2015. Chloroform values varied greatly for the different dates on which the sampling campaigns were conducted (Figure 2). The chloroform concentrations for each sampling campaign are presented.

Given the variability during the seasons, we can see the lack of a repeating pattern during hot months (December, November, April) and cold months (June, July, September), as shown by the different sampling campaigns (Figure 3).

Variability can also be seen when determining the hot and cold spots for each of the sampling campaigns (Figure 3).

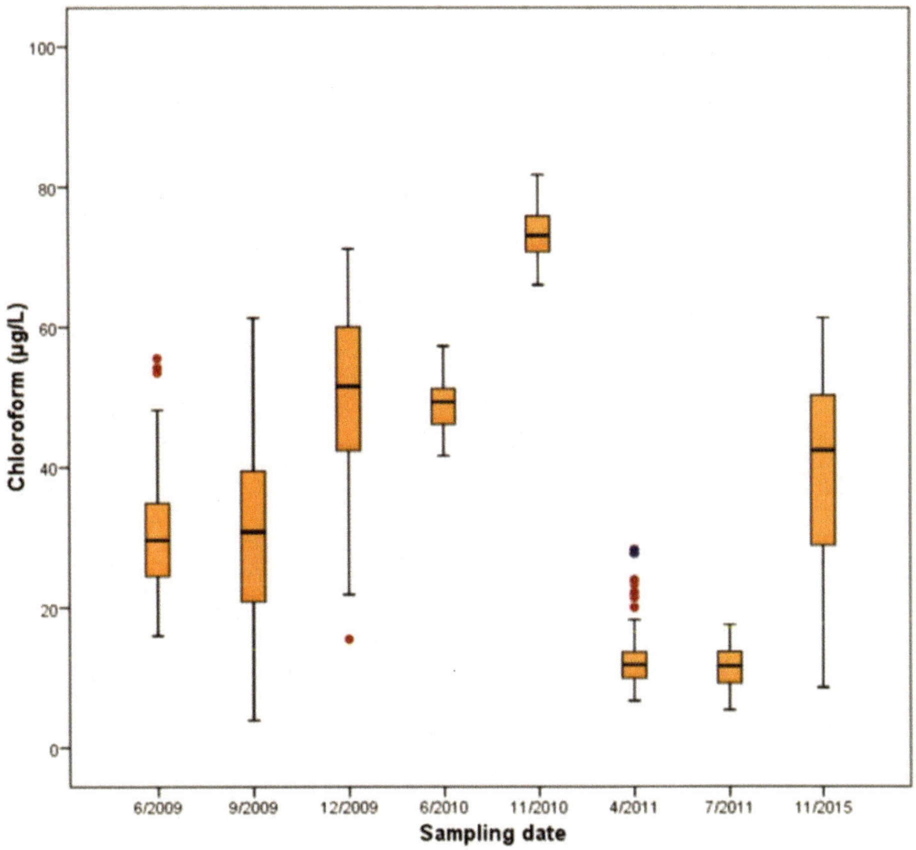

Figure 2. Box plot of chloroform concentrations for all the sampling campaigns conducted from 2009–2015, Montevideo, Uruguay.

However, when considering all the sampling campaigns as a whole (Figure 4), a definite picture emerges.

Since this figure incorporates all the results, zones where high values are likely to be found (hot spots) can be identified, as well as where low values (cold spots) are likely to be located. This makes it possible to draw conclusions about the different areas in the city and to determine which ones require greater attention on the part of the drinking water supplier.

The high values were located in a large area corresponding to the following neighborhoods: Aires Puros (28), Paso de las Duranas (32), Prado (38), Capurro (39), Aguada (40), Reducto (41), Atahualpa (42), Figurita (44), and Brazo Oriental (50). These are old and densely populated neighborhoods.

The water in this area (hot spot) is supplied through a thick branch coming from the treatment plant. A booster chlorination station is also located there, which disinfects the water again before it reaches the final consumer. The cold spots were located in the Buceo (8), Malvin (10), Malvin Norte (11), Las Canteras (18), and La Union (22) neighborhoods. No booster chlorination stations were identified near this area during the sampling period.

In order to analyze the conditions of this network, an index was constructed based on the diameters and lengths of the piping in the neighborhoods. This index reflects the density of the network in each neighborhood. It showed a good correlation with the $(G_i^*)Z$ score values (see Figure 5).

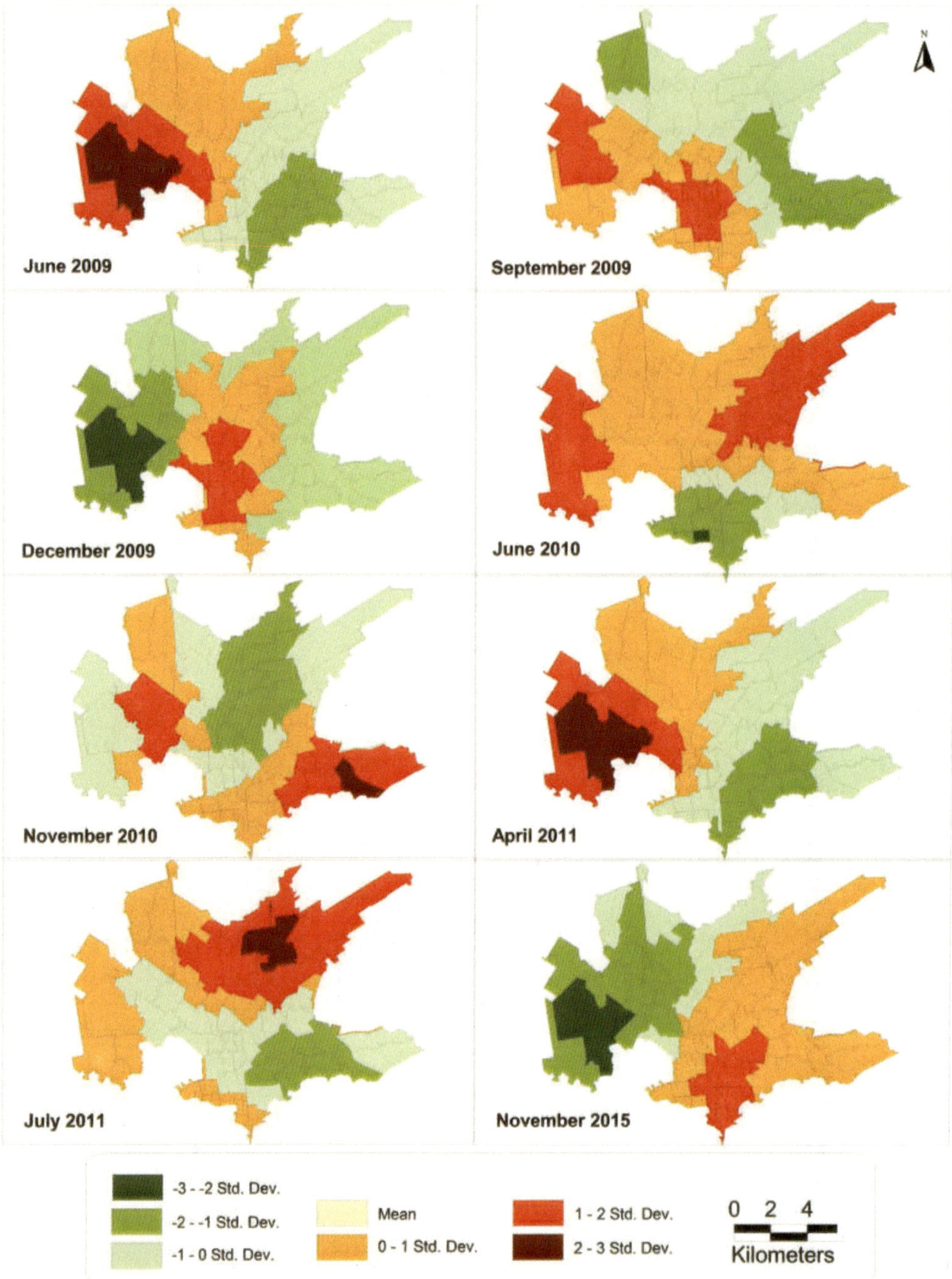

Figure 3. Chloroform hot and cold spots in the drinking water network corresponding to the different sampling campaigns.

The higher index values correspond to hot spots and the lower ones to cold spots (see Table 1). Although the construction of the index only included the length and diameter of the pipes, without considering the material or service age, it is important to note that the piping materials in the neighborhoods that appeared as hot spots were primarily iron, which represented the largest percentage, followed by fiber cement, and PVC (see Table 1).

This table shows the iron piping as a percentage of total piping in the neighborhoods, and the piping density index values for the neighborhoods, which are grouped according to expected chloroform concentrations.

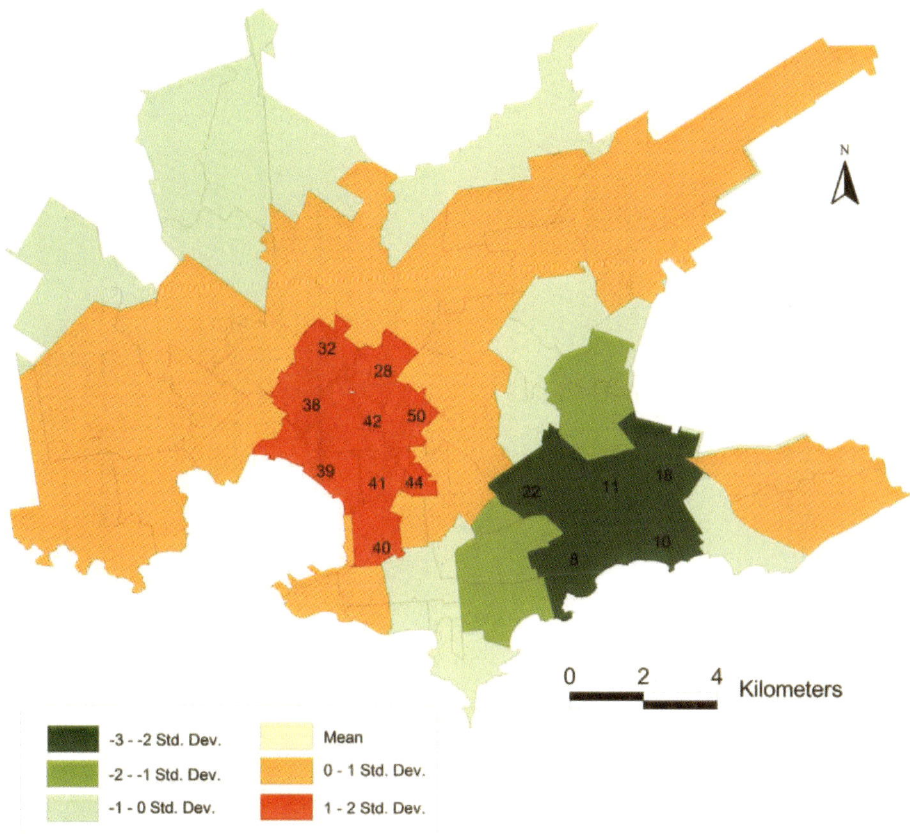

Figure 4. Chloroform hot and cold spots in the drinking water network, considering all the samples, 2009–2015, Montevideo, Uruguay.

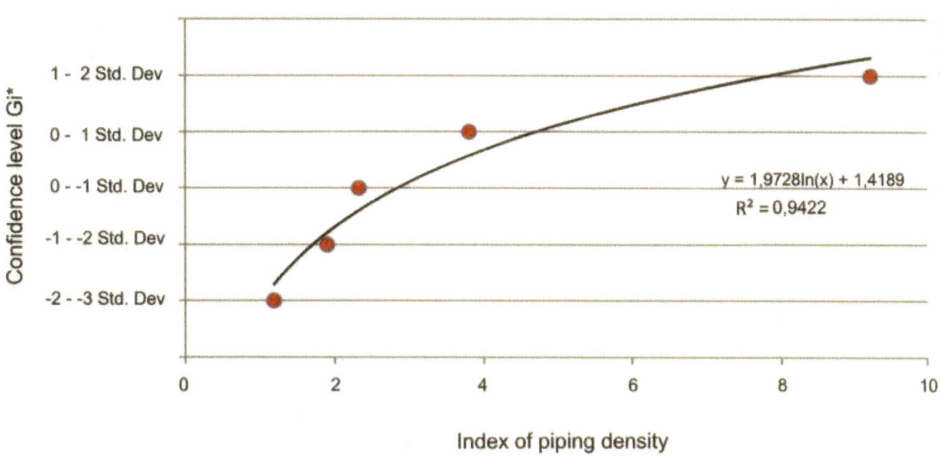

Figure 5. Correlation between the (G_i*)Z score and the piping density index for the neighborhoods.

In the neighborhoods that appeared as cold spots, the percentage of iron piping was less than 43.3% of the total piping, roughly the same percentage as fiber cement piping.

4. Discussion

The use of the geostatistic made it possible to characterize the neighborhoods in the city of Montevideo as hot spots or cold spots based on chloroform concentrations in the drinking water. Considering that a large variability in the concentrations was found when analyzing each one of

Table 1. Percentage of iron piping and density index, according to the (G_i*) Z score, for neighborhoods in Montevideo, Uruguay

Confidence Level (mean G_i* ± SD)	Neighborhood number*	Pipe wall material				Piping density index
		Iron (%)	Cement pipe	PVC	Others	
−3 to −2 SD	8, 10, 11, 18, 22	43.3	42	6.2	8.5	0.94
−2 to—1 SD	7, 9, 16, 17, 46	54.8	29.3	9.3	6.6	1.49
−1 SD to mean	3, 4, 5, 6, 12, 15, 20, 21, 23, 31, 49, 52, 56, 57, 58, 59	46.9	30.8	10.9	11.4	1.82
Mean to 1 SD	0, 1, 2, 13, 14, 19, 24, 25, 26, 27, 29, 30, 33, 34, 35, 36, 37, 43, 45, 47, 48, 51, 53, 54, 55, 60, 61, 62	50.3	31.7	8.3	9.7	2.98
1 to 2 SD	28, 32, 38, 39, 40, 41, 42, 44, 50	78.7	15.4	0.6	5.3	7.24

SD: standard deviation.

*0. Ciudad Vieja, 1. Centro, 2. Barrio Sur, 3. Cordon, 4. Palermo, 5. Parque Rodo, 6. Punta Carretas, 7. Pocitos, 8. Buceo, 9. Pque Battle, 10. Malvin, 11. Malvin Norte, 12. Punta Gorda, 13. Carrasco, 14. Carrasco Norte, 15. Bañados de Carrasco, 16. Maroñas, 17. Flor de Maroñas, 18. Las Canteras, 19. Punta Rieles, 20. Jardines del Hipodromo, 22. Union, 23. Villa Española, 24. Mercado Modelo, 25. Castro, 26. Cerrito, 27. Las Acacias, 28. Aires Puros, 29. Casavalle, 30. Piedras Blancas, 31. Manga, 32. Paso de las Duranas, 33. Peñarol, 34. Cerro, 35. Casabo, 36. La Paloma, 37. La Teja, 38. Prado, 39. Capurro, 40. Aguada, 41. Reducto, 42. Atahualpa, 43. Jacinto Vera, 44. Figurita, 45. Larrañaga, 46. La Blanqueda, 47. Villa Muñoz, 48. La Comercial, 49. Tres Cruces, 50. Brazo Oriental, 51. Sayago, 52. Conciliacion, 53. Belvedere, 54. Nuevo Paris, 55. Tres Ombues, 56. Paso de la Arena, 57. Colon, 58. Colon Centro, 59. Lezica, 60. Villa Garcia, 61. Manga, 62. Puerto.

the sampling campaigns, it can be said that the identification of hot and cold spots was possible because of the availability of numerous sampling points that were correctly distributed throughout the different geographic units of analysis (neighborhoods) in the urban space, with data covering a long time period (2009–2015).

The index proposed by this study reflects the density of the piping networks at the neighborhood level.

The study found that the standardized G_i^* levels increased as the proposed index increased. For future studies, this index can provide an early indication of points that may behave as hot spots and cold spots as long as the sampling points are selected in a manner that produces efficient measurements.

In addition, the hot spots were found to be related to a higher percentage of iron piping (Chang, Guo, Shu, Chiang, & Huang, 2010), which is consistent with previous findings showing that old cast iron pipes are one of the main causes of chlorine decay.

Larger piping diameters result in larger cross sections, which contributes to slower flows, while slower flow velocities decrease chlorine levels (due to bulk decay) (Chang et al., 2010; Ozdemir & Metin, 1998) and the wall decay rate (Hallam et al., 2002). In addition, the effect of flow conditions on the formation of THM varies and greatly depends on other factors such as residence time (which increases the formation of THM) (Hallam et al., 2002), the condition of the piping, and the piping material (iron can increase or decrease the formation of THM depending on service age) (Rossman et al., 2001).

In our case, the index increasing along with the standardized G_i^* values was related to a higher percentage of iron piping, which for the city of Montevideo suggests that piping conditions and materials are important factors in the formation of THMs. This explains why hot spots were not found at the end of the network (which would correspond to increased THM formation at points further from the water chlorination point in the treatment plant) but rather in neighborhoods that had a higher piping density and a higher percentage of iron piping (Chang et al., 2010; Guilherme & Rodriguez, 2015; Loyola-Sepulveda et al., 2009). These results are similar to findings by other studies (Al-Jasser 2007; Baird, 2012; Hallam et al., 2002; Legay et al., 2010; Rossman et al., 2001; Wang et al., 2012).

In terms of the service age of the iron piping, although information was not available, iron is known to be one of the earliest materials (OSE (National Sanitary Administration), 2017) used to build the water supply network, and therefore the service age is likely to be considerable.

This work was able to describe a relationship between chloroform levels and the conditions related to the distribution of water, specifically, the piping density as well as the type of piping material in each geographic unit (neighborhood).

While booster chlorination stations existed in the city and the quantity increased during the study period, it is worth mentioning that the analysis provided important results even without data about the presence and location of those points.

Since the hot spots were located in Montevideo's older neighborhoods, and given that iron was one of the first materials used (OSE (National Sanitary Administration), 2017) in the distribution system (Marlow, Gould, Beale, & Lane, 2015), the structure and integrity may have been affected over time (Al-Jasser 2007; Marlow et al., 2015) by factors such as corrosion, the presence of biofilms, and ruptures (unlined pipes) (Baird, 2012; Marlow et al., 2015). This reaffirms the importance of ongoing maintenance and rehabilitation programs (Baird, 2012; Marlow et al., 2015).

It would be useful to conduct other studies using this methodology in other localities where the management of the supply network varies and where water for human consumption is obtained from different sources. This can contribute to the development of a more extensive prediction system and will undoubtedly provide a more precise evaluation of the population risk of health disorders, which are still being studied.

5. Conclusions

New techniques are needed for the early detection of zones that may have unsafe levels of organic pollutants in water used for human consumption.

With the use of a geostatistical technique (Getis–Ord G_i^*), this study was able to detect sets of neighborhoods with high stable chloroform values over time. It was also able to detect sets of neighborhoods with low values.

Neighborhoods with hot spots and others with cold spots could be identified. And it was possible to associate these values with a neighborhood piping density index. The highest values corresponded to neighborhoods in the city that had a higher percentage of iron.

The results obtained by this work suggest the need for processes to transfer technology and knowledge from academic institutions to water management organizations, which can contribute to improving the efficiency of managing drinking water systems, optimizing control systems, and developing the measures needed to minimize risk from organic pollutants. A more detailed analytical process needs to be developed so that the maintenance and rehabilitation of drinking water distribution networks can be adjusted.

Acknowledgements

The authors thank the water supplier in Uruguay (Obras Sanitarias del Estado [OSE, Spanish acronym] and Dr. Jose Carlos Guerrero Antunez for his thorough reading of the manuscript and his suggestions.

Funding

The authors received no direct funding for this research.

Author details

Marcel Achkar Borras[1]
E-mail: achkar@fcien.edu.uy
Gustavo Seoane[2]
E-mail: gseoane@fq.edu.uy
Mariana Gomez-Camponovo[3]
E-mail: mgomez@cup.edu.uy
E-mail: mgomez@cup.edu
Eleuterio Umpierrez Vazquez[4]
E-mail: eumpierr@fq.edu.uy
Nicolas Perez[5]
E-mail: nico@fisica.edu.uy

[1] Department of Sciences, Territorial Organization and Sustainable Development Laboratory, University of the Republic, Montevideo, UY, Uruguay.
[2] Department of Organic Chemistry, University of the Republic, Montevideo, UY, Uruguay.
[3] Social Medicine Unit, University of the Republic, Paysandu, UY, Uruguay.
[4] Department of Chemistry's Technological Institute, Environmental, Drug & Doping Unit, University of the Republic, Montevideo, UY, Uruguay.
[5] Department of Engineering, Chemical Engineering Institute-Electric Engineering Institute, University of the Republic, Montevideo, UY, Uruguay.

References

Al-Jasser, A. O. (2007). Chlorine decay in drinking-water transmission and distribution systems: Pipe service age effect. *Water Research*, *41*, 387–396. doi:10.1016/j.watres.2006.08.032

Baird, G. M. (2012). The epidemic of corrosion, Part 2: The importance of communication. *Journal - American Water Works Association*, *104*, 20–24. doi:10.5942/jawwa.2012.104.0018

Bassey, G. I., & Egbe Jerome, G. (2016). Residual chlorine decay in water distribution network. *International Journal of Scientific Research and Engineering Studies*, *3*, 1–6. Retrieved from http://www.ijsres.com/2016/vol-3_issue-3/paper_1.pdf

Baytak, D., Sofuoglu, A., Inal, F., & Sofuoglu, S. C. (2008). Seasonal variation in drinking water concentrations of disinfection by-products in IZMIR and associated human health risks. *Science of the Total Environment*, *407*, 286–296. doi:10.1016/j.scitotenv.2008.08.019

Botton, J., Kogevinas, M., Gracia-Lavedan, E., Patelarou, E., Roumeliotaki, T., Iñiguez, C., … Villanueva, C. M. (2015). Postnatal weight growth and Trihalomethane exposure during pregnancy. *Environmental Research*, *136*, 280–288. doi:10.1016/j.envres.2014.09.035

Bove, G. E., Jr, Rogerson, P. A., & Vena, J. E. (2007a). Case-control study of the effects of Trihalomethanes on urinary bladder cancer risk. *Archives of Environmental & Occupational Health*, *62*, 39–47. Retrieved from http://www.tandfonline.com/doi/abs/10.3200/AEOH.62.1.39-47

Bove, G. E., Jr, Rogerson, P. A., & Vena, J. E. (2007b). Case control study of the geographic variability of exposure to disinfectant byproducts and risk for rectal cancer. *International Journal of Health Geographics*, *6*, 18. doi:10.1186/1476-072X-6-18

Brown, D., Bridgeman, J., & West, J. R. (2011). Predicting chlorine decay and THM formation in water supply systems. *Reviews in Environmental Science and Bio/Technology*, *10*, 79–99. doi:10.1007/s11157-011-9229-8

Buehler, C. P., Blevins, M., Ossemane, E. B., González-Calvo, L., Ndatimana, E., Vermund, S. H., … Moon, T. D. (2015). Assessing spatial patterns of HIV knowledge in rural Mozambique using geographic information systems. *Tropical Medicine and International Health, 20*, 353–364. doi:10.1111/tmi.12437

Gomez Camponovo, M., Seoane Muniz, G., Rothenberg, S. J., Umpiérrez Vazquez, E. & Marcel, A. B. (2014). Predictive model for chloroform during disinfection of water for Consumption, City of Montevideo. *Environmental Monitory and Assessment, 186*, 6711–6719. doi:10.1007/s10661-014-3884-5

Caro, J., Serrano, A., & Gallego, M. (2007). Sensitive headspace gas chromatography–mass spectrometry determination of Trihalomethanes in Urine. *Journal of Chromatography B, 848*, 277–282. doi:10.1016/j.jchromb.2006.10.034

Chang, C.-C., Shu-Chen, H., Wang, L.-Y., & Chun-Yuh, Y. (2007). Bladder cancer in Taiwan: Relationship to Trihalomethane concentrations present in drinking-water supplies. *Journal of Toxicology and Environmental Health, Part A, 70*, 1752–1757. doi:10.1080/15287390701459031

Chang, E. E., Guo, H. C., Shu, L. I., Chiang, P. C., & Huang, C. P. (2010). Modeling the formation and assessing the risk of disinfection by-products in water distribution systems. *Journal of Environmental Science and Health, Part A, 45*, 1185–1194. doi:10.1080/10934529.2010.493776

Charisiadis, P., Andra, S. S., Makris, K. C., Christophi, C. A., Skarlatos, D., Vasilis., V., … Stephanou, E. G. (2015). Spatial and seasonal variability of tap water disinfection by-products within distribution pipe networks. *Science of the Total Environment, 506–507*, 26–35. doi:10.1016/j.scitotenv.2014.10.071

Chen, K., Weiping, Y., Xinyan, M., Yao, K., & Jiang, Q. (2005). The association between drinking water source and colorectal cancer incidence in Jiashan County of ChinaR study. *European Journal of Public Health, 15*, 652–656. doi:10.1093/eurpub/cki027

Chisholm, K., Cook, A., Bower, C., & Weinstein, P. (2008). Risk of birth defects in australian communities with high levels of brominated disinfection by-products. *Environmental Health Perspectives, 116*, 1267–1273. doi:10.1289/ehp.10980

Cho, D.-H., Kong, S.-H., & Seong-Geun, O. (2003). Analysis of Trihalomethanes in drinking water using head-space-SPME technique with gas chromatography. *Water Research, 37*, 402–408. doi:10.1016/S0043-1354(02)00285-3

Cliff, A. D., & Ord, J. K. (1982). Spatial processes. Models and applications. *Population*, (37)963Retrieved fromhttp://www.persee.fr/doc/pop_0032-4663_1982_num_37_4_17407

Cool, G., Lebel, A., Sadiq, R., & Rodriguez, M. J. (2015). Modelling the regional variability of the probability of high Trihalomethane occurrence in municipal drinking water. *Environmental Monitoring and Assessment, 187*, 746. doi:10.1007/s10661-015-4969-5

Culea, M., Cozar, O., & Ristoiu, D. (2006). Methods validation for THMs determination in drinking water. *Journal of Mass Spectrometry, 41*, 1594–1597. doi:10.1002/jms.1149

Diggle, P. J. (2003). An Introduction to model-based geostatistics. Chap. 2. In J. Moller (Ed.), *Spatial statistics and computational methods. Lecture notes in statistics* (Vol. 173, pp. 43–86). New York, NY: Springer. doi:10.1007/978-0-387-21811-3_2

ESRI (Environmental Systems Research Institute, Inc) 2012. *ArcGIS resources: Hot spot analysis (Getis-Ord Gi*) (Spatial Statistics)*. Retrieved from: http://resources.arcgis.com/en/help/main/10.2/index.html#//005p00000010000000

ESRI (Environmental Systems Research Institute, Inc.) 1996. *ArcGIS 10.1 spatial analysis*.Retrieved from: http://www.esri.com/.

Getis, A., & Ord, J. K. (1992). The analysis of spatial association by use of distance statistics. *Geographical Analysis, 24*, 189–206. Retrieved from http://resources.esri.com/help/9.3/arcgisengine/java/gp_toolref/spatial_statistics_tools/how_hot_spot_analysis_colon_getis_ord_gi_star_spatial_statistics_works.htm

Goebell, P. J., Villanueva, C. M., Rettenmeier, A. W., Rübben, H., & Kogevinas, M. (2004). Environmental exposure, chlorinated drinking water, and bladder cancer. *World Journal of Urology, 21*, 424–432. doi:10.1007/s00345-003-0389-1

Grellier, J., Bennett, J., Patelarou, E., Smith, R. B., Toledano, M. B., Rushton, L., … Nieuwenhuijsen, M. J. (2010). Exposure to disinfection by-products, fetal growth, and prematurity: A systematic review and meta-analysis. *Epidemiology, 21*, 300–313. doi:10.1097/EDE.0b013e3181d61ffd

Guilherme, S., & Rodriguez, M. J. (2015). Short-term spatial and temporal variability of disinfection by-product occurrence in small drinking water systems. *Science of the Total Environment, 518-519*, 280–289. doi:10.1016/j.scitotenv.2015.02.069

Hallam, N. B., West, J. R., Forster, C. F., Powell, J. C., & Spencer, I. (2002). The decay of chlorine associated with the pipe wall in water distribution systems. *Water Research, 36*, 3479–3488. doi:10.1016/S0043-1354(02)00056-8

Hinckley, A. F., Bachand, A. M., Nuckols, J. R., & Reif, J. S. (2005). Identifying public water facilities with low spatial variability of disinfection by-products for epidemiological investigations. *Occupational and Environmental Medicine, 62*, 494–499. doi:10.1136/oem.2004.017798

IMM (Montevideo City Council) 2009. Geographic Information, City of Montevideo. Retrieved from: http://intgis.montevideo.gub.uy/.

ISO (International Organization for Standardization) 2005. *ISO/IEC 17025:2005(es). Requisitos generales para la competencia de los laboratorios de ensayo y de calibración*. Retrieved from: https://www.iso.org/obp/ui/#iso:std:iso-iec:17025:ed-2:v1:es.

Jang, H. J., Choi, Y. J., Ro, H. M., & Ka, J. O. (2012). Effects of phosphate addition on biofilm bacterial communities and water quality in annular reactors equipped with stainless steel and ductile cast iron pipes. *Journal of Microbiology, 50*, 17–28. doi:10.1007/s12275-012-1040-x

Kiene, L., Lu, W., & Levi, Y. (1998). Relative importance of the phenomena responsible for chlorine decay in drinking water distribution systems. *Water Science Technology, 38*, 219–227. doi:10.2166/wst.1998.0255

King, W. D., Marrett, L. D., & Woolcott, C. G. (2000). Case-control study of colon and rectal cancers and chlorination by-products in treated water. *Cancer Epidemiology, Biomarkers & Prevention, 9*, 813–818.

Knight, N., Watson, K., Carswell, S., Comino, E., & Shaw, G. (2011). Temporal and spatial variation of Trihalomethanes and haloacetic acids concentration in drinking water: A case study of Queensland, Australia. *Air, Soil and Water Research, 4*, 1–17. doi:10.4137/ASWR.S5618

Krasner, S. W., Weinberg, H. S., Richardson, S. D., Pastor, S. J., Chinn, R., Sclimenti, M. J., … Thruston, A. D. (2006). Occurrence of a new generation of disinfection by products. *Environmental Science & Technology, 40*, 7175–7185. doi:10.1021/es060353j

Kuo, H.-W., Chen, P.-S., Shu-Chen, H., Wang, L.-Y., & Yang, C.-Y. (2010). Trihalomethanes in drinking water and the risk of death from rectal cancer: Does hardness in drinking water matter? *Journal of Toxicology and Environmental Health, Part A, 73*, 807–818. doi:10.1080/15287391003689267

Kuo, H.-W., Tiao, M.-M., Trong-Neng, W., & Yang, C.-Y. (2009). Trihalomethanes in drinking water and the risk of death from colon cancer in Taiwan. *Journal of Toxicology and Environmental Health, Part A, 72*, 1217–1222. doi:10.1080/15287390903129176

Legay, C., Rodriguez, M. J., Sérodes, J. B., & Levallois, P. (2010). The assessment of population exposure to chlorination by-products: A study on the influence of the water distribution system. *Environmental Health, 9*, 59. doi:10.1186/1476-069X-9-59

Loyola-Sepulveda, R., Lopez-Leal, G., Jorge., M., Bravo-Linares, C., & Mudge, S. M. (2009). Trihalomethanes in the drinking water of Concepción and Talcahuano, Chile. *Water and Environment Journal, 23*, 286–292. doi:10.1111/j.1747-6593.2008.00140.x

Marlow, D., Gould, S., Beale, D., & Lane, B. (2015). Rehabilitation of small-diameter cast-iron pipe: US, UK, and Australian perspectives. *Journal - American Water Works Association, 107*, E12–21. doi:10.5942/jawwa.2015.107.0003

Rahman, Md, Bayzidur, D., & Tim, C. Christine, and Armstrong, Bruce K. (2010). Disinfection by-products in drinking water and colorectal cancer: A meta-analysis. *International Journal of Epidemiology, 39*, 733–745. doi: 10.1093/ije/dyp371

Michaud, D. S., Manolis., K., Cantor, K. P., Villanueva, C. M., Garcia-Closas, M., Nathaniel, R., ... Silverman, D. T. (2007). Total fluid and water consumption and the joint effect of exposure to disinfection by-products on risk of bladder cancer. *Environmental Health Perspectives, 115*, 1569–1572. doi:10.1289/ehp.10281

Montoya, C., Loaiza, D., Cruz, C., Torres, P., Escobar, J. C., & Delgado, L. G. (2009). Propuesta metodológica para localización de estaciones de monitoreo de calidad de agua en redes de distribución utilizando sistemas de información geográfica. *Revista Facultad De Ingeniería, 49*, 129–140. Retrieved from http://aprendeenlinea.udea.edu.co/revistas/index.php/ingenieria/article/viewFile/15967/13841

Nieuwenhuijsen, M. J., Martinez, D., James., G., Bennett, J., Best, N., Nina, I., ... Toledano, M. B. (2009). Chlorination disinfection by-products in drinking water and congenital anomalies: review and meta-analyses. *Environmental Health Perspectives, 117*, 1486–1493. doi:10.1289/EHP.0900677

Nieuwenhuijsen, M. J., Toledano, M. B., Eaton, N. E., Fawell, J., & Elliott, P. (2000). Chlorination disinfection byproducts in water and their association with adverse reproductive outcomes: A review. *Occupational and Environmental Medicine, 57*(73–85). doi:10.1136/oem.57.2.73

NIS (Uruguay National Institute of Statistic) 2014. *Uruguay en cifras 2014*. Retrieved from: http://www.ine.gub.uy/documents/10181/39317/Uruguay_en_cifras_2014.pdf/aac28208-4670-4e96-b8c1-b2abb93b5b13.

OSE (National Sanitary Administration) 2017. *Reseña Histórica*. Retrieved from: http://www.ose.com.uy/e_historia.html.

Ord, J. K., & Getis, A. (1995). Local spatial autocorrelation statistics: Distributional issues and an application. *Geographical Analysis, 27*, 286–306. doi:10.1111/j.1538-4632.1995.tb00912.x

Ozdemir, O. N., & Metin, G. A. (1998). Realistic numerical simulation of chlorine decay in pipes. *Water Research, 32*, 3307–3312. doi:10.1016/S0043-1354(98)00107-9

Ristoiu, D., von Gunten, U., Mocan, A., Chira, R., Siegfried, B., Kovacs, H., ... Sidonia. (2009). Trihalomethane formation during water disinfection in four water supplies in the Somes River Vasin in Romania. *Environmental Science and Pollution Research, 16*(supplement 1), S55–65. doi:10.1007/s11356-009-0100-1

Rodriguez, M. J., Vinette, Y., Jean-B., S., & Bouchard, C. (2003). Trihalomethanes in drinking water of greater Québec region (Canada): Occurrence, variations and modelling. *Environmental Monitoring and Assessment*, (89), 69–93. doi:10.1023/A:1025811921502

Rossman, L. A., Brown, R. A., Singer, P. C., & Nuckols, J. R. (2001). DBP formation kinetics in a simulated distribution system. *Water Research, 35*, 3483–3489. doi:10.1016/S0043-1354(01)00059-8

Sadiq, R., & Rodriguez, M. J. (2004). Disinfection by-products (DBPs) in drinking water and predictive models for their occurrence: A review. *Science of the Total Environment*, (321), 21–46. doi:10.1016/j.scitotenv.2003.05.001

Sandoval, F., Lucio, R. Z.-F., Jaime, R., León, J., & Bruno, A. (2013). Sistema para control y gestión de redes de agua potable de dos localidades de México. *Ingeniería Hidráulica y Ambiental, 34*, 112–126. Retrieved from http://riha.cujae.edu.cu/index.php/riha/issue/view/18

Tsai, P.-J., Lin, M.-L., Chu, C.-M., & Perng, C.-H. (2009). Spatial autocorrelation analysis of health care hot-spots in Taiwan in 2006. *BMC Public Health, 9*, 464. doi:10.1186/1471-2458-9-464

UNIT (Uruguay Institute for Technical Norms). 2011. *UNIT Norm 833:2008: Agua potable - Requisitos*. Retrieved from: http://www.ose.com.uy/descargas/Clientes/Reglamentos/unit_833_2008_.pdf.

Uyak, V., Ozdemir, K., & Toroz, I. (2007). Multiple linear regression modeling of disinfection by-products formation in Istanbul drinking water reservoirs. *Science of the Total Environment, 378*, 269–280. doi:10.1016/j.scitotenv.2007.02.041

Villanueva, C. M., Kogevinas, M., & Grimalt, J. O. (2001). Cloración del agua potable en España y cáncer de vejiga. Chlorination of drinking water in Spain and bladder cáncer. *Gaceta Sanitaria, 15*, 48–53. doi:10.1016/S0213-9111(01)71517-8

Villanueva, C. M., Cantor, K. P., Cordier, S., Jaakkola, J. J. K., King, W. D., Lynch, C. F., ... Kogevinas, M. (2004). Disinfection byproducts and bladder cancer: A pooled analysis. *Epidemiology, 15*, 357–367. doi:10.1097/01.ede.0000121380.02594.fc

Villanueva, C. M., Cantor, K. P., King, W. D., Jaakkola, J. J. K., Cordier, S., Lynch, C. F., ... Kogevinas, M. (2006). Total and specific fluid consumption as determinants of bladder cancer risk. *International Journal of Cancer, 118*, 2040–2047. doi:10.1002/ijc.21587

Villanueva, C. M., Gracia-Lavedan, E., Bosetti, C., Righi, E., Molina, A. J., Martin, V., ... Kogevinas, M. (2017). Colorectal cancer and long-term exposure to trihalomethanes in drinking water: A multicenter case-control study in Spain and Italy. *Environmental Health Perspectives, 125*, 56–65. doi:10.1289/EHP155

Wang, G.-S., Deng, Y.-C., & Lin, T.-F. (2007). Cancer risk assessment from Trihalomethanes in drinking water. *Science of the Total Environment, 387*, 86–95. doi:10.1016/j.scitotenv.2007.07.029

Wang, H., Masters, S., Edwards, M. A., Falkinham, J., & Pruden, A. (2014). Effect of disinfectant, water age, and pipe materials on bacterial and eukaryotic community structure in drinking water biofilm. *Environmental Science & Technology*, *48*, 1426–1435. doi:10.1021/es402636u

Wang, H., Masters, S., Hong, Y., Stallings, J., Falkinham, J. O., Edwards, M. A., & Pruden, A. (2012). Effect of disinfectant, water age, and pipe material on occurrence and persistence of Legionella, Mycobacteria, Pseudomonas Aeruginosa, and Two Amoebas.

Werneck, G. L. (2008). Georeferenced data in epidemiologic research. *Ciencia & Saude Coletiva, 13*, 1753–1766. doi:10.1590/S1413-81232008000600010

WHO (World Health Organization). 2008. *Guidelines for drinking-water quality: Fourth edition incorporating the first addendum*. Retrieved from: https://www.ncbi.nlm.nih.gov/books/NBK442372/

9

Environmental geochemistry and ecological risk assessment of potentially harmful elements in tropical semi-arid soils around the Bagassi South artisanal gold mining site, Burkina Faso

Aboubakar Sako[1*] and Mamadou Nimi[2]

*Corresponding author: Aboubakar Sako, Université de Dédougou, BP. 139, Dédougou, Burkina Faso
E-mail: aboubakar.sako@gmail.com

Reviewing editor: Arno Rein, Chair of Hydrogeology, Technical University of Munich, Germany

Abstract: This study investigates geochemistry and ecological risk of artisanal gold mining-derived potentially harmful elements in semi-arid soils in Burkina Faso. R-mode factor analysis, which reduces the variables (elements) to few factors, was applied to explain the dominant variance in the data. Three factors, which account 80% of the total variance, described differences in soil geochemistry resulting from anthropogenic and geogenic sources. High loadings of factor 1 on As, Au, Bi, Cd, Hg, Mo, Pb, Sb, Te, W and Zn suggest that the artisanal gold mining was the most important factor controlling the soil geochemistry. Factor 2 had high loadings on Al, Fe, Mn, Ti, Co, Cr, Cu, Ni, Sc, Sr, Tl and V, representing their geogenic origin. With high loadings on Ca, Mg, S and La, factor 3 describes contribution of biogeochemical cycling to the elements' abundance in the soils. Lead isotope compositions identified atmospheric deposition as the main source of Pb in farmland soils, whereas topsoil and soil profiles were primarily influenced by the mining activities. Mercury, As and, to a lesser degree, Cd posed the most serious ecological threat to the soils collected around the mining site relative to those of the farmland. Based on the findings of this study, a best pollution control plan of potentially harmful element loadings into soils is urgently required around artisanal gold mining sites across the country.

ABOUT THE AUTHORS

Aboubakar Sako is an environmental geochemist and assistant professor of environment and geosciences from University of Dédougou (Burkina Faso). He holds a PhD in Environmental Sciences from Arkansas State University (Arkansas, USA). His research involves geochemical characterization of heavy metals in surface water, groundwater and soils. Dr Sako is also interested in hydrogeochemical modeling and ecological risk assessment of industrial and artisanal gold mining sites in the West Africa region.

Mamadou Nimi is a geochemical analyst and presently head of the Geological Survey of Burkina Faso in the regional office of Bobo-Dioulasso. He gained about 13 year experience in natural water, soil and rock sample characterization and analysis for environmental assessment and mineral exploration.

PUBLIC INTEREST STATEMENT

The study describes the anthropogenic and geogenic sources of artisanal gold mining-derived potentially toxic heavy metals/metalloids and their ecological impacts on semi-arid soils in southwestern Burkina Faso. Even though artisanal gold mining contributes to poverty alleviation in rural communities, it causes several environmental problems such deforestation, soil degradation and water and air pollution. The release and accumulation of large quantities of potentially harmful elements in the surrounding soils can pose serious threat to human health and also to ecosystems. The presence of hundreds of uncontrolled artisanal gold mining sites around rural dwellings and farmlands across the country makes environmental geochemistry and ecological risk assessment of potentially harmful elements in soils highly important.

Subjects: Earth Sciences; Soil Science; Geochemistry; Environmental Change & Pollution

Keywords: potentially harmful elements; artisanal gold mining; geochemistry; ecological risk; pedogeochemical processes

1. Introduction

The Birimian greenstone belts that cross Burkina Faso are endowed with polymetallic deposits of gold (Au), zinc (Zn), manganese (Mn), copper (Cu) and molybdenum (Mo) (Huot, Sattran, & Zida, 1987). Yet, mining in Burkina Faso has been primarily based on Au (Kiethega, 1983). This is mainly due to the severe drought in late 1980s and the boom in world gold price in mid-2000s that made Au extraction attractive for both industrial investors and artisanal miners (Luning, 2008; Mégret, 2011; Werthmann, 2000). Currently, 10 industrial gold mines and about 800 artisanal gold mining (AGM) sites are active across the country. With about 36.1 tons of Au produced in 2014, gold overtook cotton as Burkina Faso's first export commodity, and the country became the fourth biggest Au producer in Africa after South Africa, Ghana and Mali (Andersson, Ola, & Niklas, 2015; Zabsonre, Agbo, Some, & Haffin, 2016). While the industrial gold mines are solely owned and operated by foreign companies, AGM activities are carried out by about 700,000 unskilled subsistence farmers (Fofana, Ouédraogo, & Zombré, 2009; Jaques, Zida, Billa, Greffié, & Thomassin, 2006; Mégret, 2011).

Although some AGM sites remain active throughout the year, this low-tech enterprise that uses rudimentary tools to extract gold from rich and near-surface deposits is largely seasonal and usually begins at the end of crop harvest (Gueye, 2001; Jaques et al., 2006). Because of low levels of agricultural mechanization, farming in Burkina Faso requires too much work for far little yields (Zhou, 2016). In order to compensate their inability to ensure self-sufficiency in food production, farmers combine both AGM and farming (Hilson, 2016). Hence, the AGM contributes to poverty alleviation and livelihood diversification of rural communities (Andersson et al., 2014). However, the unregulated AGM activities with inadequate or inexistent pollution control measures raise many environmental and human health concerns (Amegbey, Dankwa, & Al-Hasson, 1997; Davidson, 1993; Donkor, Bonzongo, Nartey, & Adotey, 2005).

The Bagassi South gold deposits are hosted in a metallogenic assemblage of quartz and sulfide minerals (Hein, 2015). These minerals are generally associated with potentially harmful elements (PHE; Plant et al., 1997; Salvarredy-Aranguren, Probst, Roulet, & Isaure, 2008), such as silver (Ag), arsenic (As), copper (Cu), lead (Pb), nickel (Ni), antimony (Sb) and zinc (Zn) that may, through mechanical removal and leaching, contaminate the surrounding soils (Cai et al., 2012). Artisanal miners also use metallic mercury (Hg^{o}) to extract gold from crude ores that could contribute to mercury (Hg) emission into the atmosphere, as well as its direct release to soils (Bose-O'Reilly et al., 2010; Donkor et al., 2005; Paruchuri et al., 2010). The interest in the PHE abundance in soils stems from the fact that many PHE have no known biological importance and can have adverse human health effects even at trace concentrations (Drasch, Böse-O'Reilly, Beinhoff, Roider, & Maydl, 2001; Duruibe, Ogwuegbu, & Egwurugwu, 2007; Weeks, 2000). Furthermore, the soil-accumulated PHE can pose long-term hazards to plants, as well as to humans that consume these plants (Agbozu, Ekweozor, & Opuene, 2007; Nagajyoti, Lee, & Sreekanth, 2010; Qian, Wang, & Tu, 1996). Pollution caused by PHE may also induce changes in size, composition and activity of soil microbial communities, and thereby affecting organic matter decomposition rates and soil structure (Giller, Witter, & McGrath, 1998; McBride, Sauvé, & Hendershot, 1997; Merrington, Rogers, & Van, 2002).

The behavior of PHE in the soil solution and its downward movement within the profile are a function of a given element's concentration and the physico-chemical properties of the soil, such as pH, grain sizes, total organic carbon content and cation exchange capacity (CEC) (Nyamangara & Mzezewa, 1999; Sposito, 1989). For example, under high pH and oxygenated conditions, PHE may be adsorbed onto the surfaces of clay particles, precipitated as sulfides, carbonate oxides or iron-

and manganese-oxyhydroxides (Moral, Gilkes, & Jordan, 2005). In contrast, under low pH and reduced conditions, previously adsorbed PHE may be released into the soil solution, and thus the soil behaves as both sink and source for PHE (McBride, 1989; McBride et al., 1997; Sauvé, Hendershot, & Allen, 2000).

Several geochemical indices, such as enrichment factor (Covelli & Fantolan, 1997), geo-accumulation index (Müller, 1981) and pollution index (Tomlinson, Wilson, Harris, & Jeffrey, 1980) have been used to differentiate between geogenic and anthropogenic sources of PHE contamination of soils. These indices are based on total concentrations of the elements in the samples, which may be affected by pedogeochemical processes (i.e., weathering and soil formation mechanisms) and regional variations. Knowing the total concentrations of PHE alone is not sufficient for a precise identification of contamination sources and the effects of PHE on the ecosystem. Stable Pb isotope compositions (^{204}Pb, ^{206}Pb, ^{207}Pb and ^{208}Pb), which are not significantly affected by physico-chemical fractionation, have been successfully used to separate anthropogenic Pb from lithogenic sources in soils (e.g., Cicchella et al., 2014; Hernandez, Probst, Probst, & Ulrich, 2003; Soderberg & Compton, 2007; Steinmann & Stille, 1997). Therefore, a combination of elemental and isotopic geochemistry of a soil can be a good indicator of AGM-derived PHE contamination of the surrounding environment. Furthermore, in the absence of toxicological study data, the adverse effects of PHE on the ecosystem can be assessed through potential ecological risk index (RI), which includes both total concentrations and toxicological-response factors of individual PHE (Håkanson, 1980).

Although some environmental investigations of soil geochemistry have been carried out on AGM sites across Sub-Saharan Africa (e.g., Ako et al., 2014; Ekengele, Danala, Zo'o, & Myung, 2016; Foli, Nude, & Apea, 2012; Manga, Neba, & Suh, 2017; Mensah et al., 2015; Rajaee et al., 2015; van Straaten, 2000; Waziri, 2014), virtually, no data are available on soils affected by AGM activities in Burkina Faso. In the present study, (1) we report elemental and isotopic geochemistry of tropical semi-arid soils, (2) discuss the factors that control dispersion of PHE around an AGM site and (3) argue that the AGM activities pose serious ecological threat to the adjacent soils.

2. Materials and methods

2.1. Environmental and geological setting

The Bagassi south AGM site was set up in 2000 following the arrival of the first wave of artisanal miners from the central plateau region of the country. This unauthorized site, located in the village of Bagassi in southwestern Burkina Faso, is part of the Bagassi exploration permits owned by Roxgold Inc. (Figure 1). The average annual rainfall of this Sudano-sahelian climatic zone is about 800 mm (Roxgold, 2017). Vegetation cover in uncultivated areas comprises savannah woodlands with dense shrubs. The landscape is characterized by a system of plains interrupted by chains of ferruginous lateritic hills (450 m above mean sea level).

The soils of the study area are made of shallow (up to 48 cm deep) and weakly developed regosolic soils, encountered on eroded valley slopes and peripheral depression (CPCS, 1967). The topsoil (0–14 cm) is dark brown (7.5YR4/4) and becomes yellowish red (5YR4/6) between 14 and 48 cm. From 48 to 110 cm, the soils are composed of unconsolidated weathered schist. Therefore, the gravelly load of the soils consists of 60–90% quartzite and schist debris. The soils contain low weatherable minerals and low silt to clay ratios. Because of their high porosity, these soils are characterized by a well-developed biological activity and high organic matter content (Roxgold, 2017). Due to the limited availability of arable land, the soils are increasing being exploited for subsistence (e.g., maize and millet) and cash crops (mainly cotton and groundnut) agriculture (Roxgold, 2017). As a result, deforestation and soil nutrient depletion are widespread in the area. In recent years, direct and indirect impacts of the AGM activities have also exacerbated the soil degradation (Roxgold, 2017).

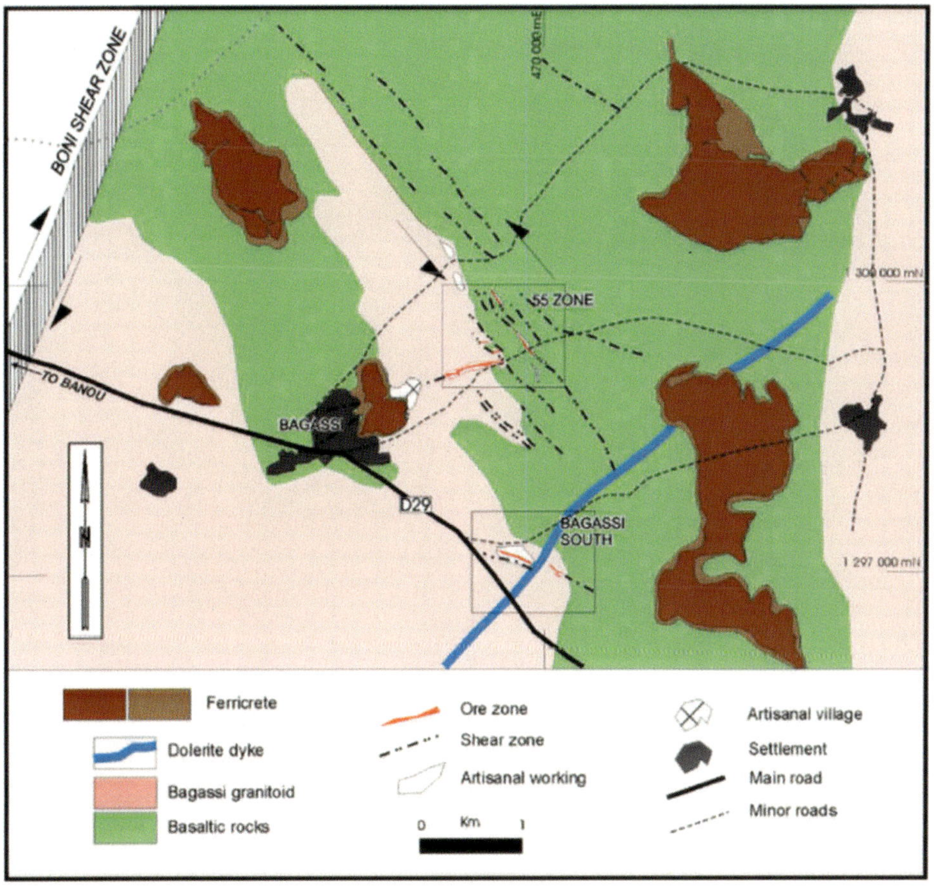

Figure 1. Geological structure of the Bagassi deposits showing the Bagassi South artisanal gold mining site (after Hein, 2015).

The Bagassi gold deposits are situated within the eastern edge of the Palaeoproterozoic Birimian Houndé greenstone belt (Lüdtke et al., 1998; Béziat et al., 2000; Figure 1). The main areas of gold mineralization are the 55 Zone and Bagassi South (Figure 1), both of which are hosted in the Bagassi tonalite-granidiorite at the margin of a metamorphosed basalt sequence composed of plagioclase, K-feldspar and quartz, with rare magmatic biotite, overprinted by chlorite, rutile, calcite, pyrite and rare pyrrhotite (Baratoux et al., 2011). These formations that occur along dextral shear zone are made up of quartz—gold ± pyrite ± pyrrholite ± chalcopyrite assemblage (Sibthorpe, 2012). The granitoid rocks are silicified, kaolinised and carbonate altered, whereas the meta-basalt units are massive and intercalated with hyaloclastite and rare pillowed basalt and sedimentary units (Hein, 2015). Both the metabasaltic rocks and the Bagassi granitoid are crosscut by a northeast trending dolerite dyke (Hein, 2015).

2.2. Sampling and analysis

A total of 11 topsoils (S1–S11) were sampled randomly around the Bagassi south AGM site on 21 May 2017 to assess lateral dispersion of the mine waste (Figure 2). Two profiles were also sampled, according to the limits of the morphological horizons, at sites S1 and S9 (0–45 cm), to investigate a possible translocation of PHE within the soil profiles. Composites of two subsamples of waste rock (WR) and tailings (TL) were also collected from the ore processing area. A representative of the host rock of the Bagassi ore (BO) was also sampled. Three topsoils (FL1–FL3) were sampled from an adjacent cotton farm and used as control samples. All samples were sealed in clean plastic bags.

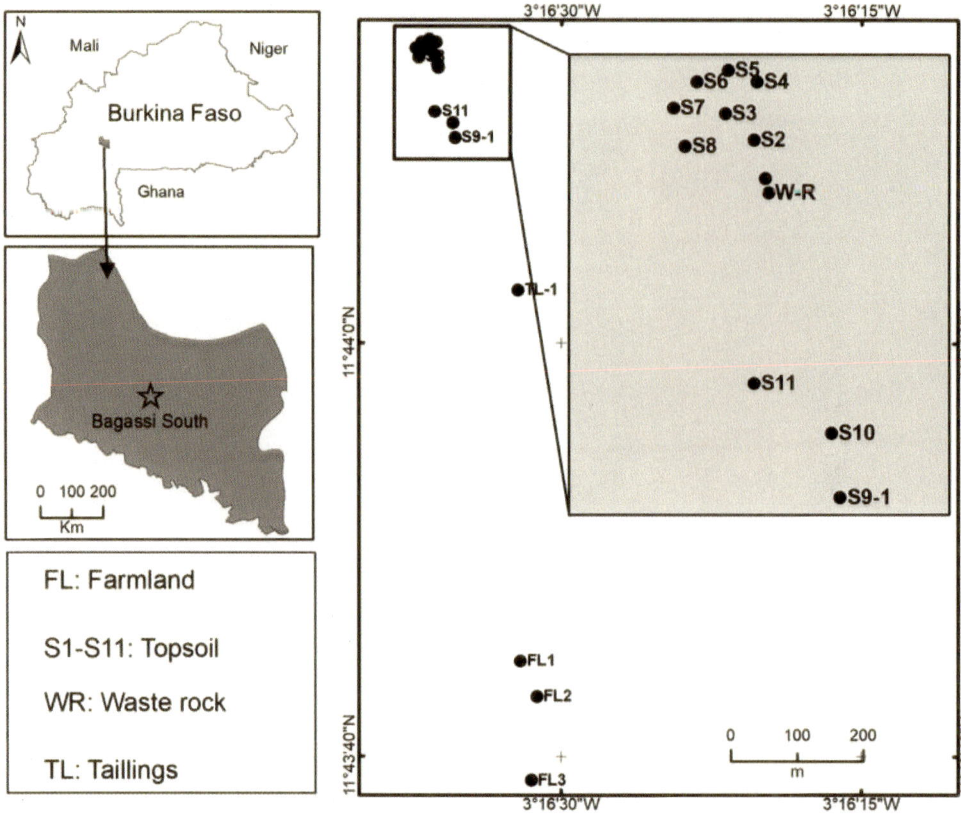

Figure 2. Location of the Bagassi South artisanal gold mining site displaying the sampling points.

Back in the laboratory, the samples were air-dried at room temperature, protected from external contamination and subsequently sieved through a stainless steel sieve in order to isolate less than the 2-mm fraction. Electrical conductivity (EC) and pH were measured in a soil to deionized water ratio of 1:2.5 mg l^{-1}. Grain size distribution (sand, silt and clay fractions) was determined using the hydrometer sedimentation method also known as the Robinson Pipette method (Rouiller et al., 1994). The carbonate content ($CaCO_3$) was estimated after loss on ignition (LOI) by combustion at 1,000°C for two hours (1974; Bengtsson & Enell, 1986).

The Ag-Thiourea (AgTu) method was used to extract basic cations (Pleysier & Juo, 1980). Extractable cations (Ca^{2+}, Mg^{2+}, Na^{+} and K^{+}) and exchangeable silver ion (Ag^{+}) concentrations were measured by a Flame Atomic Absorption Spectrophotometer (Perkin Elmer AA100). The unbuffered CEC was assumed to be the total reactive site concentrations of the solids.

A subset of finely powdered samples was analyzed at Bureau Veritas Commodities Ltd (Vancouver, Canada) for Ag, aluminum (Al), As, Au, boron (B), barium (Ba), bismuth (Bi), calcium (Ca), cadmium (Cd), cobalt (Co), chromium (Cr), Cu, iron (Fe), gallium (Ga), Hg, potassium (K), lanthanum (La), magnesium (Mg), manganese (Mn), molybdenum (Mo), sodium (Na), Ni, phosphorus (P), Pb, sulfur (S), Sb, selenium (Se), scandium (Sc), strontium (Sr), tellurium (Te), thorium (Th), titanium (Ti), thallium (Tl), uranium (U), vanadium (V), tungsten (W) and Zn concentrations with an extended package for Pb isotope analysis. The samples were analyzed using the Bureau's AQ250 package (i.e., modified aqua regia method) whereby 0.5 g of individual samples was digested in aqua regia (2:2:2 of $HCl:HNO_3:H_2O$) at 95°C for 1 h in a heating block, and the sample was made up to a final volume of 300 ml with 5% HCl. The aqua regia digestion is a "pseudo-total" concentration method for dissolution of elements bound to labile minerals, such as water soluble salts, clay particles, as well as those coated by secondary oxyhydroxide minerals (Chen & Lena, 2001). This method is adequate for determining the maximum element's concentration available to plants (Vercoutere, Fortunati, Muntau, Griepink, & Maier, 1995).

Aliquots of digested solutions were subsequently analyzed by ulitratrace Inductively Coupled-Mass Spectrometry (ICP-MS). The accuracy of the analytical method was determined by measuring a series of certified standards (SD11, NIST-981-1Y and NIST-983-1Y) and procedural blanks in each analytical set. The results were within 95% confidence limits of the recommended values for a given standard. The relative standard deviation was between 5% and 10%.

2.3. Statistical analysis

The R-mode factor analysis was used to investigate the relationships between major and trace element concentrations in the samples. The principal component analysis technique and correlation matrix were applied in the factor analysis. The number of significant factors was selected on the basis of the Kaiser criterion, and only the factors with eigenvalues greater than or equal to 1 were taken into account (Kaiser, 1961). The contribution of each variable to the principal component is considered only if the loading is greater than 0.5. Statistical analyses were carried out using IBM SPSS statistics package (version 20).

2.4. Ecological risk index

To assess the potential adverse effects of the PHE on soils, the potential ecological RI proposed by Håkanson (1980), which takes into account the sensitivity of the soil biological community to PHE (Liu et al., 2015) was used. The ecological risk (RI) estimation, usually based on Hg, Cd, As, Ni, Pb, Cr and Zn (Canli & Atli, 2003), was calculated as follows (Eqs. 1–3)

$$C_f^i = \frac{C_i}{C_n^i} \tag{1}$$

$$E_r^i = T_r^i x C_f^i \tag{2}$$

$$RI = \sum_{i=1}^{n} E_r^i \tag{3}$$

Where C^i_f is the pollution factor of an element i; C_i (mg/kg) is the measured concentration of the element i in the soil; C^i_n (mg/kg) represents the pre-industrial reference concentration of the element in soils. In the present study, the average concentrations of As, Cd, Cr, Cu, Hg, Pb and Zn (mg/kg) in farmland soils (i.e., control site) were used as background values for estimating the ecological risks posed by the AGM activities on the adjacent soils. E^i_r is the potential risk of individual element i and T^i_r is the biological toxic-response factor for an element taken as As = 10, Cd = 30, Cr = 2, Cu = 30, Hg = 40, Pb = 5 and Zn = 1 (Håkanson, 1980; Hilton, Davison, & Ochsenbein, 1985; Wang et al., 2011) and RI is the sum of potential risk of the individual metals. Håkanson defined the following five categories for E^i_r: $E_r^i < 40$, low potential risk; $40 \le E_r^i < 80$, moderate potential ecological risk; $80 \le E_r^i < 160$, considerable potential ecological risk; $160 \le E_r^i < 320$, high potential ecological risk and $E_r^i \ge 320$, very high ecological risk. Likewise, four categories describe *RI*: $RI < 150$ corresponds to low ecological risk; $150 \le RI < 300$, moderate ecological risk; $300 \le RI < 600$, ecological risk and $RI > 600$, very high ecological risk.

3. Results and discussion

3.1. Physico-chemical properties of the soils

The mobility and fate of PHE in soils is governed by their mineralogy and physico-chemical parameters, such as pH, EC, particle size, TOC and CEC (Hertz, Angehrn-Bettinazzi, & Stöckli, 1990). These factors are also climate-specific, and thus the mobility of PHE in temperate soils is significantly different from that encountered in tropical semi-arid soils. The physico-chemical parameters of the present soils reflect the semi-arid climatic conditions of the study area, as well as the pedogeochemical processes. Thus, the soil samples were predominately composed of sand, silt and, to a lesser degree, clay fraction (Table 1). This sandy-silty texture can be attributed to intense and long-term weathering, which has resulted in leaching of Si from the silicate minerals and enrichment of primary quartz and secondary Fe and Al oxide minerals (Wilcke, Kretzschmar, Bundt, & Zech, 1999). This was partially explained by low $CaCO_3$ contents (1.1–2.9%; Table 1) and predominately low pH (5.4–7.0). With slightly acidic conditions, PHE are unlikely to be retained through precipitation of carbonate minerals (Plassard, Winiarski, & Petit-Ramel, 2000). However, the co-occurrence of inner and outer sphere sorption could be the dominant geochemical processes that control PHE retention in this semi-arid soil (Appel & Ma, 2002).

The majority of the samples had CEC values in the range of that of kaolin (Halloysite: 3–57 cmolc/kg; Kabata-Pendias, 2011). The CEC also showed an increasing pattern with depth, which may be attributed to organic matter accumulation in deeper horizons (Hernandez et al., 2003). A preliminary study carried out by Roxgold (2017) in this semi-arid area reported high soil organic matter content (up to 2.8%) and C/N ratios (12–15), suggesting a well-developed biological activity and high organic matter turnover. EC of the samples ranged from 370 to 854 μS/cm. Farmland soils had the lowest EC (average ± SD = 310 ± 17 μS/cm; CV = 5.5%) followed by sample BO (370 μS/

Table 1. Physico-chemical properties of Bagassi ore (BO), waste rock (WR), tailings (TL),soil samples from a nearby farmland and topsoils (S2–S8 and S10–S11) and soil profiles (S1-1 to S1-5 and S9-1 to S9-5) collected in the close vicinity of the Bagassi South artisanal gold mining site

Sample	Depth	$pH_{(H2O)}$	EC	CEC	$CaCO_3$	Silt	Clay	Sand
	(cm)		(μS/cm)	(cmolc/kg)	(%)	(%)	(%)	(%)
BO		7.0	370		5.1	–	–	–
WR		6.8	704	6.5	1.5	9	10	82
TL		6.7	824	1.8	1.2	22	43	34
FL1	0–10	6.6	330	9.4	1.1	22	14	64
FL2	0–10	6.7	300	16.3	2.2	28	16	56
FL3	0–10	6.7	301	14.5	1.2	24	22	54
S1-1	0–10	6.9	540	12.6	1.2	16	24	60
S1-2	10–26	6.5	600	17.4	1.4	34	20	46
S1-3	26–39	6.7	680	29.3	1.6	38	20	42
S1-4	36–43	6.9	740	30.5	1.7	36	16	48
S1-5	43–45	5.9	753	28.6	1.7	38	14	48
S2	0–10	6.2	752	12.2	1.2	36	6	58
S3	0–10	5.4	783	10.6	1.2	16	28	56
S4	0–10	6.1	773	12.1	1.8	20	26	54
S5	0–10	6.7	751	13.0	1.2	18	28	54
S6	0–10	7.0	752	17.5	1.4	28	22	50
S7	0–10	6.8	767	19.8	1.4	11	13	76
S8	0–10	6.6	650	18.2	1.4	12	11	77
S9-1	0–10	6.6	894	11.1	1.2	20	22	58
S9-2	10–21	6.3	812	17.3	1.6	40	22	38
S9-3	21–28	6.7	819	24.3	2.9	44	12	44
S9-4	28–34	6.4	816	36.8	1.7	21	7	72
S9-5	34–43	6.3	815	31.6	1.7	44	8	48
S10	0–10	6.8	854	10.2	1.0	7.0	14	79
S11	0–10	6.9	781	18.0	1.4	24	26	50

cm), whereas TL and WR had the highest (824 and 704 μS/cm, respectively). EC of the soils collected around the AGM site ranged from 540 to 894 μS/cm (average ± std = 754 ± 85 μS/cm; CV = 11%), indicating the presence of high amounts of labile ions in mine waste and AGM-affected soils (Fuller, Feamebough, Mitchel, & Trueman, 1995). In light of the above physico-chemical properties, it can be concluded that kaolin, the dominant clay type in the local soils, will have less control over the PHE retention in the soils, whereas Fe and Al oxides and CEC are likely to play a central role in certain trace element immobilization.

3.2. Major and trace element distribution

Major and trace element concentrations of the soils varied between sites. All samples showed very low base cation content (Ca + Mg + Na + K: 0.35–4.37%), with the highest content observed in the less weathered sample BO and the lowest content in the farmland (Table 2 and Figure 3). The relatively low base cations in the farmland could be attributed to the absence of vegetation cover on cultivated soils compared to non-cultivated ones around the AGM site. A similar observation was observed in forested soils (tree plantation) and crop land in the Brazilian Amazon (Farella, Lucotte, Davidson, & Daigle, 2006). However, the high base cation content in BO compared to other samples clearly reflects the mineralogical compositions of the samples and the weak chemical weathering. Based on the CEC values, the soils have been identified as halloysite, which has the lowest CEC after kaolinite (Kabata-Pendias, 2011). That is, the soils had experienced an intensive chemical weathering leading to removal of the base cations (Sposito, 1989). Consequently, these soils have very low capacity to retain cations including PHE (Jordan, 1984). No depth-trends in base cation abundance were observed (Table 3). Both farmland and soil samples collected around the AGM site had similar phosphorus content (average ± standard deviation = 0.022 ± 0.009%; CV = 42%), whereas phosphorus content was slightly higher in samples BO, WR and TL. Furthermore, sulfur concentrations were below detection limit of the instrument (IDL) (except in samples BO, WR and TL), suggesting the influence of the BO on these samples. Compared to Al concentrations (average ± SD = 1.55 ± 0.63%; CV = 41%), those of Fe (average ± SD = 4.84 ± 3.1%; CV = 64%) and Ti (average ± SD 0.019 ± 0.01%; CV = 54%) varied widely across sites. This indicates that Al was the least affected by pedogeochemical processes and AGM activities among the so-called "conservative elements".

A series of trace elements, associated with gold deposits and anthropogenic pollution, had high concentrations in TL, WR and the AGM-affected soil samples relative to the farmland soils. Thus, Ag, As, Au, Bi, Cu, Cd, Hg, Sb, Te, Pb, Mo, W and Zn concentrations increased in the following order: TL> WR> topsoil> FL (Table 2). Likewise, the average concentrations of chalcophile elements, such as Ag (229 and 272 μg/kg), As (9.2 and 91 mg/kg), Au (758.9 and 342.4 μg/kg), Bi (0.38 and 1.96 mg/kg), Hg (899 and 8737 μg/kg), Pb (12.54 and 17.36 mg/kg), Mo (1.83 and 2.73 mg/kg), Sb (0.22 and 0.24 mg/kg), W (15.27 and > 100 mg/kg) and Zn (165.8 and 74.6 mg/kg) were higher in TL and WR compared to the average global soil concentrations (Kabata-Pendias, 2011). By contrast, elements, such as Co, Cr, Ni, La, Sc, Sr, Ba, Th, U, V, Ga and Tl concentrations were within the average global soil concentrations, and their distributions were fairly homogeneous across the sampling sites, suggesting that they were not significantly affected by mining activities. Au, Hg, Pb and Zn showed a decreasing pattern with depth, indicating their addition to the soil system through anthropogenic activities (Table 3 and Figure 4). Since both farmland and non-cultivated soils around the AGM site are derived from the same parent materials and exposed to similar pedogeochemical processes, it can be concluded that the presence of certain elements were enhanced by mining and mine waste dispersion through runoff and wind deposition compared to others.

3.3. Apportionment of sources of potentially harmful elements

The total concentrations of PHE in soils do not necessarily indicate anthropogenic influence, as weathering can lead to high accumulation of PHE in soils. Consequently, the R-mode factor analysis has been commonly used in environmental geochemistry to identify geogenic versus anthropogenic sources of PHE contamination (e.g., Facchinelli, Sacchi, & Mallen, 2001; Micó, Recatala, Peris, & Sanchez, 2006). The concentrations of the elements below the IDL were assigned

Table 2. Descriptive statistics of major and trace elements of topsoil samples from the Bagassi South artisanal gold mining site and the nearby farmland (n = 25)

		Farmland (n = 3)					Topsoil (n = 11)							
Element	**Unit**	**Min**	**Max**	**Med.**	**Av.**	**SD**	**Min**	**Max**	**Med.**	**Av.**	**SD**	**BO**	**WR**	**TL**
Al	%	1.02	1.56	1.16	1.25	0.28	0.87	1.47	1.21	1.21	0.18	1.02	1.21	0.52
Ca	%	0.16	0.29	0.25	0.23	0.07	0.25	0.52	0.39	0.38	0.10	2.5	0.43	0.39
Fe	%	2.33	3.57	2.56	2.82	0.66	2.41	4.21	3.54	3.47	0.54	2.65	4.99	2.53
K	%	0.04	0.07	0.06	0.06	0.02	0.06	0.12	0.08	0.08	0.02	0.09	0.09	0.06
Mg	%	0.09	0.18	0.16	0.14	0.05	0.1	0.3	0.17	0.19	0.06	1.78	0.37	0.17
Na	%	0.001	0.001	0.001	0.00	< IDL*	< IDL	0.013	0.004	0.01	0.00	0.008	0.008	0.038
P	%	0.011	0.017	0.013	0.01	0.00	0.019	0.046	0.03	0.03	0.01	0.037	0.025	0.02
S	%	< IDL	< IDL	< IDL	< IDL	< IDL	< IDL	0.02	< IDL	< IDL	< IDL	0.15	0.02	0.02
Ti	%	0.007	0.011	0.009	0.01	0.00	0.015	0.029	0.022	0.02	0.01	0.002	0.02	0.006
Ag	mg/kg	0.022	0.038	0.024	0.03	0.01	0.043	0.459	0.125	0.18	0.14	0.04	0.272	0.229
As	mg/kg	0.3	0.9	0.8	0.67	0.32	1.8	5.6	3.7	3.55	1.23	1.2	9.2	9.1
Au	mg/kg	0.0032	0.0048	0.0034	0.00	0.00	0.0156	0.0854	0.0483	0.04	0.02	0.0185	0.7589	0.3424
Ba	mg/kg	77	382	183	214	155	82	148	116	113	17	45	429	196
Bi	mg/kg	0.03	0.04	0.04	0.04	0.01	0.1	0.33	0.12	0.15	0.07	0.24	0.38	1.96
Cd	mg/kg	0.03	0.07	0.04	0.05	0.02	0.03	0.06	0.05	0.05	0.01	0.02	0.15	0.16
Co	mg/kg	16.6	64.8	30.1	37.17	24.86	23.2	35.9	26.5	27.56	3.65	15.2	50.7	18.5
Cr	mg/kg	32.7	51.8	41.7	42.07	9.56	68.6	134.8	113	107.2	22.0	23.0	96.1	86.3
Cu	mg/kg	22.14	35.69	27.65	28.49	6.81	27.71	49.33	44.96	43.22	6.61	43.73	56.87	36.28
Ga	mg/kg	4.3	5.7	4.4	4.8	0.8	4.0	6.3	5.4	5.3	0.6	2.7	5.2	2.4
Hg	mg/kg	0.021	0.039	0.03	0.03	0.01	0.084	0.146	0.107	0.11	0.02	0.025	0.899	8.737
La	mg/kg	10.3	13.3	11.8	11.80	1.5	11.7	24.1	14.5	15.27	3.71	24.4	11.7	8.8
Mn	mg/kg	557	2568	1161	1429	1032	817	1269	910	947	138	961	2282	821
Mo	mg/kg	0.17	0.38	0.3	0.28	0.11	0.42	1.06	0.62	0.65	0.19	2.83	1.83	2.73

(Continued)

Table 2. (Continued)

		Farmland (*n* = 3)					Topsoil (*n* = 11)							
Element	**Unit**	**Min**	**Max**	**Med.**	**Av.**	**SD**	**Min**	**Max**	**Med.**	**Av.**	**SD**	**BO**	**WR**	**TL**
Ni	mg/kg	19.2	54.8	30.1	34.70	18.24	23.9	52.7	34.4	36.80	8.34	41.4	57.3	20.7
Pb	mg/kg	3.07	5.24	3.74	4.02	1.11	5.03	10.56	6.76	7.22	2.15	0.95	12.54	17.36
Sb	mg/kg	0.02	0.02	0.02	0.02	0.00	0.07	0.25	0.15	0.15	0.06	0.07	0.22	0.24
Sc	mg/kg	7.0	11.7	8.6	9.1	2.4	7.4	11.4	10.5	10.1	1.2	2.5	7.9	3.3
Se	mg/kg	< IDL	< IDL	< IDL	< IDL	< IDL	0.1	0.3	0.2	0.20	0.08	< IDL	0.2	0.1
Sr	mg/kg	6.9	13.3	12.1	10.8	3.4	15.7	31.2	23.4	23.8	5.9	10.1	17.7	11.7
Te	mg/kg	0.02	0.02	0.02	0.02	< IDL	< IDL	0.14	0.05	0.06	0.04	0.02	0.28	1.23
Th	mg/kg	2.6	2.7	2.7	2.8	0.1	2.0	2.8	2.2	2.3	0.3	2.3	1.7	0.9
Tl	mg/kg	0.04	0.12	0.05	0.07	0.04	0.04	0.06	0.05	0.05	0.01	0.02	0.09	0.05
U	mg/kg	0.56	0.87	0.58	0.67	0.17	0.37	0.47	0.4	0.41	0.04	0.30	0.38	0.24
V	mg/kg	70	113	75	86.00	23.52	59	110	84	85.00	15.72	17	138	53
W	mg/kg	< IDL	< IDL	< IDL	< IDL	< IDL	0.17	1.66	0.39	0.53	0.46	0.25	15.27	> 100
Zn	mg/kg	9.9	20.8	17.1	15.93	5.54	29	68.5	46.5	47.36	11.47	58.6	165.8	74.6

*IDL = instrument detection limit.

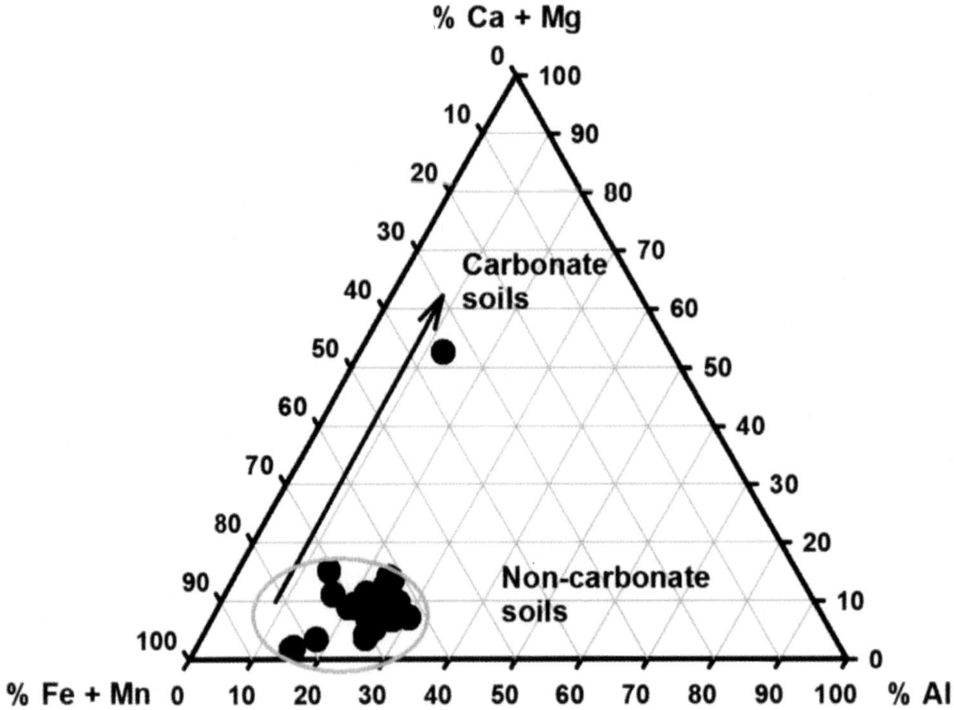

Figure 3. Ternary diagram of Bagassi South soil, illustrating major element distribution. The soil samples (except Bagassi ore) are clustered around the Fe and Mn end-members.

to 50% of the detection limits in the R-mode factor analysis (Cicchella et al., 2014). The distribution of 11 elements (Mg, Na, S, Au, Bi, Cd, Hg, Mo, Tl, U and W) was positively skewed, and thus they were log-transformed. Only the elements (30 elements) with high communalities close to unity were included in the analysis. After varimax rotation, three factors (F1, F2 and F3) that describe about 80% of the total variability were extracted (Figure 5).

The first factor (F1: 32.94 % variance) showed high loadings on Na and Au and a series of PHE (As, Bi, Cd, Hg, Mo, Pb, Sb, Te, W and Zn). The association of these PHE with Au, As, Bi, Hg and Sb, which were particularly enriched in TL and WR relative to the farmland, is a clear indication of the AGM contribution to these element loadings to the soils through dispersion and transport (Figure 5).

The anthropogenic origin of F1 was further investigated by normalizing certain elements of F1 to Sc. By definition, geochemical normalization consists of establishing correlations between PHE concentrations and the concentration of a reference element i.e., a geochemical normalizer (Donkor et al., 2005; Garcia, Barreto, Passos, & Alves, 2009). In order for an element to be a good geochemical normalizer, its abundance in the sample must be of lithogenic origin, be structurally linked to at least one of the PHE carriers in the soils and its concentration must not be significantly influenced by anthropogenic inputs (Alves, Passos, & Garcia, 2007; Loring, 1990). The most common geochemical normalizers are Al, Fe, Sc and Ti. In the present study, Al and Sc were the most homogeneously distributed in the samples with the lowest coefficients of variance (CV = 41% and 46%, respectively). Furthermore, most elements of lithogenic origin (e.g., Al, Sc, Fe, Cu, Ni and Co) exhibited significant (Sc, Al, Co and Ni) and moderate (Fe and Cu) linear relationships with CEC, a potential carrier of trace elements in the soils (Figure 6). Because Al is expressed in percent and Sc including PHE in mg/kg, Sc was used as a geochemical normalizer to assess the selected PHE enrichment in the soils and mine waste. Eight PHE (Hg, As, Ag, Mo, Cd, Pb, Sb and Zn) plus Au that showed high loadings on F1 showed weak correlations with Sc, and the soil samples most affected by AGM activities plotted outside the 95% confident interval (Figure not shown).

Table 3. Major and trace element concentrations of two soil profiles collected in the close vicinity the Bagassi South artisanal gold mining site

		Profile1					Profile 2				
		S1-1	S1-2	S1-3	S1-4	S1-5	S9-1	S9-2	S9-3	S9-4	S9-5
Element	Unit	0–10 cm	10–26 cm	26–39 cm	39–43 cm	43–45 cm	0–10 cm	10–21 cm	21–28 cm	28–34 cm	34–43 cm
Al	%	1.12	1.94	2.26	2.51	2.82	1.09	1.91	2.12	2.5	2.85
Ca	%	0.42	0.34	0.28	0.21	0.2	0.25	0.32	0.3	0.3	0.28
Fe	%	3.36	5.19	9.21	12.66	14.57	2.69	3.91	5.20	6.47	7.20
K	%	0.09	0.05	0.04	0.03	0.03	0.08	0.05	0.05	0.05	0.05
Mg	%	0.16	0.19	0.16	0.12	0.12	0.13	0.14	0.13	0.13	0.13
Na	%	0.005	0.004	0.003	0.004	0.004	0.002	0.002	< IDL	< IDL	0.003
P	%	0.032	0.014	0.015	0.017	0.018	0.019	0.014	0.015	0.014	0.014
S	%	< IDL	< IDL	< IDL	< IDL	< IDL	< IDL	< IDL	< IDL	< IDL	< IDL
Ti	%	0.019	0.019	0.025	0.039	0.044	0.015	0.013	0.011	0.011	0.012
Ag	mg/kg	0.13	0.075	0.089	0.058	0.065	0.043	0.03	0.056	0.073	0.095
As	mg/kg	4.0	5.6	9.5	12.3	16.2	2.1	2.4	3.2	4.3	4.6
Au	mg/kg	0.0156	0.012	0.0133	0.0186	0.0086	0.0478	0.0137	0.015	0.0178	0.022
Ba	mg/kg	115.6	265.7	818.8	1222.5	1135.9	106	75.9	493	694.4	989.8
Bi	mg/kg	0.11	0.11	0.16	0.15	0.18	0.12	0.18	0.18	0.2	0.22
Cd	mg/kg	0.06	0.02	0.04	0.07	0.05	0.06	0.02	0.03	0.04	0.05
Co	mg/kg	26.5	84.9	171.9	252.2	248	25.3	23.4	111.1	136.7	183.8
Cr	mg/kg	101.6	129	191.5	218.3	282.5	84.7	99.7	113.6	134.5	156.9
Cu	mg/kg	41.42	70.51	90.72	108.97	131	36.91	49.58	60.08	70.47	81.76
Ga	mg/kg	4.9	8.0	9.9	12.4	14.3	5.1	8.2	9.0	11	12.3
Hg	mg/kg	0.089	0.024	0.023	0.025	0.028	0.146	0.032	0.032	0.043	0.028

(Continued)

Table 3. (Continued)

		Profile1					Profile 2				
		S1-1	S1-2	S1-3	S1-4	S1-5	S9-1	S9-2	S9-3	S9-4	S9-5
Element	Unit	0–10 cm	10–26 cm	26–39 cm	39–43 cm	43–45 cm	0–10 cm	10–21 cm	21–28 cm	28–34 cm	34–43 cm
La	mg/kg	12.6	19.3	19.5	19.6	19.6	14.8	17.6	17.5	18.8	22.9
Mn	mg/kg	863	2905	6959	9708	9951	817	579	4342	5511	7504
Mo	mg/kg	0.53	0.80	1.56	2.08	2.44	0.55	0.68	1.11	1.26	1.37
Ni	mg/kg	33.8	64.7	131.7	211.1	196	28.7	34.3	71	92.3	120.2
Pb	mg/kg	9.84	6.57	8.53	11.12	13.27	5.16	4.58	6.51	8.03	8.88
Sb	mg/kg	0.15	0.11	0.22	0.29	0.31	0.07	0.06	0.10	0.10	0.07
Sc	mg/kg	9	16.8	19.9	22	20.5	9.9	16.5	18.2	21.2	23
Se	mg/kg	0.2	0.3	0.4	0.3	0.3	0.1	0.3	0.7	0.4	0.2
Sr	mg/kg	31.2	21.8	17.2	14.5	14.9	16.3	19	17.5	17.6	17.6
Te	mg/kg	0.03	< IDL	0.04	0.05	0.04	0.02	< IDL	< IDL	0.03	0.03
Th	mg/kg	2	3.3	3.2	2.7	3.1	2.5	3.3	3.1	3.5	4.1
Tl	mg/kg	0.04	0.12	0.24	0.32	0.33	0.06	0.08	0.34	0.43	0.53
U	mg/kg	0.37	0.53	0.53	0.58	0.74	0.47	0.53	0.53	0.55	0.69
V	mg/kg	84	116	205	275	325	64	87	117	141	151
W	mg/kg	0.17	0.06	0.17	0.18	0.15	0.51	< IDL	0.11	0.09	0.08
Zn	mg/kg	59	29.3	30.3	25	27.2	29	23.1	23.5	23.7	24

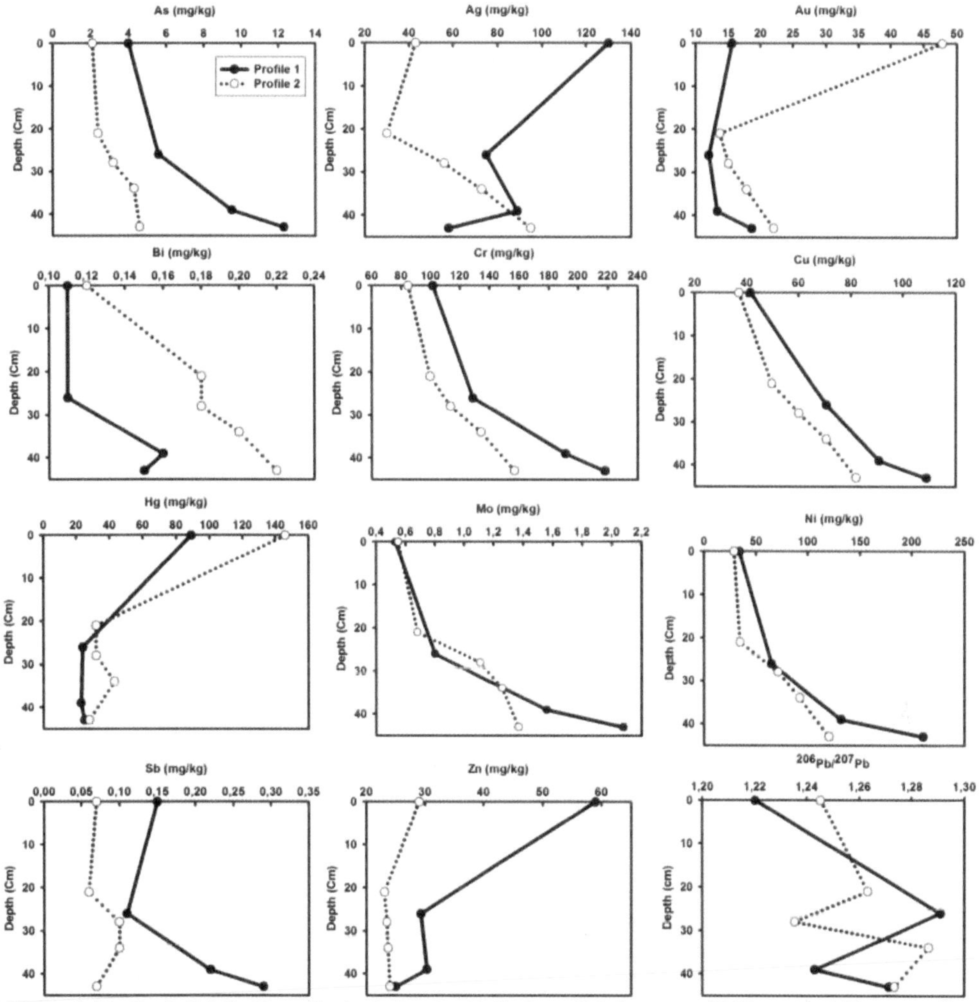

Figure 4. Vertical distribution of selected potentially harmful elements (PHE) and gold in two soil profiles.

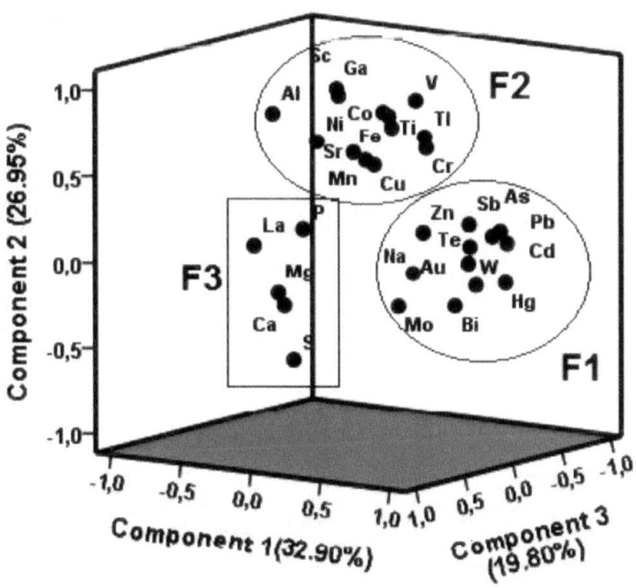

Figure 5. Loading plot of the three components of R-mode factor analysis extracted by PCA method. F1, F2 and F3 represent the influence of arti-sanal gold mining, chemical weathering and biogeochemical cycling, respectively, on the soil geochemistry.

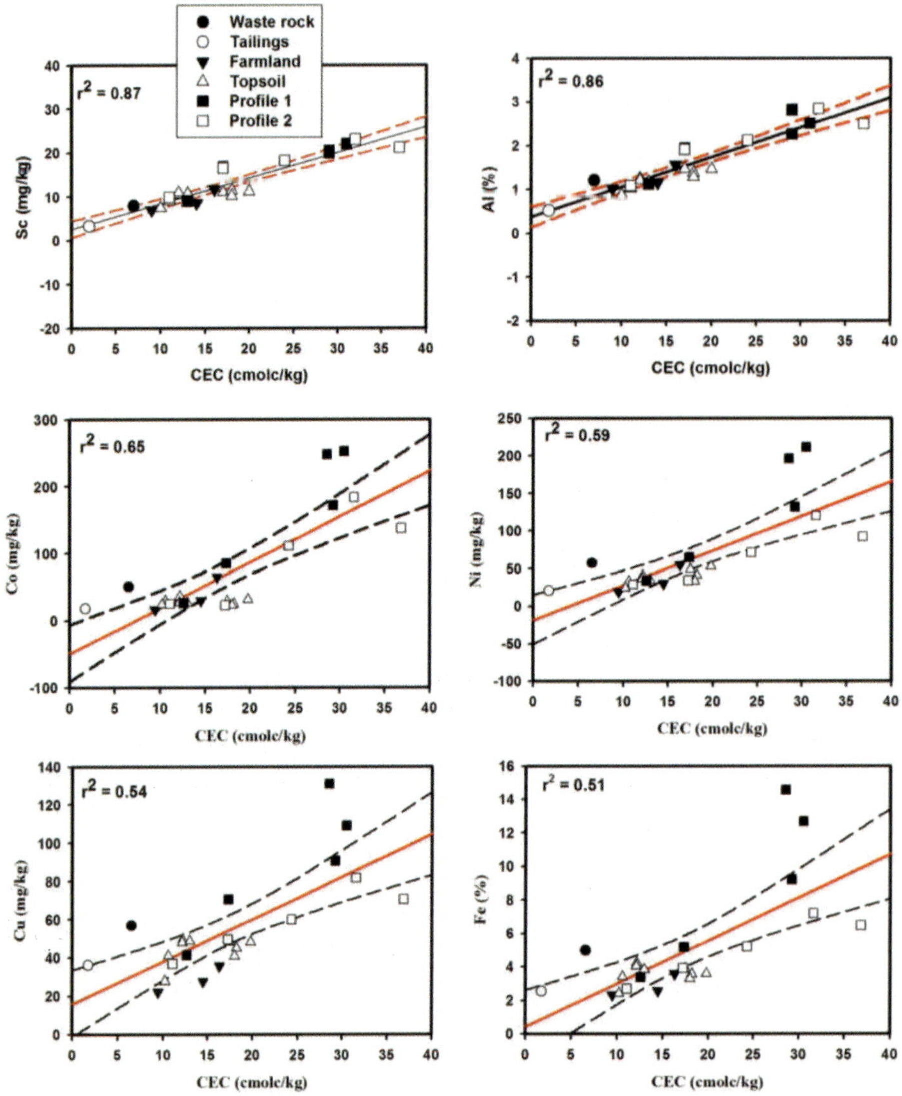

Figure 6. Relationships between cation exchange capacity (CEC) and Sc, Al, Co, Ni, Cu and Fe. The solid line represents the regression line and the dotted line the prediction interval at the 95% confidence level.

The second factor (F2: 26.95% of the variance) revealed elevated loadings on four major elements (Al, Fe, Mn and Ti) and Co, Cr, Cu, Ga, Ni, Sc, Sr, Tl and V (Figure 5). Representative elements of F2 (i.e., Co, Cu, Ni, Cr, Mn and V) co-varied with Sc (Figure 7), suggesting that the concentrations of these elements were below or close to those of the pedogeochemical background. Consequently, it can be implied that F2 represents the contribution of soil pedogeochemical processes to these element distribution.

Factor 3 (F3: 19.80% of the variance) had high loadings on Ca, Mg, P, S and La. Although rock weathering and atmospheric deposition are the major sources of Ca, Mg and P and S, respectively, these major elements are also essential to biochemical structure and metabolism functioning of organisms (Garten, 1976; Reiners, 1986). Hence, F3 could represent the biogeochemical cycling of essential elements and La in the soil system through organic matter decomposition. This is consistent with high organic matter and C/N ratios observed in the soils around the site (Roxgold, 2017).

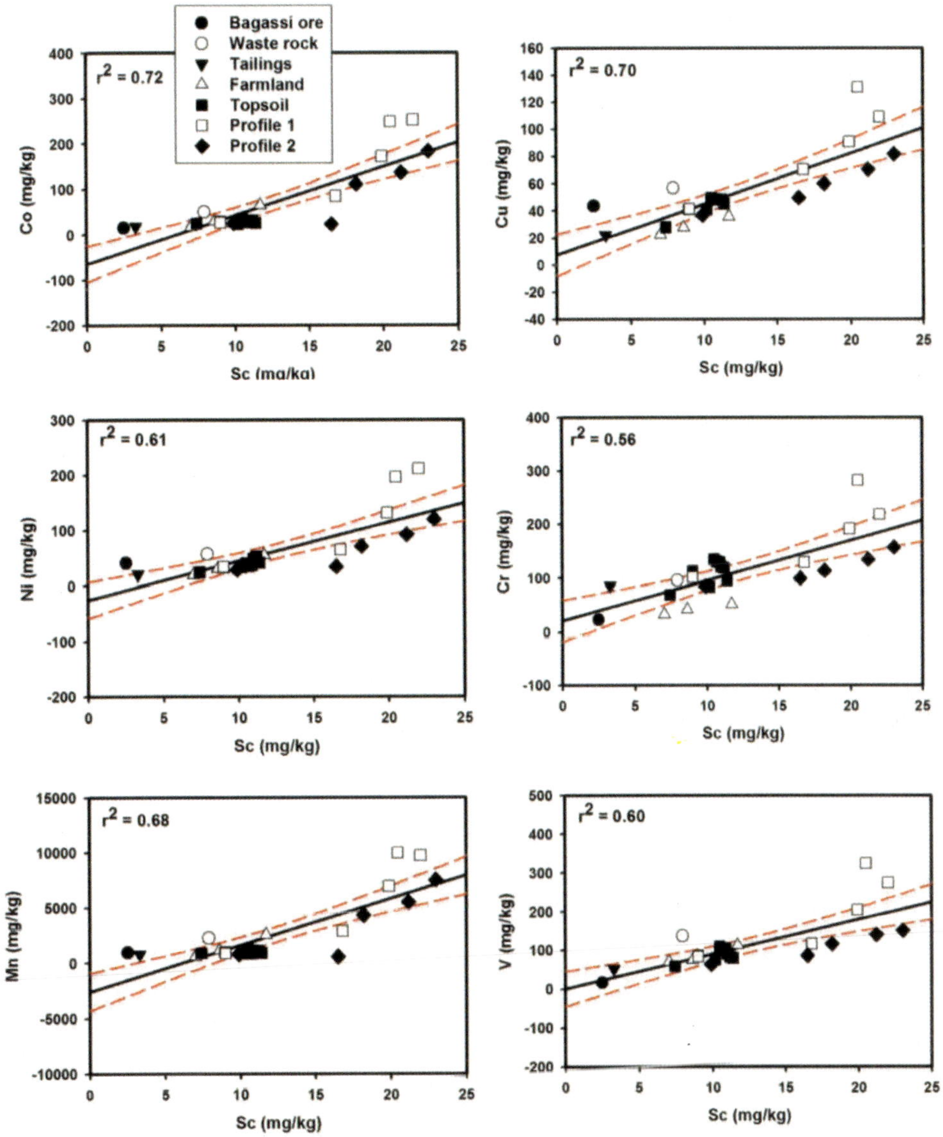

Figure 7. Strong relationships between Sc and geogenically-derived PHE with a few samples lying outside the 95% confidence interval.

In addition to AGM activities and pedogeochemical processes, other anthropogenic sources such petrol and atmospheric deposition may contribute to PHE (particularly Pb) loadings to the soils. In the present study, Pb isotope ratios (^{206}Pb/^{204}/Pb, ^{207}Pb/^{204}Pb, ^{206}Pb/^{207}Pb and ^{208}Pb/^{206}Pb) were used to evaluate the potential sources of Pb in the soil samples (Table 4).

Figure 8 highlight the relationship between ^{206}Pb/^{207}Pb and ^{208}Pb/^{206}. Although the farmland soils fall on the mixing line of two end-members, i.e., BO and petrol, they are plotted closer to the petrol isotopic compositions than to that of BO. This result suggests that the farmland soils, a relatively less polluted environment, received anthropogenic Pb contribution other than that of the local lithology (Figure 8). On the ^{206}Pb/^{207}Pb and ^{208}Pb/^{206}Pb mixing diagram, topsoil samples highly affected by AGM activities appeared to be more influenced by petrol, TL and WR (Figure 8). In contrast to the farmland soils and topsoil, isotopic ratios of TL and WR samples did not plot on the petrol and BO end-member mixing line, indicating a third source of Pb. This third source could be the portion of Pb introduced to the soil system during Au extraction and processing. It is interesting to note that soil samples from deeper horizons are clustered

Table 4. Lead isotope ratios and inverse lead concentrations in the Bagassi ore, waste rock, tailings, farmland soil, topsoil and soil profile samples collected around the Bagassi South artisanal gold mining site

Sample	$^{206}Pb/^{204}Pb$	$^{207}Pb/^{204}Pb$	$^{206}Pb/^{207}Pb$	$^{208}Pb/^{206}Pb$	1/[Pb]
BO	28.0	17.00	1.647	1.750	1.053
WR	20.7	16.63	1.244	1.937	0.080
TL	31.7	27.21	1.165	2.020	0.058
FL1	19.5	16.00	1.219	2.064	0.326
FL2	19.4	15.14	1.283	2.022	0.191
FL3	19.2	15.40	1.247	2.042	0.267
S1-1	19.6	16.08	1.220	1.988	0.102
S1-2	21.6	16.75	1.291	1.977	0.152
S1-3	20.5	16.45	1.243	1.938	0.117
S1-4	20.0	15.73	1.271	1.870	0.090
S1-5	20.0	15.22	1.314	1.875	0.075
S2	18.1	14.67	1.232	2.030	0.095
S3	18.6	15.14	1.226	2.015	0.198
S4	21.7	17.50	1.238	2.015	0.199
S5	18.5	15.57	1.188	2.100	0.097
S6	19.9	15.50	1.284	1.874	0.136
S7	18.9	14.89	1.269	2.018	0.152
S8	19.1	15.33	1.246	2.076	0.148
S9-1	18.9	15.14	1.245	2.053	0.194
S9-2	20.0	15.83	1.263	1.975	0.218
S9-3	21.0	17.00	1.235	2.018	0.154
S9-4	19.2	14.91	1.287	1.976	0.125
S9-5	21.2	16.64	1.273	1.979	0.113
S10	19.1	15.70	1.217	2.052	0.133
S11	19.1	15.29	1.252	2.067	0.190

around WR isotopic signatures (Figure 8), implying migration and accumulation of mine waste, with high Pb concentrations, within soil profiles. Anthropogenic contribution of Pb to the Bagassi south soils was determined through the regression line (r^2 = 0.88) between $^{206}Pb/^{207}Pb$ ratios and Pb concentrations. According to this plot, the anthropogenic contribution to $^{206}Pb/^{207}Pb$ corresponds to 1.190. The R-mode factor analysis, geochemical normalization and enrichment factors of PHE and Pb isotopic ratios successfully identified pedogeochemical processes, AGM activities and atmospheric deposition as the primary sources of PHE loadings to the adjacent soils.

3.4. Ecological risk assessment

The potential impact of the PHE on the ecological functions of the soils was evaluated through ecological risk assessment, proposed by Håkanson (1980). The potential ecological risk indices for single elements (E_r^i) showed that the severity of AGM-derived pollution of eight PHE decreased in the following order: Hg> As> Cd> Cu = Pb> Ni> Cr> Zn (Table 5). As expected, mercury in WR and TL samples posed very high ecological risk (E_r^i = 1,199 and 11,649, respectively). Thus, 10 topsoil samples had considerable potential ecological risk indices for Hg, whereas seven samples (WR, TL, S1-1, S1-2, S1-3, S1-4 and S1-4) exhibited considerable potential ecological risk indices for As. Although Pb and Zn showed low potential ecological risk indices at all sites, the E_r^i values were the highest for WR and TL.

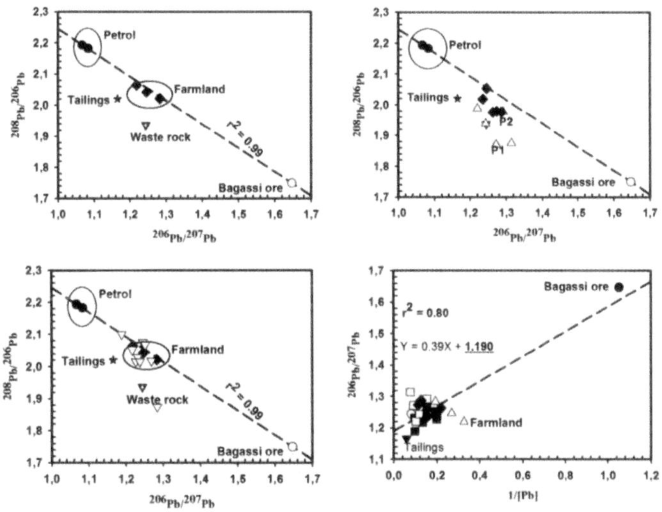

Figure 8. left panel: three-isotope diagram of $^{206}Pb/^{207}Pb$ vs. $^{208}Pb/^{206}Pb$ showing the Pb isotopic composition of farm-land (control soil), tailing, waste rock, topsoil (collected at the close vicinity of the mining site), the host rock of the Bagassi ore and European leaded gasoline (Monna, Lancelot, Croudace, Cundy, & Lewis, 1997). Right panel: three-isotope mixing diagram of $^{206}Pb/^{207}Pb$ vs. $^{208}Pb/^{206}Pb$ showing the Pb isotopic com-position of soil profiles (P1 and P2) relative to those of petrol, tailings and the host rock of the Bagassi ore, and Pb isotope diagram of $^{206}Pb/^{207}Pb$ vs. 1/[Pb]. The dotted line represents the regression through the Bagassi ore, farmland soil sam-ples and leaded gasoline.

Based on the overall ecological risk indices, only the BO had a low RI value, whereas WR and TL posed the highest ecological risk. Three sites (S1-5, S1-4 and S5) posed considerable ecological risk and the rest of the sites posed moderate ecological risk relative to the control site. According to Pb isotope ratios, all sites are vulnerable to atmospheric transport of Pb with low ecological risk, whereas the soil around the AGM site is moderately to considerably threatened by mine waste.

4. Conclusions

Trace element distribution, R-mode factor analysis results, geochemical normalization and lead isotope compositions successfully identified anthropogenic and geogenic contributions of PHE loadings to the soils. Each technique contributed important information that was capital in investigating PHE sources in the soils. The raw data showed that elements, such as Au, Hg, As, Bi, Cu, Cd, Sb, Mo and Zn that are generally associated with gold deposits had concentrations exceeding the average world's soil concentrations. The abundance of Au, As, Bi, Cu, Sb, Mo and Zn could be attributed to the gold mining, whereas Hg is added to the soils through gold extraction. By contrast, lithophile elements, such as Co, Cr, Ni, La, Sc, Ba, Th, U, V and Tl had concentrations closer to or lower than those of the average world's soil.

These results were corroborated by the R-mode factor analysis, which identified the anthropogenic activities as the main source of PHE abundance in the soils. The technique also suggested that biogeochemical cycling is partially responsible for certain major element distribution in the soils. The use of geochemical normalization indicated anthropogenic origin of Au and some PHE in the soils.

According to Pb isotope ratios, the control site appear to be impacted by atmospheric inputs of leaded gasoline, whereas topsoil samples collected around the AGM site reflect both atmospheric inputs and mine waste signature. The potential ecological risk indices identified Hg, As and Cd as the elements likely to pose serious adverse ecological effects on the soils. Although AGM is recent in the area, the study demonstrated that the soils are moderately contaminated with PHE. To protect soils and ecosystems as a whole, a sound pollution control program is urgently required for the 800 AGM sites across Burkina Faso.

Table 5. Results of Potential ecological risk indices (E_r^i) of individual potentially harmful element ecological risk indices (RI) of the samples

	Potential ecological risk (E_r^i)									
Sample	**Cu**	**Zn**	**Ni**	**As**	**Cd**	**Cr**	**Hg**	**Pb**	**Risk index (RI)**	**Pollution degree**
BO	8	4	6	18	13	1	33	1	84	Low
WR	10	10	8	138	96	5	**1,199**	16	1482	Very high
TL	6	5	3	137	103	4	**11,649**	22	11,928	Very high
S1-1	7	4	5	60	39	5	119	12	250	Moderate
S1-2	12	2	9	84	13	6	32	8	167	Moderate
S1-3	16	2	19	143	26	9	31	11	255	Moderate
S1-4	19	2	30	185	45	10	33	14	338	Considerable
S1-5	23	2	28	243	32	13	37	17	395	Considerable
S2	9	4	6	75	32	6	112	13	257	Moderate
S3	7	3	5	59	32	5	143	6	259	Moderate
S4	8	2	5	84	19	6	124	6	256	Moderate
S5	9	3	5	68	32	6	175	13	310	Considerable
S6	8	3	7	56	39	6	121	9	249	Moderate
S7	8	4	8	50	26	6	161	8	270	Moderate
S8	8	3	6	44	26	5	129	8	228	Moderate
S9-1	6	2	4	32	39	4	195	6	288	Moderate
S9-2	9	1	5	36	13	5	43	6	117	Moderate
S9-3	11	1	10	48	19	5	43	8	146	Moderate
S9-4	12	1	13	65	26	6	57	10	191	Moderate
S9-5	14	2	17	69	32	7	37	11	190	Moderate
S10	5	2	3	27	32	3	143	9	225	Moderate
S11	7	3	5	33	39	4	193	7	290	Moderate

High potential ecological risk indices are bold, whereas those of considerable potential ecological risk are underlined.

The influence of organic (soil organic matter) and inorganic ligands on trace element speciation, precipitation and adsorption could be investigated in a future study to ascertain the mobility and bioavailability of AGM-derive PHE in tropical semi-arid soils.

Acknowledgements

The authors thank Dr Sâga S. Sawadogo for compiling the sampling map. Comments and suggestions from two reviewers and Dr Arno Rein, the editor of the journal, greatly improved the early version of the manuscript.

Funding

The authors received no direct funding for this research.

Competing interests

The authors declares no competing interests.

Author details

Aboubakar Sako[1]
E-mail: aboubakar.sako@gmail.com
Mamadou Nimi[2]
E-mail: nimads2@gmail.com
[1] Université de Dédougou, BP. 139, Dédougou, Burkina Faso.
[2] BUMIGEB, Bureau des Mines et de la Géologie du Burkina Faso, Direction Régionale de Bobo, Bobo-Dioulasso, Burkina Faso.

References

Agbozu, I. E., Ekweozor, I. K. E., & Opuene, K. (2007). Survey of heavy metals in the catfish synodontis clarias. *International Journal of Environmental Science & Technology*, *4*(1), 93–98. doi:10.1007/BF03325966

Ako, T. A., Onoduku, U. S., Oke, S. A., Adamu, I. A., Ali, S. E., Mamodu, A., & Ibrahim, A. T. (2014). Environmental impact of artisanal gold mining in Luku, Minna, Niger State, North Central Nigeria. *International Journal of Geomatics and Geosciences*, *2*(1), 28–37.

Alves, J. P. H., Passos, E. A., & Garcia, C. A. B. J. (2007). Metals and acid volatile sulfide in sediment cores from the Sergipe River Estuary, Northeast, Brazil. *Journal of Brazilian Chemical Society*, *18*(4), 748–758. doi:10.1590/S0103-50532007000400013

Amegbey, N. A., Dankwa, J. B. K., & Al-Hasson, S. (1997). Small scale mining in Ghana techniques and environmental considerations. *International Journal of Surface Mining Reclamation and Environment*, *11*(11), 135–138. doi:10.1080/09208119708944077

Andersson, M., Ola, H., & Niklas, O. (2015). Mining, economic activity and remote sensing: Case studies from Burkina Faso, Ghana, Mali and Tanzania. In World Bank Team The Centre for the Study of African Economies (CSAE), Department of economics at oxford university (pp. 1–42). Malmö University Electronic Publishing.

Appel, C., & Ma, L. (2002). Concentration, pH, and surface charge effects on cadmium and lead sorption in three tropical soils. *Journal of Environmental Quality*, *31*, 581–589. doi:10.2134/jeq2002.5810

Baratoux, L., Metelka, V., Naba, S., Jessell, M. W., Grégoire, M., & Ganne, J. (2011). Juvenile Paleoproterozoic crust evolution during the Eburnean orogeny (□2.2–2.0 Ga), western Burkina Faso. *Precambrian Research*, *191*(1 – 2), 18–45. doi:10.1016/j.precamres.2011.08.010

Bengtsson, L., & Enell, M. (1986). Chemical analysis. In B. E. Berglund (Ed.), *Handbook of holocene palaeoecology and palaeohydrology* (pp. 423–451). Chichester: John Wiley and Sons Ldt.

Béziat, D., Bourges, F., Debat, P., Lompo, M., Martin, F., & Tollon, F. (2000). A Paleoproterozoic ultramafic-mafic assemblage and associated volcanic rocks of the Boromo greenstone belt: Fractionates originating from island-arc volcanic activity in the West African Craton. *Precambrian Research*, *101*, 25–47. doi:10.1016/S0301-9268(99)00085-6

Bose-O'Reilly, S., Drasch, G., Beinhoff, C., Tesha, A., Drasch, K., & Roider, G. (2010). Health assessment of artisanal gold miners in Tanzania. *Science of the Total Environment*, *408*(4), 796–805. doi:10.1016/j.scitotenv.2009.10.051

Cai, L., Xu, Z., Reu, M., Guo, Q., Hu, G., Wan, H., & Peng, P. (2012). Source identification of eight hazardous heavy metals in agricultural soils of Huizhou, Guangdong, Province, China. *Ecotoxicology and Environmental Safety*, *78*(1), 2–8. doi:10.1016/j.ecoenv.2011.07.004

Canli, M., & Atli, G. (2003). The relationships between heavy metals (Cd, Cr, Cu, Fe, Zn) levels and size of six Mediterranean fish species. *Environmental Pollution*, *121*(1), 129–136. doi:10.1016/S0269-7491(02)00194-X

Chen, M., & Lena, Q. M. (2001). Comparison of three aqua regia digestion methods for twenty Florida soils. *Soil Science Society of America Journal*, *65*(2), 491–499. doi:10.2136/sssaj2001.652491x

Cicchella, D., De Vivo, B. A., Lima, A., Albanese, S., McGill, R. A. R., & Parrish, R. R. (2014). Heavy metal pollution and Pb isotopes in urban soils of Napoli, Italy. *Geochemistry: Exploration, Environment, Analysis*, *8*(1), 103–112.

Covelli, S., & Fantolan, G. (1997). Application of normalization procedure in determining regional geochemical baseline. *Environmental Geology*, *30*(1 – 2), 34–45. doi:10.1007/s002540050130

CPCS. (1967). *Classification des sols, édition 1967* (pp. 96).

Davidson, J. (1993). The transformation and successful development of small-scale mining enterprises in developing countries. *Natural Resources Forum*, 315–326. doi:10.1111/j.1477-8947.1993.tb00192.x

Dean, W. E., Jr. (1974). Determination of carbonate and organic matter in calcareous sediments and sedimentary rocks by loss on ignition: Comparison with other methods. *Journal of Sedimentary Petrology*, *44*(1), 242–248.

Donkor, A. K., Bonzongo, J.-C. J., Nartey, V. K., & Adotey, D. K. (2005). Heavy metals in sediments of the gold mining impacted Pra River Basin, Ghana, West Africa. *Soil and Sediment Contamination: an International Journal*, *14*(6), 479–503. doi:10.1080/15320380500263675

Drasch, G. S., Böse-O'Reilly, S., Beinhoff, C., Roider, G., & Maydl, S. (2001). The Mt. Diwata study on the Philippines 1999 - assessing mercury intoxication of the population by small scale gold mining. *Science of the Total Environment*, *267*(1–3), 151–168. doi:10.1016/S0048-9697(00)00806-8

Duruibe, J. O., Ogwuegbu, M. O. C., & Egwurugwu, J. N. (2007). Heavy metal pollution and human biotoxic effects. *International Journal of Physical Sciences*, *2*(5), 112–118.

Ekengele, N. L., Danala, D. S., Zo'o, Z. P., & Myung, C. J. (2016). Physical and metals impact of traditional gold mining on soils in Kombo-Laka area (Cameroon). *International Journal of Geosciences, 7* (9), 1102–1121. doi:10.4236/ijg.2016.79084

Facchinelli, A., Sacchi, E., & Mallen, L. (2001). Multivariate statistical and GIS-based approach to identify heavy metal sources in soils. *Environmental Pollution, 114* (3), 313–324. doi:10.1016/S0269-7491(00)00243-8

Farella, N., Lucotte, M., Davidson, R., & Daigle, S. (2006). Mercury release from deforested soils triggered by base cation enrichment. *Science of the Total Environment, 368*(1), 19–29. doi:10.1016/j.scitotenv.2006.04.025

Fofana, A., Ouédraogo, D., & Zombré, B. R. (2009). Atelier sous-régional d'information despays de l'Afrique de l'Ouest Francophone sur les problèmes liés à l'orpaillage. Communication. ONUDI, Grand Hôtel de Bamako, 8–10 Décembre 2009.

Foli, G., Nude, P. M., & Apea, O. B. (2012). Geochemical characteristics of soils from selected districts in the upper east region, Ghana: Implication for trace element pollution and enrichment. *Research Journal of Environmental and Earth Sciences*, 4(2), 186–195.

Fuller, M. A., Feamebough, W., Mitchel, D., & Trueman, I. C. (1995). Desert reclamation using yellow river irrigation watr in Ningxia, china. *Soil Use and Management, 11*, 77–83. doi:10.1111/j.1475-2743.1995.tb00500.x

Garcia, C. A. B., Barreto, M. S., Passos, E. A., & Alves, J. P. H. (2009). Regional geochemical baselines and controlling factors for trace metals in sediments from the Poxim River, Northeast Brazil. *Journal of the Brazilian Chemical Society, 20*, 1334–1342. doi:10.1590/S0103-50532009000700019

Garten, C. T. (1976). Correlations between concentrations of elements in plants. *Nature, 261*, 686–688. doi:10.1038/261686a0

Giller, K. E., Witter, E., & McGrath, S. P. (1998). Toxicity of heavy metals to microorganisms and microbial processes in agricultural soils: A review. *Soil Biology & Biochemistry, 30*(10 - 11), 1389–1414. doi:10.1016/S0038-0717(97)00270-8

Gueye, D. (2001). Étude sur les Mines Artisanales et Les Exploitations Minières à Petite Échelle au Burkina Faso. IIED and WBCSD Report.

Håkanson, L. (1980). An ecological risk index for aquatic pollution control of sediment ecological approach. *Water Research, 14*(8), 975–1000. doi:10.1016/0043-1354(80)90143-8

Hein, K. A. A. (2015). The Bagassi gold deposits on the eastern margin of the Houndé greenstone belt, Burkina Faso. *Ore Geology Reviews, 78*, 660–666. doi:10.1016/j.oregeorev.2015.10.028

Hernandez, L., Probst, A., Probst, J. L., & Ulrich, E. (2003). Heavy metal distribution in some French forest soils: Evidence for atmospheric contamination. *Science of the Total Environment, 312*(1 - 3), 195–219. doi:10.1016/S0048-9697(03)00223-7

Hertz, J., Angehrn-Bettinazzi, C., & Stöckli, H. (1990). Distribution of heavy metals in various litter horizons and forest soils. *International Journal of Environmental Analytical Chemistry, 39*(1), 91–99. doi:10.1080/03067319008027685

Hilson, G. (2016). Artisanal and small-scale mining and agriculture. Exploring their limits in rural sub-Saharan Africa. Sustainable Market, Issue paper IIED, London.

Hilton, J., Davison, W., & Ochsenbein, U. (1985). A mathematical model for analysis of sediment coke data. *Chemical Geology, 48*(1 - 4), 281–291. doi:10.1016/0009-2541(85)90053-1

Huot, D., Sattran, V., & Zida, P. (1987). Gold in Birimian greenstone belts of Burkina Faso, West Africa. *Economic Geology, 82*(8), 2033–2044. doi:10.2113/gsecongeo.82.8.2033

Jaques, E., Zida, B., Billa, M., Greffié, C., & Thomassin, J.-F. (2006). Artisanal and small-scale gold mines in Burkina Faso: Today and tomorrow'. In G. Hilson (Ed.), *Small-Scale Mining, Rural Subsistence and Poverty in West Africa* (pp. 115–134). Brouton on Dunsmore, Rugby: Intermediate Technology Publications Ltd.

Jordan, C. F. (1984). Soils of the Amazon rainforest. In G. T. Prance & T. E. Lovejoy (Eds.), *Key environments Amazonia* (pp. 442). chap. 5. Oxford: Pergamon Press.

Kabata-Pendias, A. (2011). *Trace elements in soils and plants* (4th ed.). Boca Raton London New York: Taylor and Francis Group.

Kaiser, H. F. (1961). A note on Guttman's lower bound for the number of the common factors. *British Journal of Mathematical and Statistical Psychology, 14*(1), 41–72.

Kiethega, J.-B. (1983). *L'or de la Volta Noire. Exploitation traditionnelle: Histoire et archéologie*. Paris: Karthala.

Liu, J., Zhuo, Z., Sun, S. H., Ning, X., Zhao, S., Xie, W., ... Li, B. (2015). Concentrations of heavy metals in six municipal sludges in Guangzhou and their potential ecological risk assessment for agricultural land use. *Polish Journal of Environmental Studies, 24*(1), 165–175. doi:10.15244/pjoes/28348

Loring, D. H. (1990). Lithium — A new approach for the granulometric normalization of trace metal data. *Marine Chemistry, 29*, 155–168. doi:10.1016/0304-4203(90)90011-Z

Lüdtke, G., Hirdes, W., Konan, G., Koné, Y., Yao, C., Diarra, S., & Zamblé, Z. (1998). Géologie de la région Haute Comoé Nord–Feuilles Kong (4b et 4d) et Téhini-Bouna (3a à 3d), Direction de la Géologie Abidjan Bulletin, 178 pp.

Luning, S. (2008). Liberalization of the gold mining sector in Burkina Faso. *Review of African Political Economy, 35*(117)October):387–401. Accessed October 21, 2014. doi:10.1080/03056240802411016

Manga, V. E., Neba, G. N., & Suh, E. C. (2017). Environmental geochemistry of mine tailings soils in the artisanal gold mining district of Bétaré-Oya. *Cameroon. Environment and Pollution, 6*(1), 52–61. doi:10.5539/ep.v6n1p52

McBride, M. B. (1989). Reactions controlling metal heavy solubility in soils. *Advances in Soil Sciences, 10*, 1–56.

McBride, M. B., Sauvé, S., & Hendershot, W. (1997). Solubility control of Cu, Zn, Cd, and Pb in contaminated soils. *European Journal of Soil Science, 48*(2), 337–346. doi:10.1111/j.1365-2389.1997.tb00554.x

Mégret, Q. (2011). Gaining access to a globally coveted mining resource: A case study in Burkina Faso. *International Social Sciences Journal, 61*(202), 389–398.

Mensah, A. K., Mahiri, I. O., Owusu, O., Mireku, O. D., Wireko, I., Evans, A., & Kissi, E. A. (2015). Environmental impacts of mining: A study of mining communities in Ghana. *Applied Ecology and Environmental Sciences, 3*(3), 81–94. doi:10.12691/aees-3-3-3

Merrington, G., Rogers, S. L., & Van, Z. L. (2002). The potential impact of long-term copper fungicide usage on soil microbial biomass and microbial activity in an avocado orchard. *Australian Journal Soil Research, 40*(5), 749–759. doi:10.1071/SR01084

Micó, C., Recatala, L., Peris, M., & Sanchez, J. (2006). Assessing heavy metal sources in agricultural soils of a European Mediterranean area by multivariate analysis. *Chemosphere, 65*(5), 863–872. doi:10.1016/j.chemosphere.2006.04.056

Monna, F., Lancelot, J., Croudace, I. W., Cundy, A. B., & Lewis, J. T. (1997). Pb isotopic composition of airborne particulate material from France and the southern United Kingdom: Implications for Pb pollution sources in urban areas. *Environmental Technology, 31*(8), 2277–2286.

Moral, R., Gilkes, R., & Jordan, M. (2005). Distribution of heavy metals in calcareous and non calcareous soils in Spain. *Water, Air, and Soil Pollution, 162*(1 - 4), 127–142. doi:10.1007/s11270-005-5997-5

Müller, G. (1981). Die Schwermetallbelstung der sedimente des Neckars und seiner Nebenflusse: Eine Bestandsaufnahme. *Chemiker-Zeitung, 105*, 157–164.

Nagajyoti, P. C., Lee, K. D., & Sreekanth, T. V. M. (2010). Heavymetals, occurrence and toxicity for plants: A review. *Environmental Chemistry Letters, 8*(3), 199–216. doi:10.1007/s10311-010-0297-8

Nyamangara, J., & Mzezewa, J. (1999). The effects of long-term sewage sludge application on Zn, Cu, Ni and Pb levels in clay loam soil under pasture grass in Zimbabwe. *Agriculture, Ecosystems & Environment, 73*(3), 199–204. doi:10.1016/S0167-8809(99)00056-0

Paruchuri, Y., Siuniak, A., Johnson, N., Levin, E., Mitchell, K., & Goodrich, J. M. (2010). Occupational and environmental mercury exposure among small-scale gold miners in the Talensi-Nabdam District of Ghana's Upper East region. *Science of the Total Environment, 408*(24), 6079–6085. doi:10.1016/j.scitotenv.2010.08.022

Plant, J. A., Klaver, G., Locutura, J., Salminen, R., Vrana, K., & Fordyce, F. M. (1997). The forum of European geological surveys geochemistry task group inventory 1994–1996. *Journal of Geochemical Exploration, 59* (2), 123–146.

Plassard, F., Winiarski, T., & Petit-Ramel, M. (2000). Retention and distribution of three heavy metals in a carbonated soil: Comparison between batch and unsaturated column studies. *Journal of Contaminant Hydrology, 42*(2 - 4), 99–111. doi:10.1016/S0169 7722(99)00101-1

Pleysier, J. L., & Juo, A. S. R. (1980). A single-extraction method using silver-thiourea for Measuring exchangeable cations and effective CEC in soils with variable charges. *Soil Science, 129*(4), 205–211. doi:10.1097/00010694-198004000-00002

Qian, S., Wang, Z., & Tu, Q. (1996). Distribution and plant availability of heavy metals in different particle-size fractions of soils. *Science of the Total Environment, 187*(2), 131–141. doi:10.1016/0048-9697(96)05134-0

Rajaee, M., Obiri, S., Green, A., Long, R., Cobbina, S. J., Nartey, V., ... Basu, N. (2015). Integrated assessment of artisanal and small-scale gold mining in Ghana – Part 2: Natural sciences review. *International Journal of Environmental Research and Public Health, 12*(8), 8971–9011. doi:10.3390/ijerph120808971

Reiners, W. A. (1986). Complementary models for ecosystems. *American Naturalist, 127*, 57–73. doi:10.1086/284467

Rouiller, J., Souchier, B., Bruckert, S., Feller, C., Toutain, F., & Védy, J. C. (1994). Méthode d'analyses des sols. In M. Bonneau & B. Souchier (Eds.), *Pédologie 2. Constituants et propriétés du sol* (pp. 619–652). Paris: Masson.

Roxgold, (2017). Technical report for the Yaramoko gold mine, Burkina Faso. SRK Project Number 3CR016.008.

Salvarredy-Aranguren, M. M., Probst, A., Roulet, M., & Isaure, M. P. (2008). Contamination of surface waters by mining wastes in the Milluni valley (Cordillera Real, Bolivia): Mineralogical and hydrological influences. *Applied Geochemistry, 23*(5), 299–324. doi:10.1016/j.apgeochem.2007.11.019

Sauvé, S., Hendershot, W., & Allen, H. E. (2000). Solid-solution partitioning of metals in contaminated soils: Dependence on pH, total metal burden, and organic matter. *Environmental Science & Technology, 34*(7), 1125–1131. doi:10.1021/es9907764

Sibthorpe, R. (2012). Yaramoko concession, 2012 program, Investor ppt Presentation, 13 Slides.

Soderberg, K., & Compton, J. S. (2007). Dust as a nutrient source for fynbos ecosystems, South Africa. *Ecosystems, 10*(4), 550–561. doi:10.1007/s10021-007-9032-0

Sposito, G. (1989). *The chemistry of soils* (pp. 277). USA: Oxford University Press.

Steinmann, M., & Stille, P. (1997). Rare earth element behavior and Pb, Sr, Nd isotope systematics in a heavy metal contaminated soil. *Applied Geochemistry, 12*(5), 607–623. doi:10.1016/S0883-2927(97)00017-6

Sutherland, R. A. (2000). Bed sediment associated trace metals in an urban stream, Oahu, Hawaii. *Environmental Geology, 39*(6), 611–627. doi:10.1007/s002540050473

Tomlinson, D. C., Wilson, D. J., Harris, C. R., & Jeffrey, D. W. (1980). Problem in assessment of heavy metals in estuaries and the formation of pollution index. *Helgoländer Meeresunters, 33*(1 - 4), 566–575. doi:10.1007/BF02414780

van Straaten, P. (2000). Mercury contamination associated with small-scale gold mining in Tanzania and Zimbabwe. *Science of the Total Environment, 259*(1–3), 105–113. doi:10.1016/S0048-9697(00)00553-2

Vercoutere, K., Fortunati, U., Muntau, H., Griepink, B., & Maier, E. A. (1995). The certified reference materials CRM 142 R light sandy soil, CRM 143 R sewage sludge amended soil and CRM145 R sewage sludge for quality control in monitoring environmental and soil pollution. *Fresenius Journal of Analytical Chemistry, 352*(1 - 2), 197–202. doi:10.1007/BF00322326

Wang, Y., Yang, Z., Shen, Z., Tang, Z., Niu, J., & Gao, F. (2011). Assessment of heavy metals in sediments from a typical catchment of the Yangtze River, China. *Environmental Monitoring and Assessment, 172*(1 - 4), 407–417. doi:10.1007/s10661-010-1316-8

Waziri, N. M. (2014). Environmental geochemistry of soils and stream sediments from the Birnin-Gwari Artisanal Gold Mining Area, North-western Nigeria. *Universal Journal of Geoscience, 2*(1), 18–27.

Weeks, J. M. (2000). A study of the potential risks to human health from consumption of rice cultivates in paddy fields irrigated by mercury-contaminated mine waste water, Naboc River, Philippines. Report for UNIDO Contract DP/PHI/98/00511-02.

Werthmann, K. (2000). Gold rush in West Africa. The appropriation of "natural" resources: Non industrial gold mining in south-western Burkina Faso. *Sociologus, 50*(1), 90–104.

Wilcke, W., Kretzschmar, S., Bundt, M., & Zech, W. (1999). Metal concentrations in aggregate interiors, exteriors, whole aggregates and bulk of Costa Rican soils. *Soil Science Society of America Journal, 63*(5), 1244–1249. doi:10.2136/sssaj1999.6351244x

Zabsonre, A., Agbo, M., Some, J., & Haffin, I. (2016). Gold exploitation and income disparities: Case of Burkina Faso. Conference economique africaine, Abuja, Nigeria.

Zhou, Y. (2016). *Agricultural mechanization in West Africa* (pp. 1–11). Syngenta Foundation for Agriculture.

10

Microbial quality of ablution kettles used at mosques in the Tolon District

A. Waris Abu-Safianu[1], S. J. Cobbina[1] and E. Yeleliere[1*]

*Corresponding author: E. Yeleliere, University for Development Studies-Tamale, Ghana
E-mail: enochyeleliere.ye@gmail.com; zaroukimoro@yahoo.com

Reviewing editor: Geun Young Yun, Kyung Hee University, Republic of Korea

Abstract: The ubiquitous nature of microorganism and its ability of being transmitted have raised serious concerns, particularly items/objects used by multiple people. The aim of this study is to assess the microbial quality of water-containing vessels/containers (kettles) for ablution in Nyankpala in the northern region of Ghana. Thirty ablution kettles from 10 mosques were sampled by taking 90 swab samples which were taken from the external surfaces of these kettles for laboratory analyses for four bacteria (*Staphylococcus* spp., *Salmonella* spp., *Shigella* spp. and *Escherichia coli*) known to majorly cause infectious diseases pertaining to hygiene and sanitation. Swabbed samples were spread-plated on various appropriate media and incubated. Viable colonies were then counted and recorded as colony-forming units (cfu/cm^2). Confirmations of bacteria were done using citrate, urease, Triple Sugar Iron and catalase tests. A questionnaire was also administered to assess the knowledge of mosque management regarding bacterial contamination of surfaces as well as mosque management practices that influence it. Also, 70% of the mosque management in Nyankpala had no idea about microbial contamination of the ablution kettle, while 30% had fair knowledge about it. The study indicated that *Staphylococcus* spp. was the highest in occurrence (79.33%), followed by *Shigella*

ABOUT THE AUTHORS

Mr. Abu-Safianu A. Waris holds a BSc Renewable Natural Resources (majored in Ecotourism and Environmental Management).

Dr. Cobbina S. J is a Senior Lecturer and Vice Dean of the Faculty of Natural Resources and Environment of the University for Development Studies. He has a PhD in Environmental Science and Engineering from Jiangsu University and an MPhil in Environmental Science. He has been the principal researcher in various projects such as the UNICEF-funded studies on hand washing with soap (HWWS) in basic schools in the northern region of Ghana and the CIDA-funded FARMER project on the quality of water in dugouts in northern Ghana. He is currently a member of a research team assessing the environmental and human health impacts of oil and gas activities on frontline communities in Ghana.

Mr. Yeleliere E has a BSc Renewable Natural Resources and majored in Environmental management.

PUBLIC INTEREST STATEMENT

The incidence of hygiene and sanitation has become a nationwide concern which Ghana in no exception. However, efforts have not been made with regard to sustainable strategies to curb and mitigate it. The article seeks to create awareness to make conscious efforts to prevent microbial contamination. Also, the work is to notify Government of Ghana and other stakeholders to take the issue serious. Consequently, this work sought to make available comprehensive information about the extent of microbial contamination of ablution kettles in selected mosque in the Nyankpala community. Also, 70% of mosque management in Nyankpala had no idea about microbial contamination of the ablution kettle, while 30% had fair knowledge about it. The study indicated that *Staphylococcus* spp. was the highest in occurrence (79.33%), followed by *Shigella* spp. (10.76%), *E. coli* (9.17%) and *Salmonella* spp. (0.75%). The study recommends cautiousness in the use of ablution kettles in public places.

spp. (10.76%), *E. coli* (9.17%) and *Salmonella* spp. (0.75%). It was recommended that proper hand washing among ablution kettle users should be practiced as most contaminations were observed from contact with palms of handlers and avenues be created by stakeholders of hygiene and public health for educating mosque managements on the risk of microbial contamination as well.

Subjects: Environmental Management; Environment & Resources; Environment & Health

Keywords: ablution kettle; microbial quality; bacteria; mosque; Nyankpala (Tolon District)

1. Introduction

Ablution is a general term used to describe ritual purification. Basically, ritual purification is a standard set by the Islamic religion to cleanse oneself prior to performing Islamic prayers or worship. It is a recognized ritual among at least four major religions of the world: Islam, Christianity, Judaism and Shintoism. In Islam, ablution ("Wuḍu" in Arabic) mainly denotes the Islamic established method for washing parts of the body. *Wudu* is an item of indulgence which has become an everyday inevitability for all Muslims as it is typically done in preparation for formal prayers (*salat*) and also before handling and reading the Qur'an (C. Glasse 2001).

The vessels/containers in which water can be held for use for ablution according to the Islamic tradition have not been specified. But evolutions in technology and civilization clearly have had tremendous impact on them when comparison is made between vessels used in recent times and what were conventionally used in the early ages of Islam. In the early Islam, quite apart from the use of water in their natural environments such as rivers, dams, springs, among others, it was much more common to find the use of bowls, gourds and buckets as containing vessels for ablution water. In recent times, due to technological advancement and changes in the social lives of Muslims, much more sophisticated ways have been developed to make performing ablution easier. Running taps, ablution bowls and water bottles are examples of such improvements in the developed Muslim world such as Saudi Arabia, United Arab Emirates, Qatar and most portions of the Middle East. In West Africa, however, as of now, such levels of advancement have not been reached on a large scale, but a few innovative trends have obviously taken place pertaining vessels for water for performing ablution in recent times as compared to the early days of Islam.

The ablution kettle "booter" has become the most pervasive water container for ablution in West Africa. It, therefore, comes as no surprise to see that many Ghanaian Muslims make optimum use of the ablution kettle in performing their religious rituals. The quality of hygiene exercised over these kettles is, however, always brought to question, especially with emphasis on the consequent effects on the health of its users.

Infectious diseases affect the world unequally. Sixty-two percent (62%) of all deaths in Africa and 31% of all deaths in Southeast Asia are caused by infection diseases (Global Health Council, 2005). About half of all the infant mortalities each year are due to diarrhea and acute respiratory infections, both of which are transmitted from person to person during everyday interaction, through droplet and airborne spread, through skin contact and through contamination of the environment (World Health Organization [WHO], 2002). However, many of these diseases are preventable as it is estimated that the global disease burden could be lessened by one-tenth simply by improving water quality, sanitation and personal hygiene (World Health Organization [WHO], 2014).

The worst perpetrators of infections on humans are microorganisms which are deemed ubiquitous; hence, despite the unique role of the ablution kettle in the daily lives of its users, can serve as reservoirs for microbes. They can help spread infectious diseases by recurrent contact with hands

(L. H. Kilic, L. H. Kilic, et al. 2009), and as a matter of fact, several epidemiological studies have corroborated that surfaces play major roles in the spread of infectious diseases (J. Nobel 2001). By way of a disclaimer, however, the hygienic conditions of the vessels for ablution water such as the ablution kettle have not been specifically prescribed by the Islamic tradition. Hence, the hygiene exercised over these kettles varies from place to place, society to society, culture to culture and even from person to person. Again, there are varying levels of hygiene and sanitation ablution kettle users in the urban and rurul ureas or less endowed areas both in the field of knowledge and that of income.

Wudu involves washing the hands, mouth, nostrils, arms, head and feet with water and is an important part of ritual purity in Islam. However, all these serve as effective routes for disease transmission (Glasse, 2001). The Muslims of West Africa have taken up the use of ablution kettles to perform this all important ritual from the Islamic perspective. However, the major challenge is the quality of hygiene and the status of the ablution kettle and its consequent effects on the quality of water put in them as well as the health of its users. From the Muslim tradition, it is strongly encouraged to use water when one attends the restroom or the urinal to ease him/herself. Hence, most Muslims attend the restroom with water in ablution kettles which are some of the most contaminated areas where infectious diseases can be contracted (ESFJ, 2014).

Also, there has being limited research on the hygienic status of water-containing vessels for ablution among its users. Hence, this study therefore sought to assess the microbial quality of water-containing vessels/containers (kettles) for ablution in Nyankpala in the northern region of Ghana by detecting for *Staphylococcus* spp., *Salmonella* spp., *Shigella* spp. and *Escherichia coli*, which are known for causing infectious diseases.

2. Methodology

2.1. Study area

This study was conducted in Nyankpala, a suburb of the Tolon District in the northern region of Ghana. It is a community with about 2,000 natives and an influx of about a similar population coming into the University for Development Studies (UDS) situated in its environs. The dominant ethnic group in this area is Dagomba with the Dagbani language as the most widely spoken dialect. Being one of the communities of the Tolon District known to have about 94.1% of its inhabitants as Muslims, most of the people there as such make use of the ablution kettles mostly at the various mosques for their ritual purification (Ghana Statistical Service [GSS], 2014).

There are a total of about 29 public mosques in Nyankpala, about 18 of which provide ablution kettles for public use. Out of these, 10 were selected at random for this study (Figure 1).

2.2. Sampling procedure

First of all, the study area (Nyankpala) was zoned into three sections, namely the eastern zone, western zone and the UDS campus area. The eastern and the western zones were determined by using the main road that leads into the township towards the Nyankpala market. Within each cluster/zone, mosques with ablution kettles were purposively targeted. Eighteen mosques in total (including mosques on the UDS campus) were discovered to have ablution kettles for public use at the mosques. Hence, using simple random sampling, 10 mosques were chosen to have samples taken from their ablution kettles for the laboratory analysis. The modal number of ablution kettles found in the mosques in Nyankpala was about 5 with the highest number of kettles in a mosque being about 17 and the least number of kettles being 3. Hence, simple random sampling was again employed to select three ablution kettles at each chosen mosque. A total of 30 ablution kettles were selected for the swab samples to be taken, and in the end, 90 swab samples were taken for laboratory analyses of the four strains of bacteria selected. This is because that one of the media for culturing the

Figure 1. Map showing the study area.

Source: GSS, (2014).

bacteria (Salmonella–Shigella Agar) can accommodate two strains of bacteria on the same plate (*Salmonella* spp. and *Shigella* spp.), and so instead of 120 swabs, 90 swabs were enough to accommodate the four strains.

In addition, a semi-structured questionnaire was designed and administered to key informants (mosque management) for information on how ablution kettles were treated, the various demographics of mosques and the knowledge level of mosque management as far as microbial contamination of ablution kettles is concerned.

Plate 1. Taking swabs from ablution kettles.

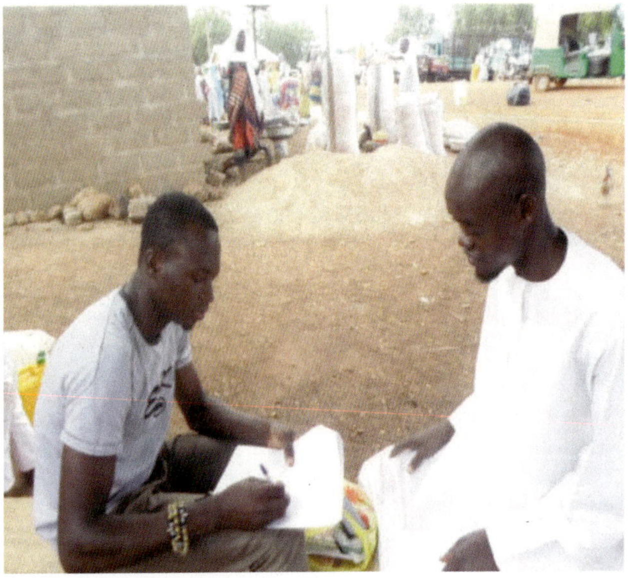

Plate 2. Qualitative data collection.

2.3. Study design

The research was based on experimentation, observation and a field survey. Laboratory experiments were conducted at the Spanish Laboratory Complex in Ghana UDS Nyankpala campus while data concerning environmental, hygienic and sanitary conditions as well as the knowledge level of users of the ablution kettles about microbial contamination were taken by administering questionnaires and personal observation.

The experiment was done by culturing the various microorganisms (*Staphylococcus* spp., *Salmonella* spp., *Shigella* spp. and *E. coli*) using various appropriate media. Thus,

- MacConkey Agar was used for culturing *E. coli*,
- Mannitol Salt Agar for culturing *Staphylococcus* spp. and
- Salmonella-Shigella Agar for culturing both *Salmonella* spp. and *Shigella* spp.).

Samples were picked up using swab sticks. Three swabs were used to swab the handles and external surfaces of each of the ablution kettles selected. Each of the swabs was, respectively, used to culture for the aforementioned bacteria. The swabs, however, under sanitized conditions were processed into a moist form by using (0.9%) saline solution so as to ease the process of swabbing. The swabbed samples were then transported to the lab in an ice chest filled with ice cubes to keep the tip of the swab moist.

The various culture media were prepared following the manufacturers' instructions. Petri dishes, some of the culture media and other necessary glassware and equipment needed were sterilized by autoclaving at 121°C for about 15 min. After cooling the media, they were poured into 15–20-ml petri dishes. The swabs were spread-plated on the various media and incubated at temperatures based on the requirements of the bacteria cultured for. After the incubation period, plates were counted for colonies of bacteria. Confirmatory tests were done for all the colony-forming units (cfu/cm^2) obtained.

2.4. Microbiological analyses

Microbial analyses with the aid of Laminar flow/Hood were carried out using the total plate count method. The total level of microbial contamination was determined. The media used were MacConkey agar, Salmonella–Shigella agar and Mannitol Salt agar for the culturing of *E. coli* spp., *Salmonella* spp., *Shigella* spp. *and Staphylococcus* spp., respectively, which were prepared in the laboratory according to the prescribed standard. Biochemical analysis for confirmation of the various bacteria were done using urease, citrate and Triple Sugar Iron agars as well as (H_2O_2) for catalase tests.

2.5. Statistical analyses

Statistical analyses of results were conducted using Genstat Discovery version 4.10. 3, IBM SPSS Statistics version 23 and Microsoft Excel to find means, standard deviations, frequencies and percentages and to produce tables and charts for the final results and conclusions.

3. Results and discussion

3.1. Results from questionnaire

Figure 2 illustrates the educational status of respondents in the various mosques studied. The mosque management who responded to the questionnaire issued have gone through formal

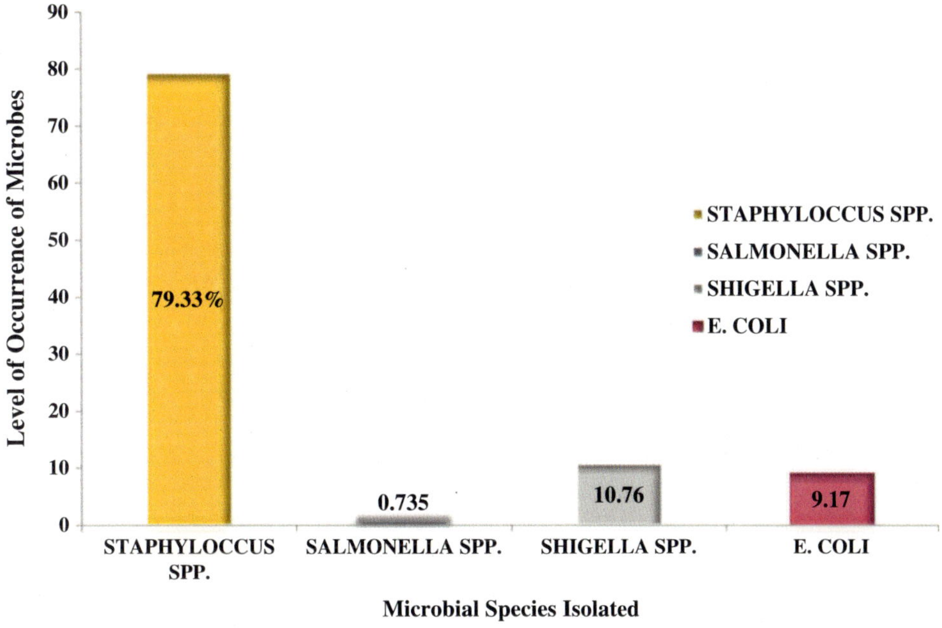

Figure 2. Level of education of respondents.

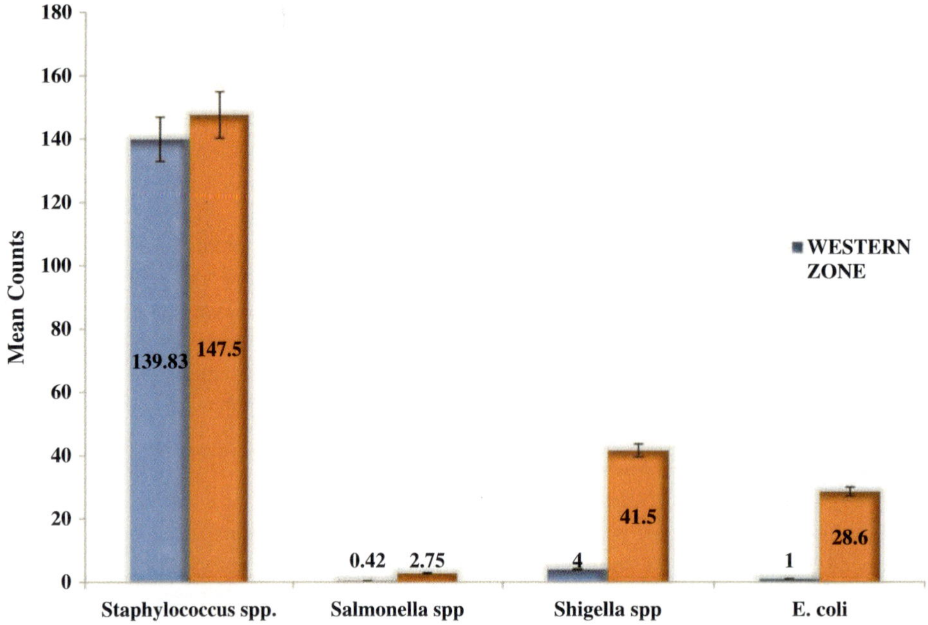

Figure 3. Knowledge of mosque management about bacterial contamination.

education to up to tertiary level, 30 (30%) of the respondents, however, have not in any way had the opportunity to go through formal education and 10 (10%) each of the respondents have been through formal education up to secondary education, respectively.

Figure 2 illustrates the level of awareness of mosque management about bacterial contamination of surfaces. We observed that 70% of the respondents were unaware of ablution kettle contamination by bacteria, while 30% responded positively.

Figure 3 displayed the knowledge of mosque management to prevent bacterial contamination.

Interestingly, only 40% and 30% of the respondents clean their ablution kettles weekly and three times a year, respectively (Figure 4).

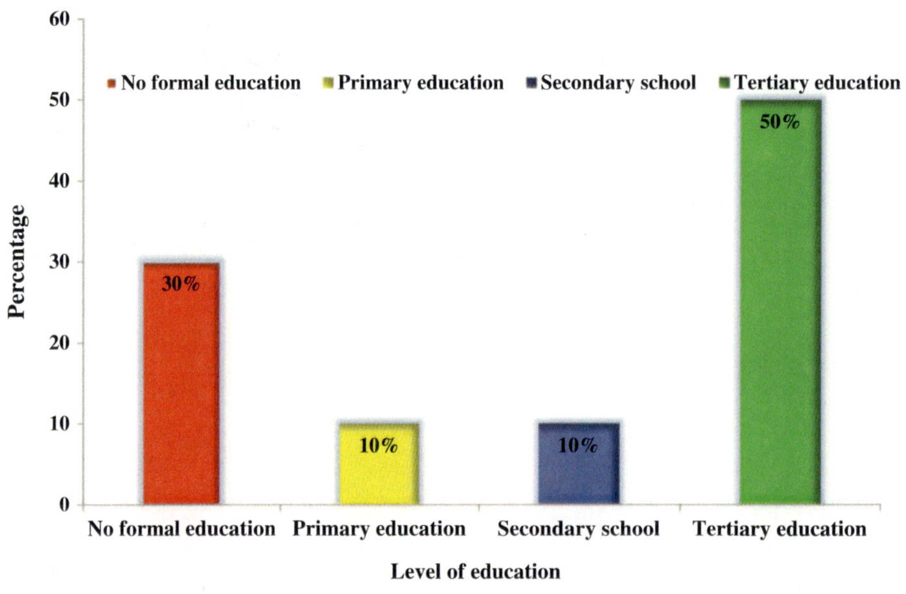

Figure 4. How often mosque management wash their ablution kettles.

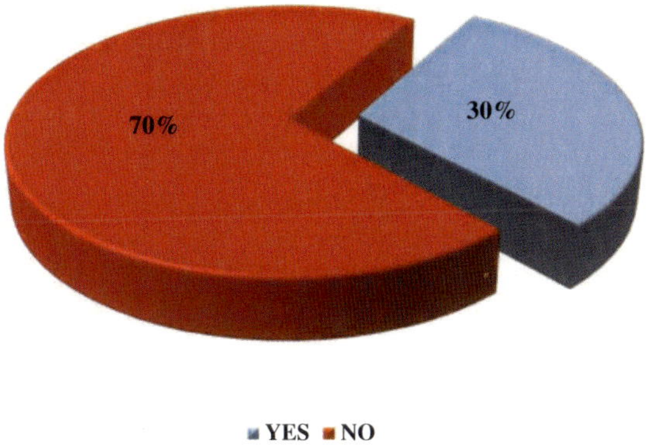

Figure 5. Storage of ablution kettles.

Figure 5 shows the percentage of mosques with racks to keep their ablution kettles or not. From the table, 80% of the mosques did not have racks, while 20% did.

Table 1 illustrates the main sources of water for ablution. Thirty percent of the mosques make optimum use of pipe-borne water for ablution, while 70% mostly make use of dam water. None of the mosques, however, made use of borehole or well water sources.

3.2. Knowledge level of mosque management about microbial contamination

Contaminated surfaces when brought even into relatively brief contact with the fingers, palms or inanimate surfaces, significant number of organisms can be transferred (E. Scott and S. F. Bloomfield 2008). Inadequate knowledge about where germs/microbes lurk could result in the spread of health disorders such as intestinal tract infections as a result of the unhygienic use of artifacts such as the ablution kettle (P. Nancy, P. Nancy, et al. 2008). Awareness on contamination of surfaces and the likely microbial contamination of ablution kettle surfaces was very low on the part of mosque management in Nyankpala. This trend could be a very significant contributing factor to the ineffective management strategies put up in the mosques to maintain sanity, especially of the ablution kettles.

3.3. Microbial analyses

Different species of bacteria were isolated from the various samples that were collected from the various mosques. The bacteria genera that were isolated include *Staphylococcus* spp., *Salmonella* spp., *Shigella* spp. *and E. coli*. Table 2 shows the various bacteria isolated from the ablution kettles of each of the sampled mosques within Nyankpala. *Staphylococcus* spp. and *Shigella* spp. were both identified in all the mosques. *Salmonella* spp. were found only in mosques *A, F, G* and *H*. ***E. coli***, on the other hand, was found present in all the mosques except mosques *A, B* and *J*.

Figure 6 displayed the percentage contribution of various bacteria isolated from sampled ablution kettles. *Staphylococcus* spp. (79.33%) was the most dominant, next was *Shigella* spp. (10.76%), *E. coli* (9.17%) and *Salmonella* spp. (0.75%).

Table 1. Sources of water for ablution in Nyankpala mosques

Responses	Frequency	Percentage (%)
Pipe borne	3	30
Borehole	0	0
Well	1	10
Dam	6	60
Total	10	100

Table 2. The genera of bacteria isolated from the ablution kettles of the various mosques

Mosque	Types of microbes isolated
A	*Staphylococcus* spp., *Shigella* spp. and *Salmonella* spp.
B	*Staphylococcus* spp. and *Shigella* spp.
C	*Staphylococcus* spp., *Shigella* spp. and *E. coli*
D	*Staphylococcus* spp., *Shigella* spp. and *E. coli*
E	*Staphylococcus* spp., *Shigella* spp. and *E. coli*
F	*Staphylococcus* spp., *Shigella* spp., *Salmonella* spp. and *E. coli*
G	*Staphylococcus* spp., *Shigella* spp., *Salmonella* spp. and *E. coli*
H	*Staphylococcus* spp., *Shigella* spp., *Salmonella* spp and *E. coli*
I	*Staphylococcus* spp., *Shigella* spp., and *E. coli*
J	*Staphylococcus* spp. and *Shigella* spp.

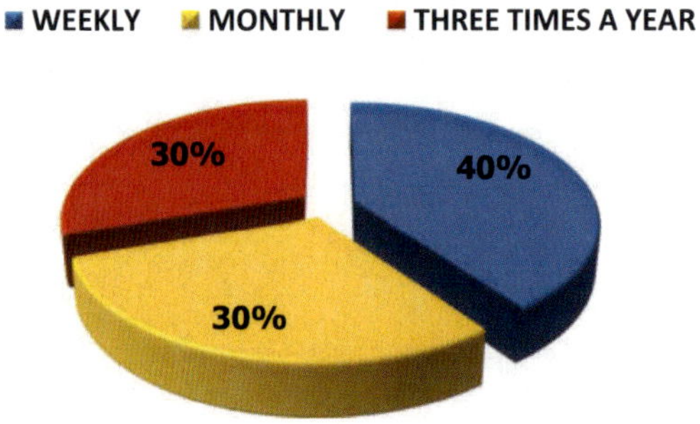

Figure 6. Bacteria isolated from the ablution kettles at various mosques in Nyankpala.

However, the analyses of variance (ANOVA) of occurrence of the various bacteria species in the various mosques in Nyankpala were not statistically significant in occurrence of any of the various bacteria between the various treatments (mosques) and the replications (sampled ablution kettles).

Figure 7 illustrates a comparison of the mean level of occurrence of the various bacteria (*Staphylococcus* spp., *Salmonella* spp., *Shigella* spp. and *E. coli*) on ablution kettles of mosques within two zones of Nyankpala: the western and the eastern zones. The western plots of the various bacteria are represented by light blue bars, while the eastern plots are represented by orange bars. It was observed that the mean occurrence of *Staphylococcus* spp. in the eastern zone (147.5 cfu/cm^2) was higher than the western zone (139.83 cfu/cm^2). Also, for *Salmonella* spp., the eastern zone (2.75 cfu/cm^2) yet again recorded a higher occurrence than in the western zone (0. cfu/cm^2). Also, for *Shigella* spp., the eastern zone (41.5 cfu/cm^2) once again had a higher recording than that of the western zone (4.00 cfu/cm^2), and finally, for *E. coli*, the eastern zone (28.6 cfu/cm^2) once again recorded the highest count than in the western zone (1).

Table 3 shows the total aerobic counts of the various bacteria on the surfaces of ablution kettles under study. The total aerobic counts ranged from 1.12×10^2 cfu/cm^2 to 3.79×10^2 cfu/cm^2 for the various mosques. The order of the viable bacteria were 3.79×10^2 cfu/cm^2 (*F*) > mosque 2.36×10^2 cfu/cm^2 (*E*) >

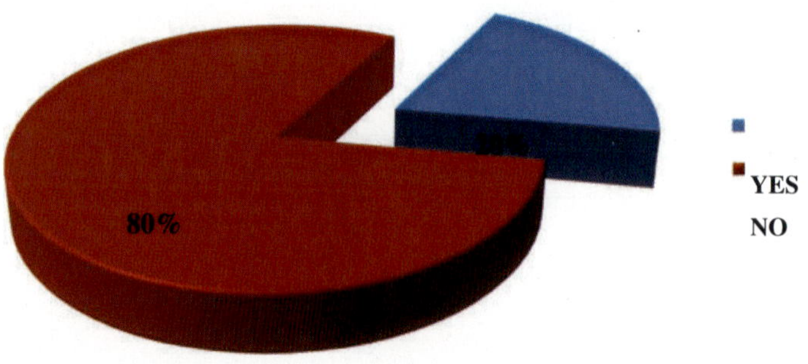

Figure 7. Comparison of bacteria counts between the east-ern and the western zones of Nyankpala.

Table 3. Mean viable count of bacteria in the various mosques in Nyankpala

Mosque	Mean viable count (cfu/cm^2)
(Sample ID)	Total
A	1.24×10^2
B	1.68×10^2
C	1.41×10^2
D	1.47×10^2
E	2.36×10^2
F	3.79×10^2
G	1.20×10^2
H	1.46×10^2
I	1.48×10^2
J	1.12×10^2

1.68×10^2 cfu/cm^2 (*B*) > 1.48×10^2 cfu/cm^2 (*I*) > 1.47×10^2 cfu/cm^2 (*D*) > 1.46×10^2 cfu/cm^2 (*H*) > 1.41×10^2 cfu/cm^2 (*C*) > 1.24×10^2 cfu/cm^2 (*A*) > 1.20×10^2 cfu/cm^2 (*G*) > 1.12×10^2 cfu/cm^2 (*J*).

3.4. Microbial load on ablution kettles

The level of microbial contamination of ablution kettles in mosques at Nyankpala was examined in this study. It was observed from the study that the ablution kettles were contaminated with *Staphylococcus* spp., *Salmonella* spp., *Shigella* spp. and *E. coli.* Using a keyboard in his study, R. Marsden 2009 posited that anytime an object, such as the ablution kettle in this case, is shared among two or more people, and it poses a risk of the spread of infectious agents.

Staphylococcus spp. was the most occurring species of bacteria observed on the ablution kettles with a percentage of 79.33% as shown above. This may be attributed to palm/hand contact with users. The human skin is a good habitat for *Staphylococcus* spp. because users of the ablution kettles have direct contact with its handle or upper surface (E. A. Grice, E. A. Grice, et al. 2009). Users of the ablution kettles' palms/hands are in direct contact with the handle or upper surface of the ablution kettle. The palm is moist to varying degrees due to perspiration, the result of which makes the palms to be basic which is known for sustaining the growth of halophytic bacteria such as *Staphylococcus* spp. (T. Elliot, M. Hastings and U. Desselberger 2001).

Shigella spp. was the second most occurring bacteria species with 10.76% in the various mosques. This largely is attributed to the handling by users as *Shigella* spp. are found in the human intestines, thereby suggesting fecal sources of contamination of the ablution kettles. This is supported by the observation made in the mosque which recorded the highest count of this species (*mosque F*) having a toilet facility sited close to it and the locals making use of it by taking some of the ablution kettles to the facility to ease themselves.

The presence of *E. coli* being the third highest in occurrence (9.17%) indicated recent fecal contamination of ablution kettles. This may be attributed to contamination with feces after visiting the toilet, human sewage or contaminated water, fecal matter of animals and improper/inadequate hand washing as stated (D. H. Tambekar, S. D. Shirsat and S. B. Suradkar 2007). The source of contamination of *E. coli* is closely related to *Shigella* spp. as this species is known to be abundant in the gut of mammals. The mosque which recorded the highest count of this species (*mosque E*) was situated close to the abattoir as well as the market of the town and could be the source of contamination. Also, *mosque F* which recorded the second highest number of this species as already indicated had a toilet facility close by and could be the reason for the contamination. It was also observed in *mosque I* that bats were hanging up on a tree under which their ablution kettles were kept. These bats could drop their fecal matter on these ablution kettles which might have accounted for the presence of the species at that mosque. The presence of *E. coli* on the ablution kettle samples is an indication of fresh fecal contamination.

Salmonella spp. was the least because it thrives in a protein-rich medium which many of the respondents could not afford regularly. Nonetheless, *mosque H and F* located at the center of the town recorded the highest count of this species and could be attributed to the fact that the mosque were sited close to a protein source and the place was concentrated with food products. Also there are higher numbers of people with different levels of hygiene around the market who might be making use of the ablution kettle.

Not all the above-mentioned microbial species were found in each mosque. *E. coli* for instance was absent in mosques *A, B* and ***J***. *Salmonella* spp. was also found to be absent in mosques *B, C, D, E, I* and *J*. It is worth noting that the level of microbial contamination varied from one mosque to the other with some recording high microbial presence while others recorded low microbial presence.

3.5. Factors that influenced microbial load and contamination of ablution kettles

All the microbial contamination of ablution kettles can be attributed to factors such as the location of the mosques, the mosque management strategies and also the demographics of the mosque attendees. Some of the mosques were located within the township, especially the market areas as well as the UDS campus, and are therefore easily accessed by numerous users, while others were found in very obscure locations where only few people had access to them. Generally speaking, the more accessible these mosques are, the more diverse the types of microbes present and hence the higher the level of microbial contamination (A. K. Al-Ghamdi, A. K. Al-Ghamdi, et al. 2011). In situations where there are a lot of people using the ablution kettles in mosques, there is likely to be a good number of sick people or people with varying hygienic conditions. Through them, new bacteria would settle on the ablution kettle, get release into the air or from physical contact (V. Lam, V. Lam, et al. 2011). This could account for the differences in microbial types and numbers among the different mosques that was observed. Also, the environmental conditions in the areas where these mosques were located played a very important role. Some of the mosques were located around areas where the environment was highly polluted with fecal matter, litter, other forms of excreta and some waste disposal areas. For instance, some of the mosques were located around the abattoir where the gut of slaughtered animals could serve as convenient sources of these bacteria strains, others around drainage gutters where liquid waste and others were discarded and others around toilet facilities. Others, however, were found in less polluted environments. J. Fleming 2007 posited that a variety of reservoirs in the environment can serve as sources of contamination.

Predominantly, hygiene levels influence the quantity of microbes on surfaces and thus the chance of transmission during handling of ablution kettles. Antimicrobial resistance has been observed to steadily be on the rise (H. Gedik, H. Gedik, et al. 2010), and contaminated surfaces such as the ablution kettle contribute to the spread of these multi-drug-resistant microbes among the public. One factor that had a significant influence on the microbial contamination of the ablution kettle in mosques is the management strategies put into play by mosque management. Generally, it was observed that microbial contamination in mosques which regularly washed their ablution kettles (e.g. mosques *A, I* and *J*) was found to be lower as compared to mosques which did not as regularly wash their ablution kettles (e.g. mosques *E, F* and *H*). Also, the fact that none of the mosques provided soap and antiseptic for use by the public after visiting urinals or toilets could be an accounting factor to the levels of microbial contamination on ablution kettles in mosques. Human hands and artifacts of public use/contact are prime sources of disease transmission, and the ablution kettle may serve as a key vehicle in the broadcast of diseases or disease-causing organisms (T. K. C. Pope, P. T. Ender and W. K. Woek 2002). Another contributing factor to microbial contamination of ablution kettles noticed in mosques due to mosque management strategies could be the inability to provide racks and proper storage facilities for keeping them, there by leaving them to the mercy of stray animals, rodents and other unfavorable environmental conditions.

3.6. Potential health hazards of contamination of ablution kettles

Staphylococcus spp. is a strain of bacteria usually found in large quantities in the environment and also found on the human body. Some strains of this species can be pathogenic and can cause serious infections in immunocompromised persons varying from urinary tract infections to osteomyelitis and valvular endocarditis (C. Mims, C. Mims, et al. 2004). Other popular infections caused by *Staphylococcus* spp. are boils, meningitis, bumble foo, arthritis, pneumonia, black pox, bronchitis, carbuncle, cystitis and scaled skin (B. Stuart 2005). Therefore, the high occurrence of *Staphylococcus* spp. on ablution kettles in mosques within the Nyankpala community is a major cause for concern.

Salmonella spp. belongs to the family Enterobacteriaceae. *Salmonella* spp. is facultative anaerobic, gram-negative, oxidase-negative and rod-shaped bacteria. *Salmonellae* cause diseases in both humans and animals. The serovar *S. Typhi* and most *S. Paratyphi* strains cause serious systemic infections in humans and are specific human pathogens. *Salmonella* cause a wide range of symptoms from diarrhea, cramping, vomiting and nausea to more serious complications and illness (P. N. Acha and B. Szyfres 1987).

E. coli is a gram-negative, facultative anaerobic and rod-shaped strain of bacteria which is usually found in the intestines of warm-blooded organisms. Most *E. coli* strains are known to be harmless, but some serotypes of which are also known to be pathogenic. *E. coli* (0157:H7) is a strain of this particular bacteria which can cause serious anemia or kidney failure and can eventually lead to death (www.webmd.com/a-to-Z guide/e. coli). Several laboratories have therefore characterized them as causing self-limiting diarrhea, hemorrhagic colitis, hemolytic uremic syndrome and thrombotic thrombocytopenic purpura in children and other susceptible groups of individuals (J. Meng, J. Meng, et al. 2001).

Shigella spp. is a genus of gram-negative, nonspore-forming, facultative anaerobic, nonmotile and rod-shaped bacteria which genetically has been observed to be closely related to another strain of bacteria known as *E. coli* (E. Yabuuchi 2002). *Shigella* is causative agent of human shigellosis; it causes disease in primate, but not in other mammals (K. J. Ryan and C. G. Ray 2004)

On a global scale, *Shigella* is one of the leading bacterial causes of diarrhea, resulting in an estimated 80–165 million cases (A. Bowen 2016). Annually, the number of deaths it causes is estimated between 74,000 and 600,000 deaths (Bowen, 2016; S. Mani, T. Wierzba and R. I. Walker 2016). It counts as part of the top four pathogens that are responsible for moderate-to-severe diarrhea in African and South Asian children (K. L. Kotlogg, J. P. Nataro and W. C. Blackwelder 2013).

The illustration shows the levels of contamination that are deemed acceptable as well as levels that are deemed contaminated according to BC CDC (2010). From the above, swab samples which have <45 cfu/cm^2 of bacteria on them are deemed clean enough for use by humans. However, even more specifically, if there is a specified area of 25 cm^2 that is swabbed, then it is required that the number of cfu/cm^2 of bacteria found should be <5. Contaminated levels are also said to range between 140 and 260 cfu/cm^2, while any number >260 cfu/cm^2 is deemed highly contaminated. From the experiment on swabs taken from ablution kettles at the various mosques as shown in Table 4, the mean total aerobic counts ranged from 1.12×10^2 to 3.79×10^2 cfu/cm^2 which in essence showed that all the kettles from the mosques were either *contaminated* or *very contaminated* according to BC CDC (2010) standards of swabbed surfaces.

4. Conclusion and recommendation

The study assessed the microbial quality of ablution kettles used at mosques in Nyankpala by examining the level of occurrence of four strains of bacteria, namely *Staphylococcus* spp.,

Table 4. BC Centers for Disease Control (CDC) standard for swabbed surfaces

Interpretation	Sponges or swabs (total cfu)	Log conversion	Counts on surface based on 25 cm^2 surface area sampled
Clean	<45	<1.65	Less than 5 cfu per cm^2
Contaminated	140–260	2.15–2.41	~5 to 10 cfu per cm^2
Very Contaminated	>260	>2.41	Greater than 10 cfu per cm^2

Source: BC Centers for Disease Control (BC CDC, 2010).

Salmonella spp., *Shigella* spp. and *E. coli*. The study revealed the abundance of *Staphylococcus* spp. contamination of ablution kettles recorded 79.33%, *Shigella* spp. 10.76%, *E. coli* 9.17% and *Salmonella* spp. 0.75%. Ablution kettles no doubt have matchless importance to its users, but its multi-user nature poses a risk serving as reservoirs and vehicles for microbial contamination and transmission. Slight differences exists among the concentration of the bacteria studied; however, sources of contamination could be discovered to range from individuals who used these ablution kettles, the environment in which these ablution kettles were located as well as the various mosque management activities that were carried out in these mosques.

In general, the prevalent means of bacterial contamination are through palm contact with the kettles. It is therefore imperative that proper hand washing practices and decontamination of kettles should be adopted by mosque users.

The study therefore recommend that,

- Avenues for education of mosque management and other ablution kettle users about microbial contamination should be taken up by stakeholders of hygiene and environmental safety.

Competing Interest

The authors declare no competing interest.

Author details

A. Waris Abu-Safianu[1]

E-mail: abusafianuabdulwaris@yahoo.com

S. J. Cobbina[1]

E-mail: scobbina@uds.ed.gh

E. Yeleliere[1]

E-mail: enochyeleliere.ye@gmail.com

[1] Department of Ecotourism and Environmental Management, Faculty of Natural Resources and Environment, University for Development Studies, Tamale, Ghana.

References

Acha, P. N., & Szyfres, B. (1987). Zoonotic tuberculosis. *Zoonoses and communicable diseases common to man and animals* (2nd ed.). Washington, DC: Pan American Health Organization (PAHO)/World Health Organization, 503.

Al-Ghamdi, A. K., Abdelmalec, S. M. A., Ashshi, A. M.,). Faidah, H., Shukri, H., & Jiman-Fatani, A. A. (2011 Bacterial contamination of computer keyboards and mice, elevator buttons and shopping carts. *African Journal of Microbiology Research, 5*(23), 3998–4003.

BC Centre for Disease Control. (2010). *Environmental hygiene monitoring. A guide for environmental health officers* (3 ed., pp. 5–6).

Bowen, A. (2016). Chapter 3: Infectious diseases related to travel. In *The yellow book: Health information for international travel*. CDC. ISBN 978-0-19-937915-6. Retrieved 16th May, 2017.

Elliot, T., Hastings, M., & Desselberger, U. (2001). *Lecture notes on medical microbiology* (3rd ed.). Oxford: Blackwell Publishing, 20 [Google Scholar].

Fleming, J. (2007). *Computer keyboards, a new reservoir for infection transmission. Managing infection control.*

Gedik, H., Yahyaoglu, M., Yörük, G., & Fincanci, M. (2010). Extended-spectrum beta-lactamase production rates of *Klebsiella spp.* and *Escherichia coli* strains isolated from infections and faecal samples of healthy people. *Infectious Diseases in Clinical Practice, 18*, 104–106.

Ghana Statistical Service. (2014). 2010 Population and housing census: District analytical report. Tolon District.

Glasse, C. (2001). *The new Encyclopedia of Islam* (pp. 477). Walnut Creek, CA: Altamira Press.

Global Health Council. (2005). Global health news. Initiative of University of Uta. Retrieved 10 June, 2017, from www.old.globalHealth.org.new/past-sear

Grice, E. A., Kong, H. H., Conlan, S., Deming, C. B., Davis, J., Young, A. C., ... Murray, P. R. (2009). Topographical and temporal diversity of the human skin microbiome, *Science, 324*(5931), 1190–1192. PMC 2805064. PMID 19478181.

Kilic, L. H., Ozaslan, M., Karagoz, I. D., Zer, Y., & Davutoglu. (2009). The microbial colonization of mobile phones used by healthcare staff. *Pak Journal Biologic Sciences, 12*(11), 882–884.

Kotlogg, K. L., Nataro, J. P., Blackwelder, W. C., Nasirin, D., Farag, T. H., Panchalingam, S., ... Breiman, R. F. (2013). Burden and aetiology of diarrhoeal disease in infants and young children in developing countries (the Global Enteric Multicenter Study, GEMS): A prospective, case-control study, *The Lancet, 382*(9888), 209–222. PMID 23680352.

Lam, V., Chia-Wei, C., Aquino, E., & Han, Y. (2011). Computer keyboard. Retrieved 9 May, 2017, from http://microbewiki.kenyon.edu/index.php/ComputerKeyboard

Mani, S., Wierzba, T., & Walker, R. I. (2016). Status of vaccine research and development for Shigella, *Vaccine, 34*(26), 2887–2894. PMID 23680352.

Marsden, R. (2009). *A solid-surfaced infection control computer keyboard* (pp. 1–5). P Eng Clean key Inc.

Meng, J., Doyle, M. P., Zhao, T., & Zhao, S. (2001). Enterohemorrhagic *Escherichia coli* O157: H7. In M. P. Doyle, L. R. Beuchat, & T. J. Montville (Eds.), *Food microbiology: Fundamentals and frontiers* (2nd ed., pp. 193–213). Washington, D.C: ASM Press.

Mims, C., Dockrell, H., Goering, R., Roitt, I., Wakelin, D., & Zuckerman, M. (2004). *Medical microbiology,* Elsevier Mosby updated (3rd ed., pp. 287).

Nancy, P., Moreno, P. D., Barbara, Z., Tharp, M. S., Deanne, B., Erdmann, M. S., ... Denk, M. A. (2008). The science of microbes, *microbes are everywhere* ISBN 13: 978-1-888997-54-5ISBN-10: 1-888997-54-0.

Nobel, J. (2001). *Text book of primary care medicine* (3rd ed., pp. 8). St Louis, Mo: Mosby.

Pope, T. K. C., Ender, P. T., Woek, W. K., et al. (2002). Bacterial contamination of paper currency. *Southern Medica Journal, 95*, 1408-1410.

Ryan, K. J., & Ray, C. G. (2004). *Sherries medical microbiology; an introduction to infectious diseases* (4th ed.). Chapter 48, Candida albicans, pp. 661-663.

Scott, E., & Bloomfield, S. F. (2008). The survival and transfer of microbial contamination via cloths, hands and utensil. *Journal of Applied Microbiology, 68*, 271–278.

Stuart, B. (2005). *Infrared spectroscopy.* John Wiley & Sons, Inc. Kirk-Othmer Encyclopedia of Chemical Technology.

Tambekar, D. H., Shirsat, S. D., & Suradkar, S. B. (2007). Prevention of transmission of infectious disease: Studies on hand hygiene in health-care among students in. *Continental Journal of Biochemical Science, 1*, 6–10.

WHO. (2002). World Health Report. Geneva. doi:10.1044/1059-0889(2002/er01)

WHO. (2014). Water. World Health Organization. Retrieved 17 May, 2017, from http://www.who.int/topics/water/en/

Yabuuchi, E. (2002). Bacillus dysentericus (sic) 1897 was the first taxonomic rather than Bacillus dysenteriae 1898, *International Journal of Systematic and Evolutionary Microbiology, 52*(Pt 3), 1041. PDMID 12054222.

11

Characterization of aerosol loading in urban and suburban locations: Impact on atmospheric extinction

Nnaemeka Dom Onyeuwaoma[1,2*], Theo Chidiezie Chineke[2], Okeychukwu Kelechi Nwofor[2], Ian Crandell[3], Olushina Olawale Awe[4], Ajileye Olasumbo[1], Alex I Opara[5], Nwachukwu Pius[6], Mba Tochukwu[7] and Nwachukwu Joy[6]

*Corresponding author: Nnaemeka Dom Onyeuwaoma, Cooperative Information Network, NASRDA, PMB, 022, Obafemi Awolowo University, Ile-Ife, Nigeria; Atmospheric Physics Group, Department of Physics, Imo State University, Owerri, Nigeria
E-mail: emekadonn@gmail.com

Reviewing Editor: M. Manzurul Hassan, Jahangirnagar University, Bangladesh

Abstract: The study of atmospheric aerosols remains a very strong factor in the understanding of the Earth's climate and predicting the air quality within any given environment. In view of this, analysis of aerosol characteristics measured using AERONET sun photometer, MODIS, Cloud-Aerosol Lidar with Orthogonal Polarization, AIRS and TOMS/OMI sensors, over six distinctly different environments in Nigeria for a period of 10 years was carried out. The results revealed significant regional and temporal variation. It showed that different aerosol types impact differently on atmospheric extinction properties. The result further revealed that aerosol loading peaks in the south and middle belt regions during the dry season (November–March), while the reverse was the case in the north during the rainy season (April–October). It was found out that coarse aerosols prevail all year round within these regions, with the rate of occurrence varying with season. The range of the ratio of coarse-to-fine aerosol was found to be as high as 19:1 during the rainy season in Sokoto and Maiduguri, and 3:1 during the dry season. In Abuja and Ilorin, it ranged from 13:1 to 2:1 in both seasons. The atmospheric extinction values plotted against relative humidity (RH) showed that extinction grows at very low RH of about 45% in a less anthropogenic polluted environment (north), while a RH of 60% and above is

ABOUT THE AUTHOR

Onyeuwaoma Nnaemeka holds a PhD in atmospheric physics from Imo State University, Owerri. His research interest is in aerosols with specific interest in air quality and aerosol direct radiative forcing. Neaemeka is also interested in topics on climate change and cloud seeding.This present study is on the role aerosols play in atmospheric extinction growth. The authors found out that there are no uniform values of relative humidity for extinction growth in Nigeria. This implies that atmospheric extinction is dependent on the prevailing aerosol specie within an environment. Subsequently, it was discovered from this research that the air quality in Port Harcourt and Ikeja is reducing as the year goes by and needs concerted efforts from the government to alleviate in other to forestall an outbreak of air-quality related epidemic.

PUBLIC INTEREST STATEMENT

This research article is on aerosols and how it affects the atmospheric environment. We discovered that the urban settlements of Ikeja and Port Harcourt (PH) are more polluted than the other locations studied. We also realized that in Ikeja and PH, extinction requires a lot of water vapour to grow significantly in the atmosphere, while in the other places, it requires less water vapour. This research also shows that in spite of the aerosols in the atmosphere, rainfall is increasing over the years studied.

Also, air quality in Nigeria is on the decrease in the south as a result of industrialization and fossil-fuel combustion, while in the north, desert dust and biomass burning is responsible for the phenomena. Therefore, the rate of acid rain, pollution-related ailment and contamination of soil is expected to be on the increase in the south.

required for growth to take place in a more polluted environment in the south. This implies that the use of rain water for drinking and other domestic purposes should be done with caution.

Subjects: Environmental Issues; Environment and the City; Environmental Change and Pollution

Keywords: AOD; air quality; Sahel; carbonaceous aerosols; biomass burning

1. Introduction

Aerosols in the atmosphere comprising smoke from biomass burning (BB), sea salts and dust particles originating from deserts and bare surfaces are among the atmospheric particles that affect air quality and the Earth's meteorological cycle. The understanding of their meteorology, source and effects on the environment needs to be properly studied. This calls for continuous and consistent study aimed at understanding aerosols through measurements from ground, space and air-bound platforms. But ground-based measurements cannot comprehensively monitor aerosols because these particles can be transported and deposited over long distances remote from their source and monitoring stations. Meanwhile, air-bound platforms are very expensive and therefore not available to the larger research community especially in developing countries such as Nigeria. The shortcomings of the ground and air-bound platforms underscore the choice of satellite as the platform of interest for data acquisition in this study. Winker et al. (2010) reported that satellite aerosol remote sensing has been applied widely in recent researches in atmospheric studies; this was exploited in order to arrive at a holistic overview of the aerosol situation in the country. Before now, most aerosol researches in Nigeria had applied various techniques to study aerosol dynamism but most studies made use of data from synoptic stations and aerosol measuring platforms. Anuforom (2007) for example applied data from synoptic stations observed in Nigeria to determine that aerosol particle size decreases from north to south during Harmattan, while Nwofor (2010) observed that extinction growth is enhanced by increased relative humidity (RH) in Ilorin. Elsewhere, Skupin, Ansmann, Engelmann, Baars and Müller (2014) observed that Aerosol Optical Depth (AOD) also increases with RH as a result of particle size (growth) enhancement and chemical composition (Winkler, 1988; Okuda et al., 2004; Sun et al., 2013). The increase in RH results in increased aerosol particle extinction (Cheng et al., 2015; Liu et al., 2012; Malm & Day, 2001; Sun et al., 2013), which implies that soluble particles take up water vapour, increasing the particle size in the atmosphere and inhibiting visibility (Liu et al., 2012; Malm & Day, 2001) causing air pollution (Goyal and Sidhartha, 2003; Tsai, 2005). The increase in particle growth is mostly as a result of increased urbanization due to vehicular emission, fossil-fuel combustion and industrialization (see Chen et al., 2012; Goyal and Sidhartha, 2003; Tsai, 2005; Zhang, Wang, Sheng, Kanai, & Ohta, 2004), depending on the meteorological condition (Zujev & Krekov, 1986). Zhou et al. (2014) and Qu, Wang, Zhang, Wang and Sheng (2015) further reported that increased atmospheric extinction as a result of RH also depends on different particle sizes, while Zhang and Mcmurry (1992), Tang (1996), Mcmurry (2000), Tang (1996) and Zhang, Shen, Li, Zhu and Yu (2017) opined that extinction can increase by more than 100% when RH exceeds 70–80%. Despite all these studies, the comprehensive knowledge about aerosols in Nigeria are limited to where NASA AERONET station is located and some academic environments where *in-situ* measurements are carried out as the need arises. Therefore, this research paper is the first comprehensive study of aerosols in Nigeria using satellite remote sensing, AERONET and meteorological data to create models which can be replicated in other parts of the country to segregate between coarse and fine aerosols. Hence, in this work, aerosol optical of the entire country was mapped and its relationship with atmospheric extinction and RH was also examined.

2. Materials and methods

2.1. Study location

Nigeria is located in West Africa on the Gulf of Guinea and has a total area of 923,768 km^2 (356,669 sq. mi). It lies between latitudes 4° and 14°N, and longitudes 2° and 15°E. The far south is defined by its tropical rainforest climate, where annual rainfall is 60–80 in (1,524–2,032 mm) a year. In the southeast stands the Obudu Plateau. Coastal plains are found in both the southwest and the southeast. This forest zone's most southerly portion is defined as salt water swamp, also known as a mangrove swamp because of the large amount of mangroves in the area. Everything in between the far south and the far north is savannah and rainfall is between 20 and 60 in (508 and 1,524 mm) per year. Also, the amount of rainfall decreases as one moves northwards. It is also made up saline water towards the coastal region and fresh water at the hinterlands. The southern region of Nigeria experiences a double rainfall maxima characterized by two high rainfall peaks, with a short dry season and a longer dry season falling between and after each peaks. The first rainy season begins around March and lasts to the end of July with a peak in June. This rainy season is followed by a short dry break in August known as the August break which is a short dry season lasting for 2–3 weeks in August. This break is broken by the short rainy season starting around early September and lasting to mid-October with a peak period at the end of September. The ending of the short rainy season in October is followed by Long Dry Season. This period starts from late October and lasts till early March with peak dry conditions between early December and late February. Nigeria has a population of 160 million people according to the 2006 census (NPC, 2009).

The study sample locations are Sokoto, Maiduguri, Abuja, Ilorin, Ikeja and Port Harcourt (PH) (see Figure 1). These locations were selected based on their characteristic economic and geographical vantage locations. Sokoto and Maiduguri are located within the Sahel, but locations are less industrialized and mostly agrarian. Abuja and Ilorin are located within the middle belt of Nigeria which is the transition belt between the savannah and rain forest. The vegetation is characterized mainly by scanty forest and tall grasses known as guinea savannah. PH and Ikeja are located within the fringes of the Atlantic Ocean and are characterized by huge industries (such as oil companies, fertilizer, paint manufacturing, food process, pesticides, wood processing, petrochemicals plants etc.), which are sources of anthropogenic aerosols. Generally, meteorological seasons

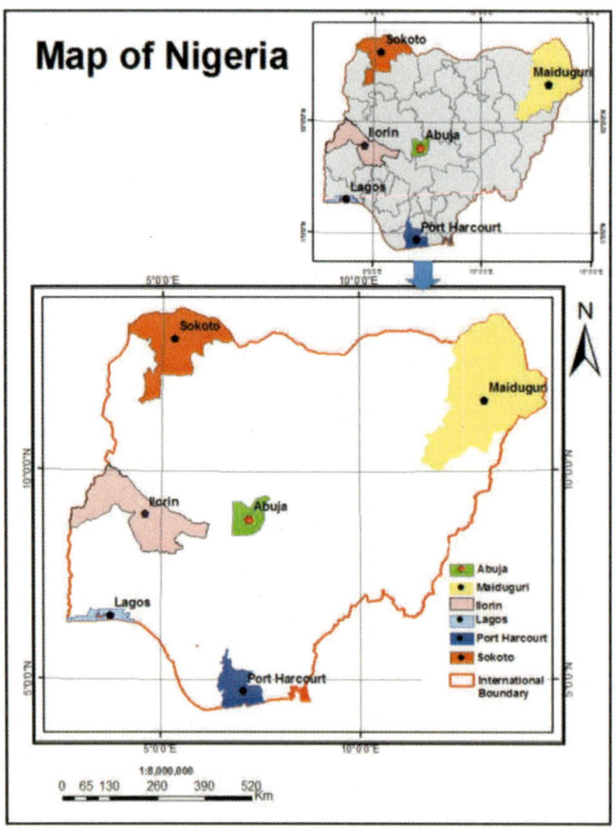

Figure 1. Study area map.

in Nigeria are categorized into dry and rainy season. The dry season begins in November and lasts till March each year, while the raining season starts in April and lasts till October in most part of the country. The rainy season starts earlier in the middle belt and south than in the Sahel. RH in the south and middle belt is large in both seasons compared to the upper moist and dry Sahel areas. Low humidity of <60% is experienced during the Harmattan, when dust is transported southwards from the Sahara and Bodele depression (Nwofor, 2010).

2.2. Data set and instruments

Version 2 aerosol optical depth retrievals from AERONET Ilorin measured with CIMEL sun photometer (quality assured Level 2 monthly averaged data) (Holben et al., 1998) were used for this study, which was also correlated with MODIS Level 2 data over the same area at NASA AEROSTAT web page. Also, the AERONET fine AOD data were used together with AIRS carbon monoxide (CO) to ascertain the ratio of coarse-to-fine aerosol loading.

MODIS Terra monthly mean AOD at a spatial resolution of 1°× 1 was obtained for the period 2003–2012 from NASA at http://disc.sci.gsfc.nasa.gov/giovanni/. These data were used to study the annual, seasonal and cumulative aerosol loading over the entire country.

This study also used observations from Cloud-Aerosol Lidar with Orthogonal Polarization, Level 2 versions 3.01 and 3.02 profile product consisting of aerosol subtype and vertical mask feature. These data were used to identify the various aerosol subtypes of dust, polluted dust (PD), polluted continental (PC) and smoke. Further literature on CALIPSO instrument configuration and sensor characteristics can be found in Winker, Hunt and McGill (2007) and Winker, Pelon and McCormick (2003).

The data used in this study are monthly values of meteorological visibility (at 9 km altitude) for 10 years (2003–2012) and RH (%) measured at the PH, Ikeja, Abuja, Ilorin, Maiduguri and Sokoto NIMET stations. The following mathematical expressions were used to computing the statistics in Section 3.

$$\text{Average} = \frac{\text{total AOD}}{\text{no years}} \tag{1}$$

$$\%\ \text{difference} = \frac{T_{rf} - T_n}{T_{rf} + T_n} \times \frac{100}{1} \tag{2}$$

T_{rf} is the total loading for reference location and T_n is the total loading for other location n. Equation (2) was replicated in order to calculate the results in Tables 2.

$$\text{Cumulative loading} = \text{total sum annual values} \tag{3}$$

Furthermore, Figure 2 shows the schema of how the various data were integrated to arrive at the results. Finally, all the data sets used in this study are monthly averages which took care of several gaps inherent in daily data acquisitions.

3. Results and discussions

3.1. Cumulative aerosol loading

The result shows that the central locations of Abuja and Ilorin had CO values of 24.8×10^{18} and 24.35×10^{18} molecules/cm^2, respectively, which implies that these locations had the highest volume of cumulative carbonaceous aerosols for the entire study period when compared with the other locations sampled, followed by PH and Maiduguri at 23.48×10^{18} and 23.23×10^{18} molecules/cm^2, respectively, then 22.1×10^{18} molecules/cm^2 for Ikeja and 22.33×10^{18} molecules/cm^2 for Sokoto, respectively. It was observed that locations with the highest CO values recorded lower aerosol absorption index (AI) and AOD values; these high values are attributable to the type of anthropogenic activities within those locations. The carbonaceous aerosol in Abuja can be attributed to atmospheric transport of particulates (see Abiye et al., 2014) from BB and industrial activities from

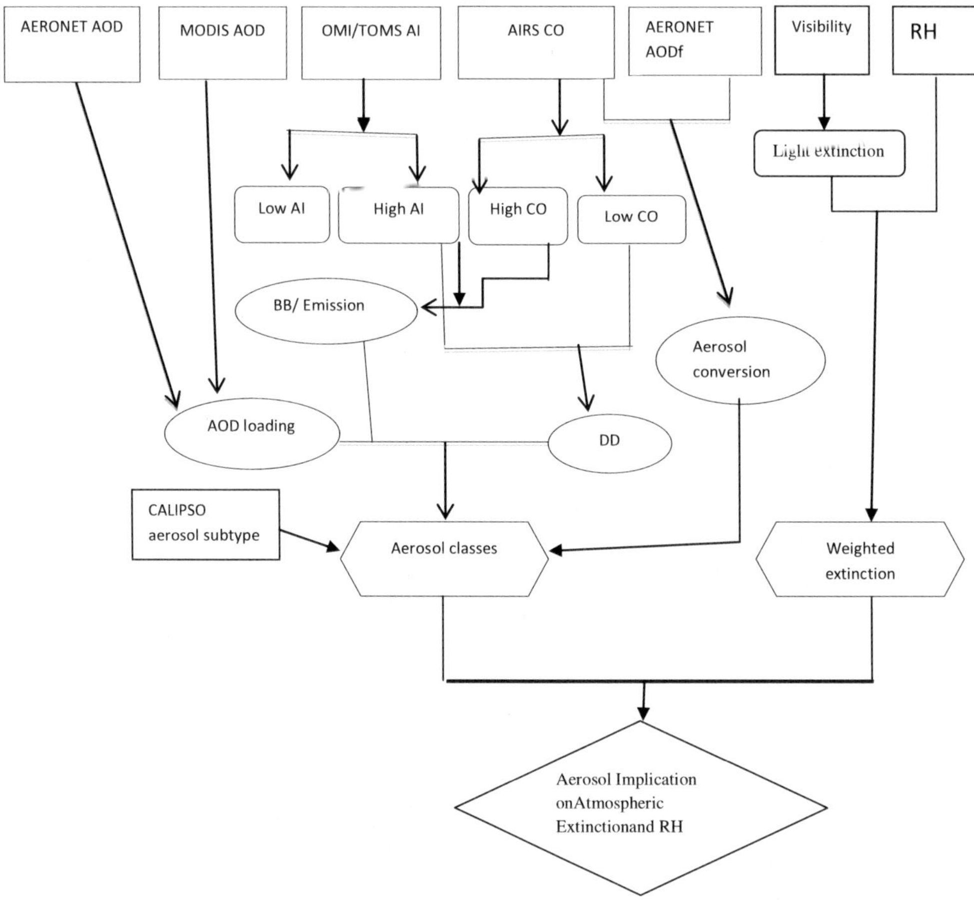

Figure 2. Work flowchart for data integration.

the surrounding localities, which were higher than the value obtained for Lagos by Eze et al. (2012). Ikeja which had the least CO value was observed to have recorded the highest total value of 5.2 for the AOD followed by PH at 4.91, while Ilorin and Abuja recorded 4.4 and 4.25, respectively. Subsequently, Sokoto and Maiduguri had AOD values of 3.47 and 3.99, respectively.

3.2. Aerosol loading trend

The results from scatter plots show that the aerosol loading in all the study sites for the period under review was on the decrease (table not shown). This is indicated by the negative values of the slopes of the individual parameters. The result obtained from the calculated annual mean shows a decreasing trend in AOD, absorbing and carbonaceous aerosols in all the study locations except for PH where the AOD showed a positive trend. This indicates that absorbing aerosols are not the only contributor to the aerosols loading in PH. By implication, non-absorbing aerosols contribute substantially to the aerosol loading in PH, as a result of the various industrial activities in PH such as shipping (which crates bubbles that bust and emit sea salt particles into the atmosphere) and gas flaring. For instance, Nigeria flared a total of 575 billion cubic feet (BCF) of gas in 2007 which was reduced to 515 BCF in 2011 (EIA, 2015), which makes Nigeria the second largest gas flaring nation in the world after Russia. This was a supporting factor for decreasing aerosol trend in Nigeria. This is in line with the trend (extract from slope = S) obtained from AERONET observation at Ilorin (for AOD [AERONET −0.002, satellite −0.010] and fine AOD [AERONET −0.0016, satellite (CO) −0.011]). Therefore, we considered the use of CO to infer the presence of fine mode aerosols which have sources from biomass, fossil combustion and other industrial activities. Furthermore, the aerosol loading in the various locations during the dry season for the period under study was compared using Ikeja as the reference location. The results obtained showed that the aerosol loading in Ikeja was greater than those in other locations by varying degrees with the exception of CO values in Ilorin and Abuja which were higher by 1.31% and 3.57%, respectively, despite that the AOD and AI in Ikeja still remain higher by13.37% (AOD) and 2.57% (AI) in Ilorin and 15.77% (AOD) and 3.22% (AI) in Abuja. A breakdown of loading shows that in PH, it varied by 2.68% for AOD, 1.58% for AI and 2.13% for CO. In Sokoto and Maiduguri, the aerosol loading variations were very obvious, 42.05% (AOD), 11.09% (AI) and 6.75% (CO) in Sokoto and 33.63% (AOD), 9.09% (AI) and 6.75% (CO) in Maiduguri (see Table 1 for % variations with respect to the other locations).

The reasons for the spatial variation of aerosol can be attributed to the prevailing wind regime (see Figures 3 and 4), in which during the dry season, fine dust aerosols are transported from Sahara Desert and Bodele depression down to the Gulf of Guinea. The coarse particles are therefore deposited mainly in the north (Sokoto and Maiduguri). It should be noted that the dry season is the period when dust and smoke from BB from parts of Sahara and West African Sahel are transported downwind by the north-east trade wind (NETW) over Nigeria to the Gulf of Guinea.

During the rainy season, it was observed that the pattern of aerosol distribution between the northern and southern parts of the country was reversed (see Table 2). The result indicates that Sokoto and Maiduguri had more cumulative aerosol (AOT) and absorbing aerosol (AI) than any other location studied but less carbonaceous aerosols. The reason for the high AI and AOT, in the north compared to the south, was a result of non-intrusion of dusts from the deserts by the counteraction of the NETW by the south-west trade wind (SWTW) as was observed in Figure 5. (It should be noted that the SWTW brings in moisture from the Atlantic Ocean which wets the soil and at the same time reduces the rate at which dusts are buoyance in the south.) The constant carbonaceous aerosols loading in the south indicate continue industrial activity all year round.

3.3. Seasonal variation of aerosol loading

The result on seasonal analysis of aerosol loading analysis is (see table not shown) that it displays the cumulative seasonal aerosol loading difference between the dry and rainy seasons for each location. The results imply that aerosol loading increases from south to north during the dry season, while the loading decreases from south to north during the rainy season. The magnitude of the seasonal variations indicates that PH, Ikeja, Ilorin and Abuja had decreased AOD by 26.55%, 25.52%, 12.39% and 10.16%, respectively, while it increased by 20.75% in Maiduguri and 24.29% in Sokoto. Subsequently, the values for AI reduced significantly, for PH by 24.12%, Ikeja 20.73%, Ilorin 9.39% and 12.44% for Abuja, while it increased in Maiduguri and Sokoto by 11.88% and 12.46%, respectively. Meanwhile, CO values did not show any significant decrease, this was either as a result of long residence time of CO compared to aerosols or the continuous emissions of CO from the sources such as factories, automobile combustion, generating sets and gas flaring. This implies that these figures may not necessarily mean increased CO emission but do not rule out CO contributions from the continuous emission source irrespective of the season. In Maiduguri and Sokoto, where CO increased by 2.60% and 3.64% respectively could be as a result of biomass emission between May and July hence the northern part of the country experience late onset of rain. Furthermore, this could also be as a result of aerosol transport from Egypt, Libya, Chad, Europe and the Atlantic (see Figure 4).

Subsequently, the restriction of dust particles to Maiduguri and Sokoto was result of the influence of the tropical maritime air mass (TMAM) which counteracts the tropical continental air mass (TCAM) around latitude 11°N as shown in Figure 5, thereby creating a wind flow in perpendicular direction. This gave rise to the concentration of both coarse and fine dust particles to Maiduguri and Sokoto, which conversely increased the aerosol loading and decreased the moisture-laden TMAM for a longer period. In the south and the central parts of Nigeria, this interaction between the TMAM and the TCAM ensures that dust particles from the active dust sources in the Sahel are restricted to northern fringes of the country. The trend of aerosol loading extracted from the scatter plot for this season also showed that AOD increased in PH, Ikeja and Ilorin. Generally, this result implies that there is an enormous difference in the aerosol loading between the

Table 1. Percentage difference of aerosol loading in the various locations during the dry season

Ikeja		% Diff	Port Harcourt		% Diff	Ilorin		% Diff	Abuja		% Diff	Sokoto		% Diff
Ilorin	AI	2.57	Ilorin	AI	0.99	Abuja	AI	0.64	Maiduguri	AI	5.88	Maiduguri	AI	−2.02
	CO	−1.31		CO	−3.45		CO	−2.26		CO	9.67		CO	−0.62
	AOT	13.37		AOT	10.73		AOT	2.44		AOT	18.86		AOT	−9.81
Port Harcourt	AI	1.58	Abuja	AI	1.63	Maiduguri	AI	6.52	Sokoto	AI	7.90			
	CO	2.13		CO	−5.70		CO	7.43		CO	10.30			
	AOT	2.68		AOT	13.14		AOT	21.21		AOT	28.15			
Abuja	AI	3.22	Maiduguri	AI	7.51	Sokoto	AI	8.54						
	CO	−3.57		CO	3.99		CO	8.05						
	AOT	15.77		AOT	31.23		AOT	30.39						
Maiduguri	AI	9.09	Sokoto	AI	9.52									
	CO	6.12		CO	4.61									
	AOT	33.63		AOT	39.82									
Sokoto	AI	11.09												
	CO	6.75												
	AOT	42.05												

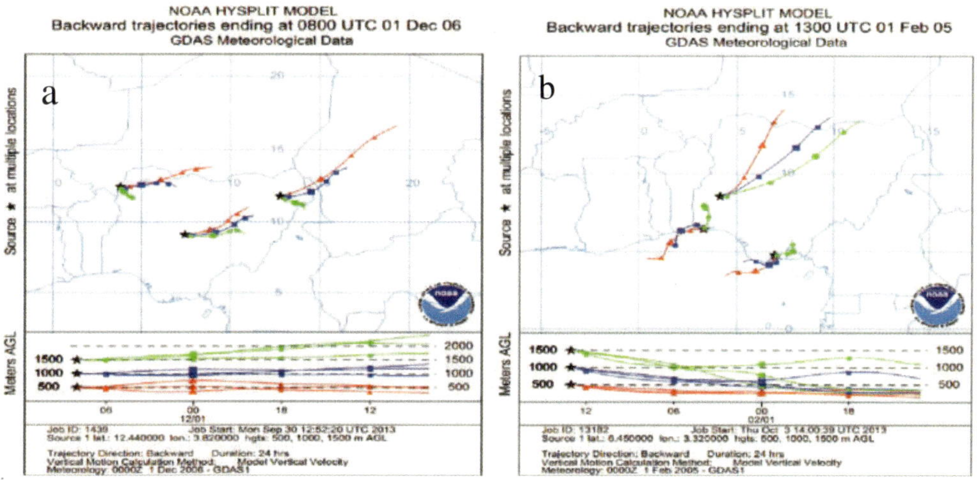

Figure 3. HYSPLIT 7 days back trajectory for the dry season (a) Abuja, Maiduguri, Sokoto (b) Ikeja, Port Harcourt, Ilorin.

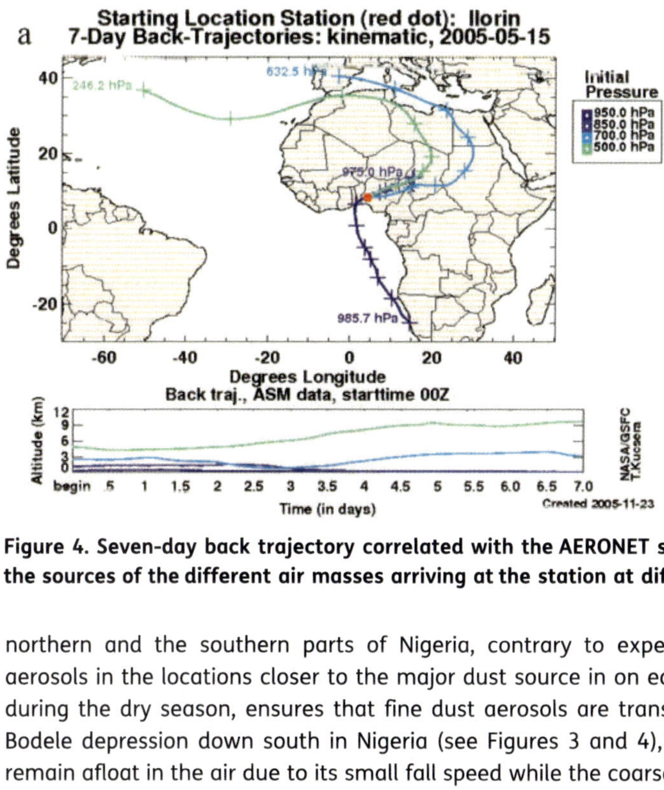

Figure 4. Seven-day back trajectory correlated with the AERONET station at Ilorin (red dot) showing the sources of the different air masses arriving at the station at different isobars in May 2005.

northern and the southern parts of Nigeria, contrary to expectation that there would more aerosols in the locations closer to the major dust source in on earth. The prevailing wind regime, during the dry season, ensures that fine dust aerosols are transported from Sahara Desert and Bodele depression down south in Nigeria (see Figures 3 and 4), in which the fine dust particles remain afloat in the air due to its small fall speed while the coarse particles are deposited close to locations in the north. Anuforom, Akeh, Okeke and Opara (2007) further noted that the dry season is the period when dust and smoke from BB from parts of Sahara and West African Sahel are transported downwind of the NETW over Nigeria to the Gulf of Guinea. Furthermore, it is pertinent to note that AOD loading in Nigeria has been reported to be the highest compared to the other adjoining locations within the Sahel such as Cape Verde, Banizoumbou and Agoufou. Oluleye, Ogunjobi, Bernard, Ajayi and Akinsanola (2012) reported aerosol loading from different platforms for various locations in Africa for a period ranging from 2005 to 2009, Mail (Agoufou) (AERONET: 0.52, TOMS: 0.58, MODIS Terra: 0.47), Niger (Banizoumbou) (AERONET: 0.57, TOMS: 0.85, MODIS Terra: 0.69), Cape Verde (AERONET: 0.37, TOMS: 0.84, MODIS Terra: 0.37), Nigeria (Ilorin) (AERONET: 0.75, TOMS: 1.07, MODIS Terra: 0.88). The high AOD loading in Nigeria is attributable to the following reasons.

Table 2. Percentage difference of rainy season aerosol loading for the different locations

Ikeja		% Diff	Port Harcourt		% Diff	Ilorin		% Diff	Abuja		% Diff	Maiduguri		% Diff
Ilorin	AI	-9.0	Ilorin	AI	−14.09	Abuja	AI	3.73	Maiduguri	AI	−18.35	Sokoto	AI	1.43
	CO	1.13		CO	−0.53		CO	0.25		CO	0.69		CO	−0.41
	AOT	−0.18		AOT	−3.97		AOT	0.18		AOT	−12.1		AOT	6.10
Port Harcourt	AI	5.16	Abuja	AI	−10.42	Maiduguri	AI	−14.72	Sokoto	AI	−16.96			
	CO	1.66		CO	−0.27		CO	0.95		CO	0.27			
	AOT	3.78		AOT	−3.78		AOT	−11.92		AOT	-6.04			
Abuja	AI	−5.28	Maiduguri	AI	−28.23	Sokoto	AI	−13.31						
	CO	1.39		CO	0.41		CO	0.53						
	AOT	−8.41E−15		AOT	−15.82		AOT	−5.86						
Maiduguri	AI	−23.41	Sokoto	AI	−26.90									
	CO	2.08		CO	5.59E−15									
	AOT	−12.10		AOT	−9.81									
Sokoto	AI	−22.05												
	CO	1.66												
	AOT	-6.04												

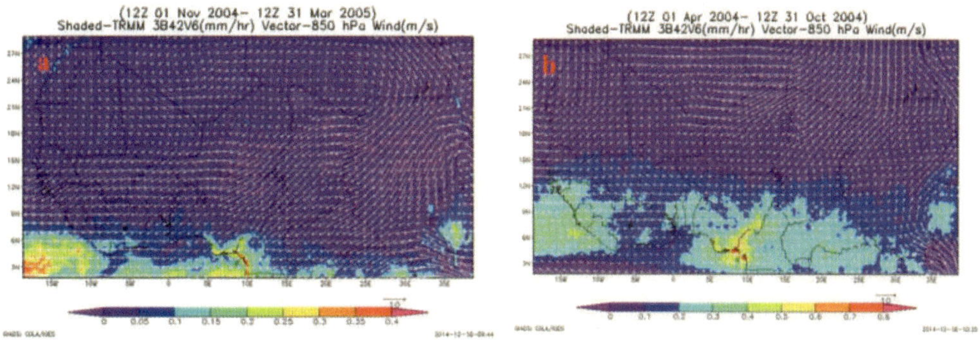

Figure 5. Wind speed and direction generated from NASA Hurricane Analysis Tool, the vectors shows the wind direc-tion while the length of the arrow stands for the magnitude of the wind speed during the (a) dry season (b) rainy season.

Large quantities of dust and plant debris are easily lifted locally from the surface in the dry season; this could also account for the differential aerosol loading within Nigeria compared with the neighbouring countries.

The high CO despite less BB implies that the influence of accumulation mode aerosols in the seasonal flux patterns is supported by the high industrialization and traffic density of Ikeja and PH.

The prevailing wind pattern in West Africa as shown in Figure 4 shows that the wind speed reduced drastically when in Nigeria, giving rise to a unique wind pattern in Nigeria different from the neighbouring countries. D'Almeida (1986) observed that about 60% by mass of the dust produced in the Sahara and the surrounding region is transported southward to the Gulf of Guinea, 28% westward to the equatorial North Atlantic Ocean and 12% northward to Europe. This implies that dust particles lifted from the Bodele depression that passes through Nigeria are more or less deposited in the country. This phenomenon could probably be responsible for the continuous high AOD loading in Nigeria reported by Oluleye et al. (2012).

Increase population and land-use intensification in the zone due to oil and gas exploration. It is also possible that anthropogenic activities which affect land surface conditions may also contribute to high aerosol loading in Nigeria. Onyeuwaoma et al. (2015) further showed that high population density results in increased aerosol emission in southern Nigeria.

3.4. Relationship between CO, fine aerosol and coarse aerosol

The fine AOD from AERONET and CO from AIRS for Ilorin were used to create a relationship between CO and fine aerosols (AODf), at a correlation of 0.67. A linear regression analysis was performed to determine how CO levels may predict fine aerosol concentrations. Once this relationship was established, the prediction function was applied to the original (aerosol data) data set to transform the CO readings into fine aerosol readings. Furthermore, two models (see Equations 4 and 5) were created, one to assess the relationship between CO and fine aerosols and the other predict fine aerosol levels for Ilorin.

The model equations are

$$f = 0.0627 + 0.047 \times \mathrm{CO} + \beta_{\mathrm{month}} \tag{4}$$

$$f = -0.322 + 0.234 \times \mathrm{CO} \tag{5}$$

The predicted fine aerosol was compared with the AERONET fine aerosol and the result is shown in Figure 6. The black line is the raw CO data, the red line is the raw fine aerosol data and the blue line represents the prediction. The closer the red to the blue lines, the better the prediction.

Furthermore, Equation (4) was used to generate the ratio of fine-to-coarse aerosol for all the locations as shown in Figure 7. The ratio of the coarse-to-fine aerosols shows that coarse aerosols

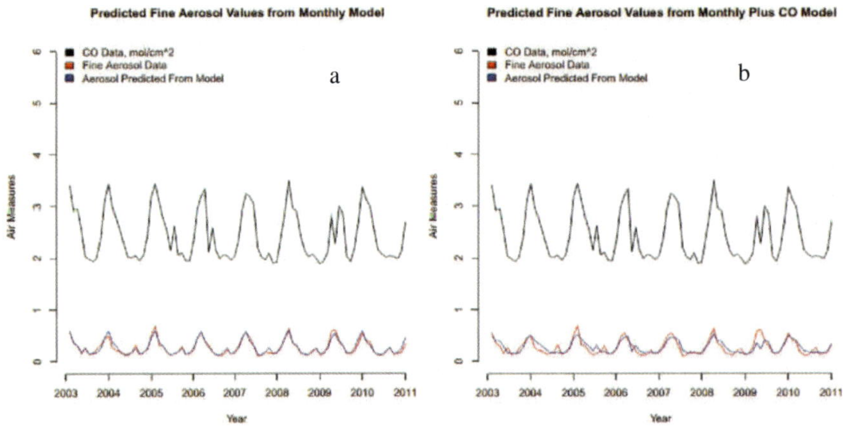

Figure 6. The results from two different models used in con-verting CO to fine aerosols. The models were run from data acquired from AERONET Ilorin and CO from AIRS for the same location.

prevail more in the north (Maiduguri and Sokoto) ranging from 7:1 to 14:1 during the rainy seasons and reducing to 3:1 during the dry season. In Abuja and Ilorin, the ratio of coarse-to-fine aerosols ranged from 11:1 to 2:1. Meanwhile, the situation in PH and Ikeja shows that the coarse-to-fine aerosol ranged from 11:1 to 1.5:1. This result indicates that coarse aerosols dominate the air space in Nigeria, but the magnitude of the dominance decreases from north to south.

The trend line for Ikeja shows an ever-increasing fine aerosol loading which was a result of increased industrialization, fossil energy consumption from generators, vehicles and domestic cooking. This is because Ikeja has the highest volume of industries in Nigeria, population density and vehicular activities. The curve for Abuja maintains a relative constant ratio between the coarse and fine aerosol loading indicated by the steady trend. The trend line also shows that coarse aerosols increased from 2003 to 2004 and then drop in 2005 while it maintained a steady flow thereafter, from 2005 to 2012. The fine aerosols are aerosol originate majorly from local construction, vehicular emission and biomass from burning from neighbouring locations during the dry season (see Abiye et al., 2014). The curve for Ilorin maintained the same pattern with Abuja, but with BB as the major source of the fine aerosol; hence, Ilorin is not industrialized but mostly agrarian. The curves for Sokoto and Maiduguri show that the volume of coarse aerosol increased between 2003 and 2004 and maintains a steady level thereafter.

Subsequently, the CALIPSO image (figure not shown) shows that in all the locations sampled, the characteristic aerosols are basically dust, PD and smoke (BB aerosol), while the PC occurred in PH and Ikeja. Ezeh et al. (2014) and Nie et al. (2014) reported that PDs occur as a result of anthropogenic activities which may have resulted from industrial or vehicular emissions. Meanwhile, Abuja and Maiduguri recorded the absence of smoke during the rainy season. Generally, the vertical height of the aerosols drastically decreases during the rainy seasons as a result of wetness of the soil, vegetation (which are not easily ignited during the rainy season) and reduced influence of NETW in the country during the rainy season. Interestingly, there is no diurnal variation in the aerosols captured during the day and night overpass by the CALIPSO. This implies that the night and day overpass captures the same aerosol types for the same location at the same period.

3.5. Implications of findings on atmospheric extinction and RH

The analytical relation between humidity and light extinction was evaluated to make more deductions on the evaluated seasonal visibility averages. In order to improve the correlation, the visibility data were first converted to atmospheric extinction values using the equation by Middleton (1952) (see Equation (6)).

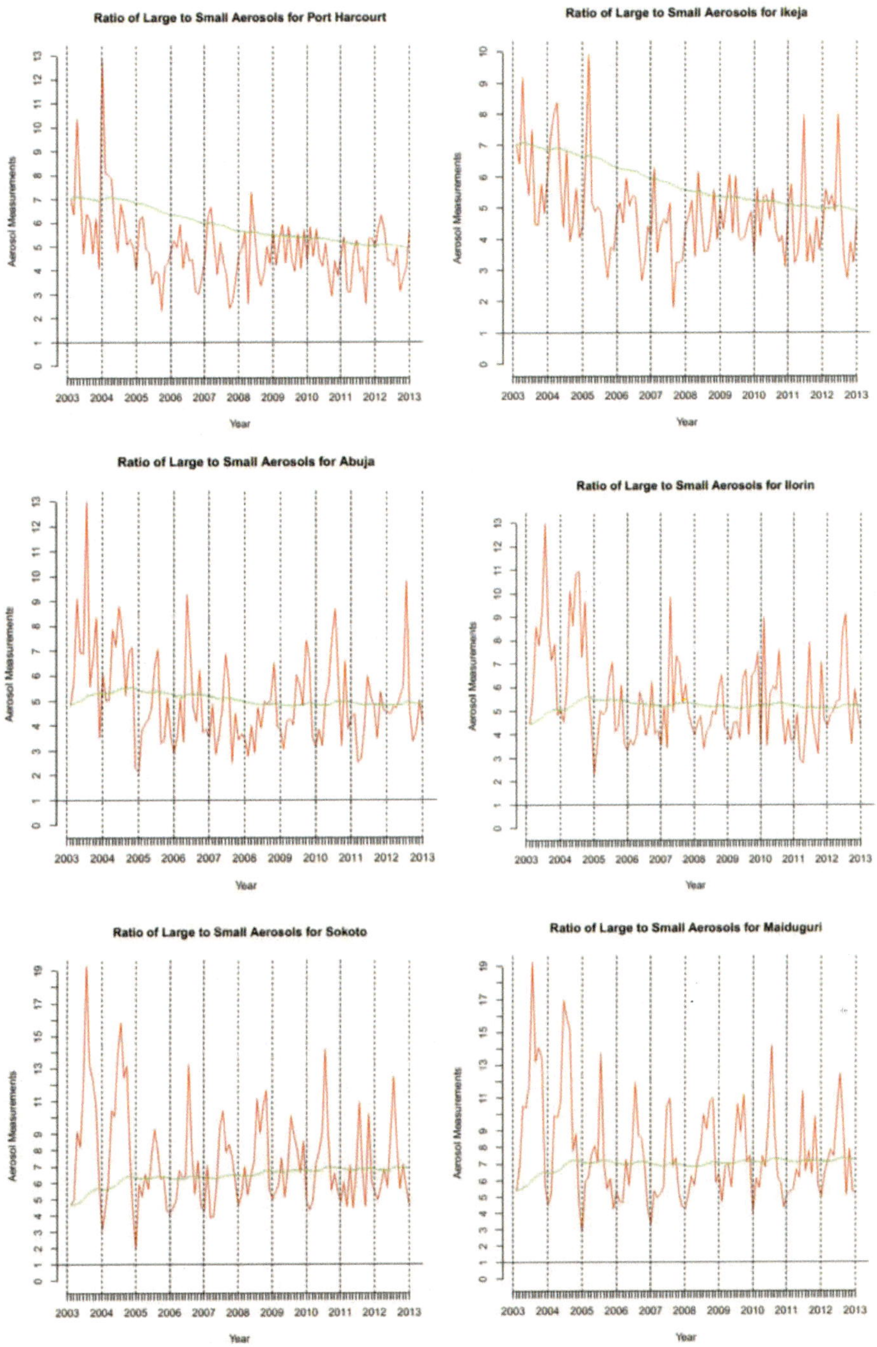

Figure 7. Curve showing the ratio of coarse-to-fine aerosol in Ilorin, which indicates the volume of the prevailing aerosol.

$$\mathrm{Rv} = \frac{3.912}{\sigma} \tag{6}$$

where Rv is the visibility averages and σ is the atmospheric extinction.

These extinction values were then correlated with RH data for the same period as shown in the correlation between the extinction parameter (σ) and the RH was made more sensitive by weighting the extinction (σ_e) values with the RH according to the expression in Nwofor (2010) (see Equation (7)).

$$\sigma_e = \frac{\sigma}{\mathrm{RH}} \tag{7}$$

σ_e values were then plotted against the aerosol modal fraction (figure not shown). The deductions made from the atmospheric extinction are as follows:

In the Sahel area, σ grows at low RH (Maiduguri RH <45%, Abuja RH >70%), Sokoto (RH >50%), compared with the southern and middle belt areas, (PH RH >88%, Ikeja >80%, Ilorin RH >80%). This shows that particles become hygroscopic based on their meteorology, location and back ground economic activity. The back trajectory in Figure 3 shows that SWTW and local sources prevail over PHC and Ikeja during both season, which suggests that there is the likelihood of sea salt influx from the Atlantic Ocean and local pollution. For the Sahel and middle belt, the NETW from the Sahara plays more critical role in bring in aerosol than the weak SWTW. This suggests that Sahara dust prevails over the north with little influence in the south.

In Maiduguri, it was observed that σ increases during the dry season at 24% RH, when dust particles dominate in the atmosphere.

The plots in Figure 8 also show a rapid decay of σ for every part of the country except for Sokoto, because particles in Sokoto tend to be more hygroscopic than in other parts of the country. Generally, σ grows more in the wet season presumably due to the influence of RH on particles in the atmosphere.

Areas where σ grows at very high RH are the most polluted due to industrial emission as was observed (table not shown). It suggests that dust particles increased in Maiduguri and Sokoto during the rainy season by 20% and 24%, respectively, while carbonaceous aerosols increased by 2% and 3% as well. Meanwhile, dust reduced in PHC and Ikeja by 26% and 25% and carbonaceous by 0.97% and 1.44%, respectively. Furthermore, Ilorin and Abuja recorded reduced dust by 12%

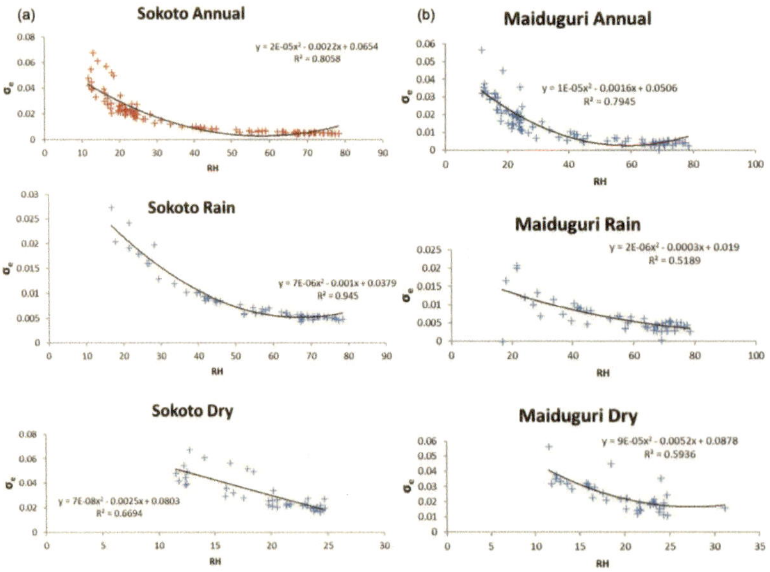

Figure 8. A cross-section of some of the CALIPSO aerosol subtype images used in con-firming the various aerosol types occurring in the different parts of the country during the various seasons. (a) Graph of the atmospheric extinction parameter versus the relative humidity (%) for Sokoto. (b) Graph of the atmospheric extinction parameter versus the relative humidity (%) for Maiduguri. (c) Graph of the atmospheric extinction para-meter versus the relative humidity (%) for Abuja. (d) Graph of the atmospheric extinction parameter versus the relative humidity (%) for Ilorin. (e) Graph of the atmo-spheric extinction parameter versus the relative humidity (%) for Port Harcourt (PHC). (f) Graph of the atmospheric extinction parameter versus the relative humidity (%) for Ikeja.

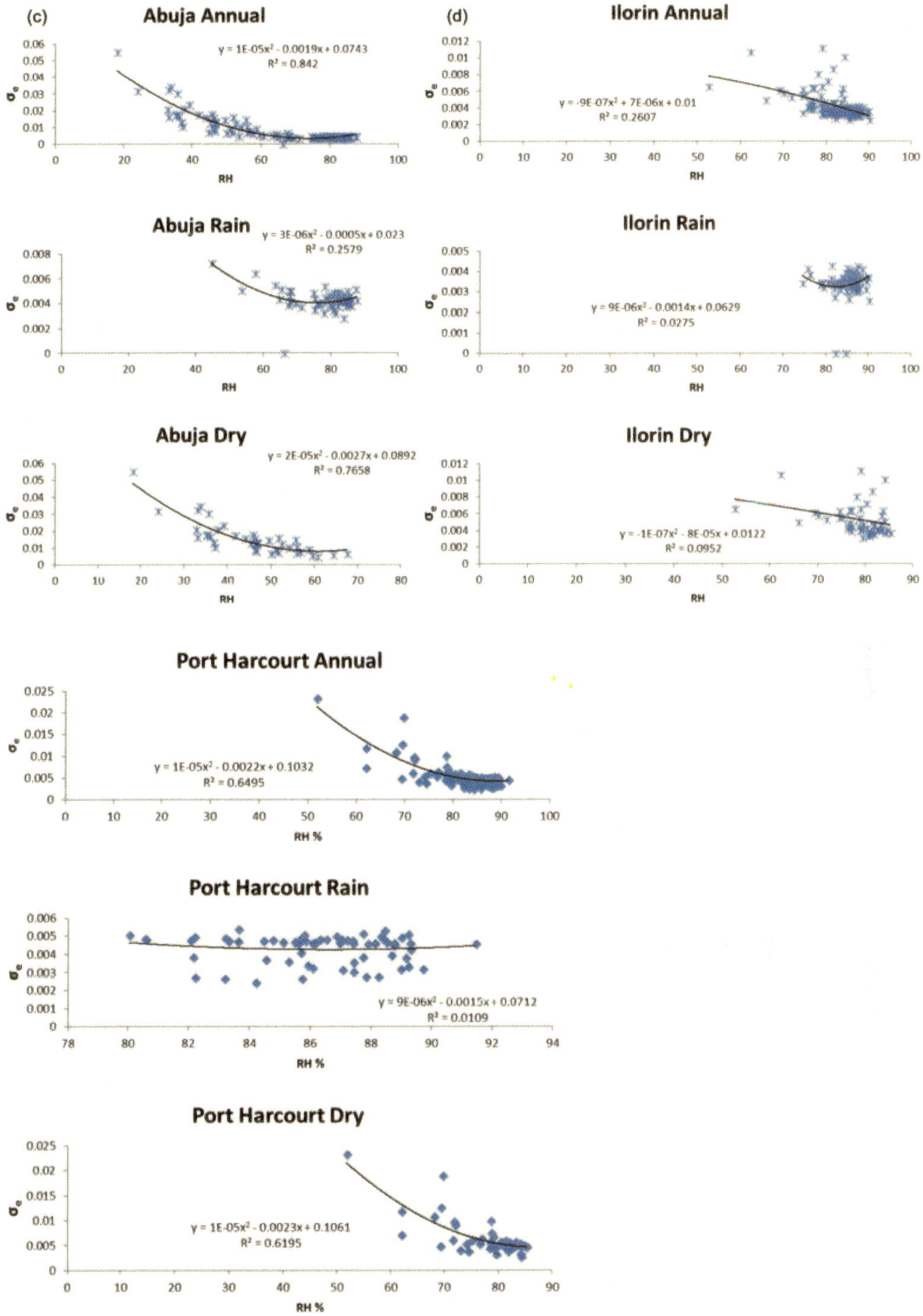

Figure 8. Continued

and 10% and reduced carbonaceous aerosol by 3% and 6%, respectively. When these results are integrated with the deductions from Figure 8, it implies that areas with more dusts and less carbonaceous aerosols do not require a lot of RH for extinction to take place.

Also, σ grows at very low RH when pure dusts are in the atmosphere during the rainy season especially within the Sahel. This can be confirmed with Figure 5a, which shows that the wind-laden dust from the Sahara does not penetrate beyond latitude 9.5°N.

3.6. Extractions from the relationship between coarse aerosol, fine aerosol and atmospheric weighted extinction

Figure 9 shows the graphs of the relationship between σ_e and fine aerosols, and σ_e and coarse aerosols. The results from the analysis as shown in Figure 9a and b indicate that during the rainy season, when the coarse aerosols optical depth is high, it corresponds to very low extinction between 0.005 and 0.025 at 0.1–0.9 coarse optical depth while the same σ_e value is elicited by low fine aerosols optical value of 0.12–0.18 for the same season. During the dry season, both the coarse and fine aerosols had the same effect on particle extinction (0.02–0.056) but at different optical ranges (0.1–0.74 for coarse and 0.25–0.55 for fine aerosols). Furthermore, the annual plots for both the coarse and fine aerosols show that fine particles impact more on σ_e than coarse aerosols at a lower optical depth.

Figure 9c and d shows that annually both aerosol species had the same effect on σ_e, which grows with decreasing optical depth though this is more pronounced with the fine aerosols. During both the seasons, increased coarse AOD results in increased σ_e while decreased fine AOD results in increased σ_e.

In PH, the annual plots (see Figure 9e and f) for both the fine and coarse aerosols show that both particles impact σ_e equally (0.002–0.023) but at different optical depths (coarse [0.1–0.9] and fine [0.13–0.57]). This implies that at low optical depth, fine aerosols have more impact on atmospheric extinction than coarse aerosols even when the optical depth of the later is higher. Seasonally, coarse aerosols have significant impacts on σ_e from an optical depth >0.4 at a σ_e ranging from 0.004 to 0.023. The difference between the extinction by fine and coarse aerosols during both seasons is that it takes a smaller volume of fine aerosols to scatter equal amount of radiation by coarse aerosols. A comparison of the results from both seasons shows that aerosols scatter more

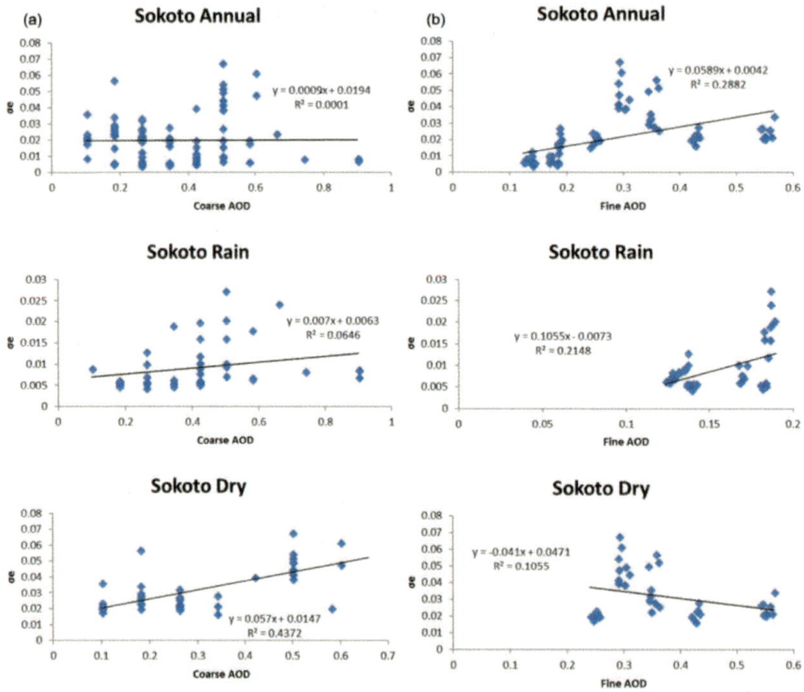

Figure 9. (a) Coarse aerosol versus atmospheric weighted extinction for Sokoto. (b) Fine aerosol versus atmospheric weighted extinction for Sokoto.(c) Coarse aerosol versus atmospheric weighted extinc-tion for Maiduguri. (d) Fine aerosol versus atmospheric weighted extinction for Maiduguri. (e) Coarse aerosol versus atmospheric weighted extinction for Port Harcourt. (f) Fine aerosol versus atmo-spheric weighted extinction for Port Harcourt. (g) Coarse aero-sol versus atmospheric weighted extinction for Ikeja.(h) Fine aerosol versus atmo-spheric weighted extinction for Ikeja. (i) Coarse aerosol versus atmospheric weighted extinc-tion for Ilorin. (j) Fine aerosol versus atmospheric weighted extinction for Ilorin. (k) Coarse aerosol versus atmospheric weighted extinction for Abuja.(l) Fine aerosol versus atmo-spheric weighted extinction for Abuja.

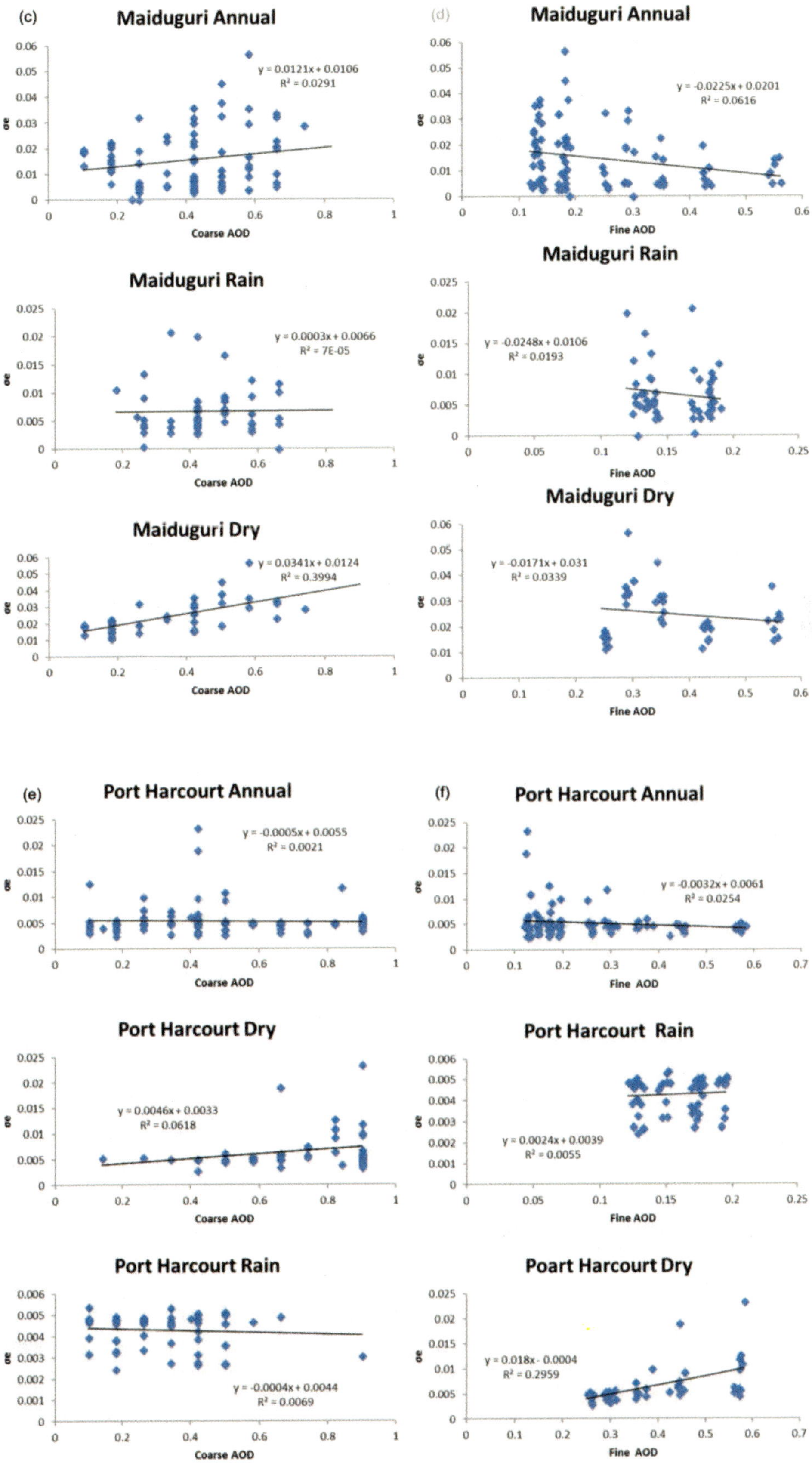

Figure 9. Continued

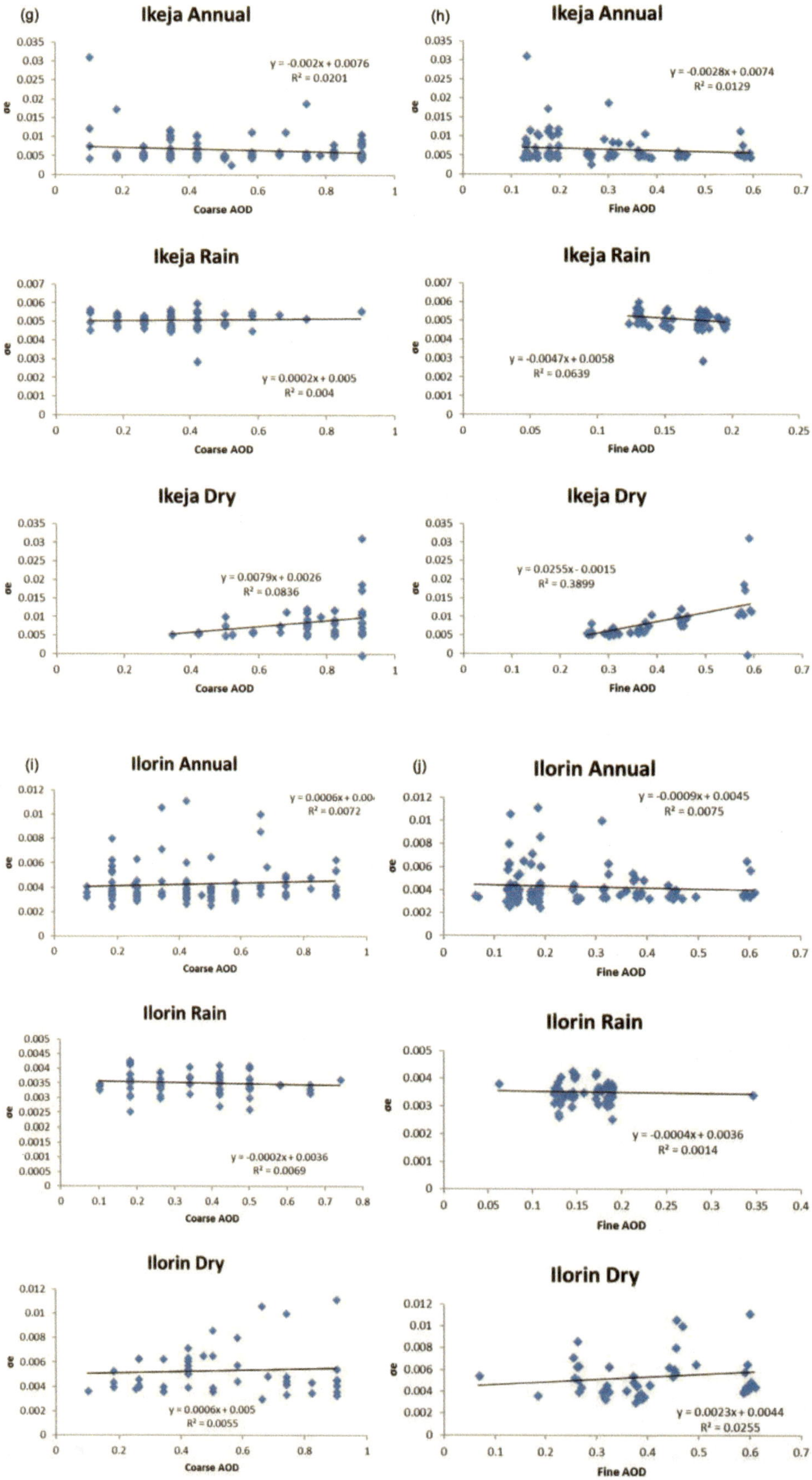

Figure 9. Continued

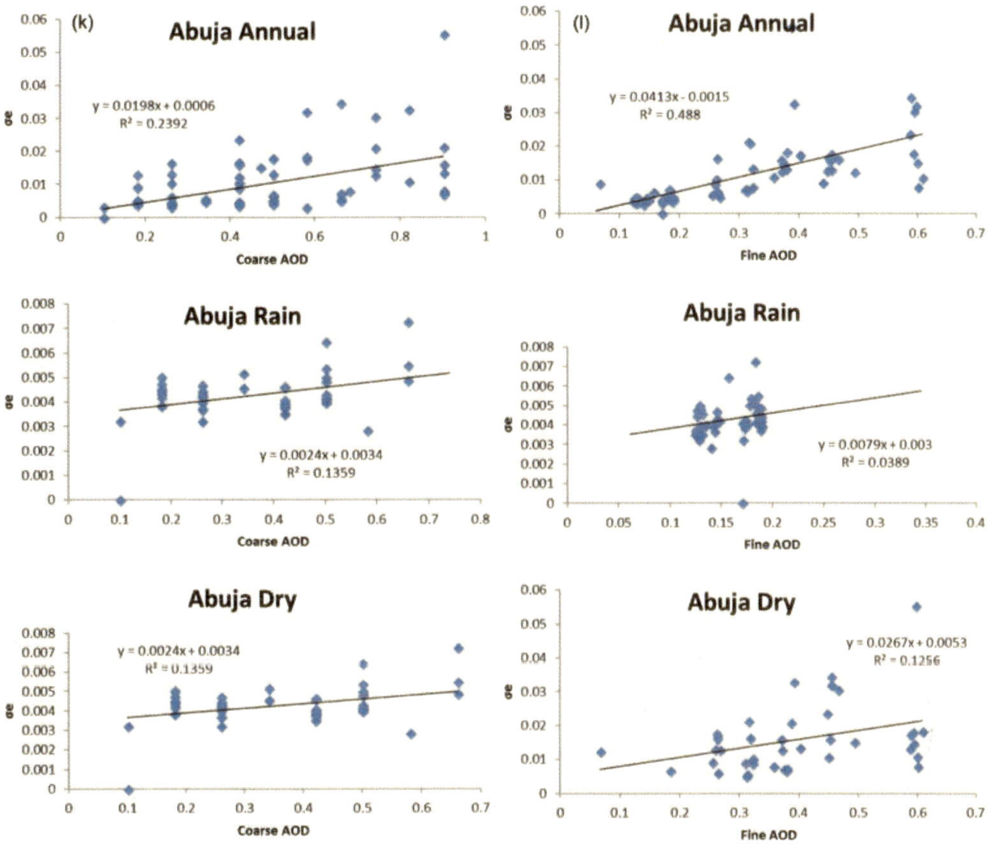

Figure 9. Continued

radiation during the rainy season than during the dry season, as shown by the low σ_e values recorded during the rainy season for both the coarse and fine aerosols.

In Ikeja, the annual coarse and fine aerosol plots (see Figure 9g and h) are such that at an optical range (0.1–0.9), a σ_e ranging from 0.004 to 0.031 was elicited. (Coarse aerosol grew exponentially, from an optical depth >0.6.) Meanwhile at lower fine optical depth, σ_e growth was pronounced. Furthermore, the role of coarse aerosol on σ_e growth is such that at low optical depth >0.4, a σ_e of 0.005–0.031 was observed. During the dry season, the coarse optical range was between 0.34 and 0.9 with the major AOD impact on σ_e at an optical depth ranging from 0.7 to 0.9, resulting in a σ_e of 0.005–0.031. The fine aerosols on the other hand increase with a decreasing σ_e during the rainy season and increase with AOD during the dry season.

The situation in Ilorin shows that σ_e grows with increasing AOD, except that during the rainy season, increased AOD (coarse and fine) results in decreased σ_e. Subsequently, in Abuja (Figure 9k and l), increased optical depth corresponds to increase in σ_e. The difference between the σ_e as a result of fine and coarse aerosols is that coarse aerosols require large optical depth to achieve the same σ_e effect at low fine optical depth.

Furthermore, the rainfall trend was plotted as shown in Figure 10. The analysis shows that over the entire study period, rainfall has been on the increase, which implies that particle growth enhances rainfall. This implies that the particles prevalent in Nigeria are good cloud condensation nuclei, which enhances rain formation at varying rates depending on their chemical composition as indicated (see also Figure 8).

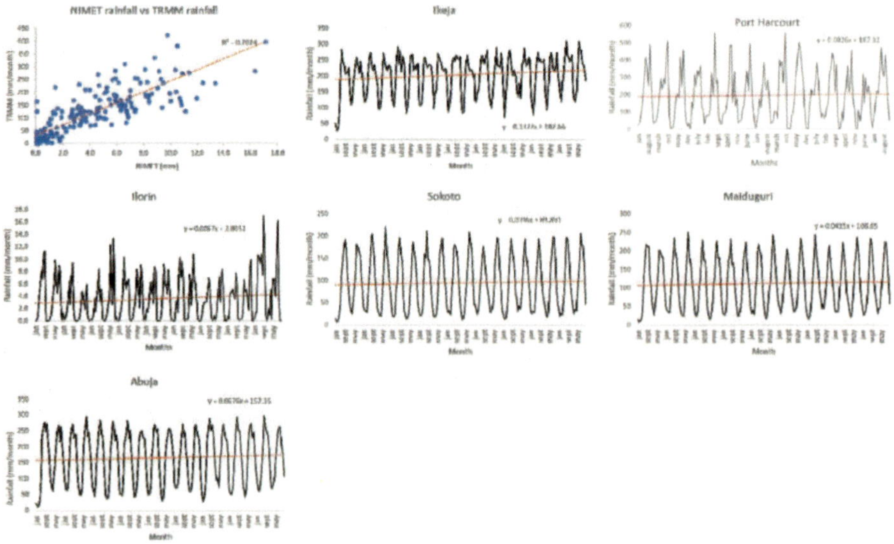

Figure 10. Correlation of ground-based data from Nigerian Meteorological Agency (NIMET) and satellite rainfall data from Tropical Rain Measuring Mission (TRMM), and rainfall trend over the entire study period. The top left graph in Figure 10 shows that ground data (NIMET) correlate well with the rainfall data acquired from TRMM at a correlation of 0.702.

4. Conclusion

This study has shown that the air quality in Nigeria is on the decrease; in the south, this is a result of industrialization and fossil-fuel combustion, while in the north, desert dust and BB is responsible for the phenomena. Therefore, it is expected that the rate of acid rain, pollution-related ailment and contamination of soil is expected to be on the increase in the south. Furthermore, this phenomenon has the likelihood to threaten food security in Nigeria; hence, increased aerosol loading affects the photosynthetically available radiation by reducing the ability of plants to engage in photosynthesis because the particles settle on the leaves and shade them from receiving enough energy from sun.

Acknowledgements

We wish to express our thanks to NASA for the provision of the various data used for this research on their various platforms; GIOVANNI, AERONET and CALIPSO. Also, we want to thank the Laboratory for Interdisciplinary Statistical Analysis and Collaboration (LISAC), O.A.U, Ile-Ife for data analysis and interpretation.

Funding

The authors received no direct funding for this research.

Competing interests

The authors declare no competing interests.

Author details

Nnaemeka Dom Onyeuwaoma[1,2]
E-mail: emekadonn@gmail.com
ORCID ID: http://orcid.org/0000-0003-0213-6166
Theo Chidiezie Chineke[2]
E-mail: chidiezie@gmail.com
Okeychukwu Kelechi Nwofor[2]
E-mail: okeynwofor@gmail.com
Ian Crandell[3]
E-mail: ian85@vt.edu
Olushina Olawale Awe[4]
E-mail: ajileyeseun@rocketmail.com
Ajileye Olasumbo[1]
E-mail: oparazanda2001@yahoo.com
Alex I Opara[5]
E-mail: olawaleawe@gmail.com
Nwachukwu Pius[6]
E-mail: piusnwachukwu@hotmail.com
Mba Tochukwu[7]
E-mail: mbatochukwu27@yahoo.com
Nwachukwu Joy[6]
E-mail: felicitas4real@yahoo.com

[1] Cooperative Information Network, NASRDA, PMB, 022, Obafemi Awolowo University, Ile-Ife, Nigeria.
[2] Atmospheric Physics Group, Department of Physics, Imo State University, Owerri, Nigeria.
[3] Department of Mathematics, Virginia Tech Institute, Blacksburg, VA, USA.
[4] Department of Mathematics, Obafemi Awolowo University, Ile-Ife, Nigeria.
[5] Department of Geology, Federal University of Technology, Owerri, Nigeria.
[6] Advanced Space Applications Laboratory, University of Uyo, Uyo, Nigeria.
[7] Regional Center for Space Science Technology Education in English, Obafemi Awolowo University, Ile-Ife, Nigeria.

References

Abiye, O. E., Obioh, I. B., Ezeh, G. C., Alfa, A., Ojo, E. O., & Ganiyu, A. K. (2014). Receptor modeling of atmospheric aerosols in Federal Capital Territory (FCT), Nigeria. *Ife Journal of Sciences, 16*(1), 107–118.

Anuforom, A. C. (2007). Spatial distribution and temporal variability of Harmattan dust haze in sub-Sahel West Africa. *Atmospheric Environment, 41*, 9079–9090. doi:10.1016/j.atmosenv.2007.08.003

Anuforom, A. C., Akeh, L. E., Okeke, P. N., & Opara, F. E. (2007). Inter-annual variability and long-term trend of UV-absorbing aerosols during Harmattan season in sub-Saharan West Africa. *Atmospheric Environment, 41*, 1550–1559. doi:10.1016/j.atmosenv.2006.08.024

Chen, J., Zhao, C. S., Ma, N., Liu, P. F., Göbel, T., Hallbauer, E., ... Wiedensohler, A. (2012). A parameterization of low visibilities for hazy days in the North China Plain. *Atmos Chemical Physical, 12*, 4935–4950. doi:10.5194/acp-12-4935-2012

Cheng, Y. C., Liang, C. S., Syu, J. Y., Chang, Y. Y., Yan, Y. L., Chen, S. J., ... Lin, W. Y. (2015). Characteristics of aerosol extinction coefficient in Taipei Metropolitan Atmosphere. *Aerosol Air Qualitative Researcher, 15*, 1823–1835.

D`Almeida, G. A. (1986). A model for Saharan dust transport. *Journal of Applied Meteorology, 25*, 903–916.

EIA (U.S Energy Information Administration). (2015) Country analysis brief: Nigeria. Retrieved March 10, 2015 from http://www.eia.gov/countries/analysis briefs/Nigeria/nigeria.pdf

Ezeh, G. C., Obioh, I. B., Asubiojo, O. I., & Abiye, O. E. (2012). PIXE characterization of PM10 and PM2.5 particulates sizes collected in Ikoyi Lagos, Nigeria. *Toxicologyand Environmental Chemistry, 94*, 884–894. doi:10.1080/02772248.2012.674133

Ezeh, G. C., Obioh, I. B., Asubiojo, O. I., Chiari, M., Nava, S., Calzolai, G., ... Nuviadenu, C. K. (2014). Elemental compositions of PM10–2.5 and PM2.5 aerosols of a Nigerian urban city using ion beam analytical techniques. *Nuclear Instruments and Methods in Physics Research, 334*, 28–33. doi:10.1016/j.nimb.2014.04.022

Goyal, P., & Sidhartha. (2003). Present scenario of air quality in Delhi: A case study of CNG implementation. *Atmospheric Environment, 37*, 5423–5431. doi:10.1016/j.atmosenv.2003.09.005

Holben, B. N., Eck, T. F., Slutsker, I., Tanré, D., Buis, J. P., Setzer, A., ... Smirnov, A. (1998). AERONET - A federated instrument network and data archive for aerosol characterization. *Remote Sensing of Environment, 66*, 1–16. doi:10.1016/S0034-4257(98)00031-5

Malm, W. C., & Day, D. E. (2001). Estimates of aerosol species scattering characteristics as a function of relative humidity. *Atmospheric Environment, 35*, 2845–2860. doi:10.1016/S1352-2310(01)00077-2

Mcmurry, P. H. (2000). A review of atmospheric aerosol measurements. *Atmospheric Environment, 34*, 1959–1999. doi:10.1016/S1352-2310(99)00455-0

Middleton, W. E. (1952). *Vision through the atmosphere.* Toronto: University of Toronto Press.

National Population Commission (NPC) (2009). 2006 Population and Housing Census of the Federal Republic of Nigeria, Priority Tables, Abuja, 347 pages.

Nie, W., Ding, A., Wang, T., Kerminen, V.-M., George, C., Xue, L., ... Kulmala, M. (2014). Polluted dust promotes new particle formation and growth. *Scientific Reports, 4*, 6634. doi:10.1038/srep06634

Nwofor, O. K. (2010). Seasonal levels of meteorological visibility at Port Harcourt Nigeria and possible links to aerosol loading and humidification. *Pacific Journal of Sciences and Technical 11*(2), 544–551.

Okuda, T., Kato, J., Mori, J., TENMOKU, M., SUDA, Y., TANAKA, S., ... YU, X. (2004). Daily concentrations of trace metals in aerosols in Beijing, China, determined by using inductively coupled plasma mass spectrometry equipped with laser ablation analysis, and source identification of aerosols. *Science of the Total Environment, 330*, 145–158. doi:10.1016/j.scitotenv.2004.04.010

Oluleye, A., Ogunjobi, K. O., Bernard, A., Ajayi, V. O., & Akinsanola, A. A. (2012). Multiyear analysis of ground-based sunphotometer (AERONET) aerosol optical properties and its comparison with satellite observations over west. *Africa Global Journal of Human Social Sciences Geology & Environment Geological Science, 12*, 31–48.

Onyeuwaoma, N. D., Nwofor, O. K., Chineke, T. C., Eguaroje, E. O., & Dike, V. N. (2015). Implications of MODIS impression of aerosol loading over urban and rural settlements in Nigeria: Possible links to energy consumption patterns in the country. *Atmospheric Pollution Research, 6*, 484–494. doi:10.5094/APR.2015.054

Qu, W. J., Wang, J., Zhang, X. Y., Wang, D., & Sheng, L. F. (2015). Influence of relative humidity on aerosol composition: Impacts on light extinction and visibility impairment at two sites in Coastal Area of China. *Atmospheric Research, 153*, 500–511. doi:10.1016/j.atmosres.2014.10.009

Skupin, A., Ansmann, A., Engelmann, R., Baars, H., & Müller, T. (2014). The Spectral Aerosol Extinction Monitoring System (SÆMS): Setup, observational products, and comparisons. *Atmos Measurement Technical, 7*, 701–712. doi:10.5194/amt-7-701-2014

Sun, Y. L., Wang, Z. F., Fu, P. Q., Jiang, Q., Yang, T., Li, J., & Ge, X. L. (2013). The impact of relative humidity on aerosol composition and evolution processes during wintertime in Beijing, China. *Atmospheric Environment, 77*, 927–934. doi:10.1016/j.atmosenv.2013.06.019

Tang, I. N. (1996). Chemical and size effects of hygroscopic aerosols on light scattering coefficients. *Journal of Geophysical Research, 101*, 19245–19250. doi:10.1029/96JD03003

Tsai, Y. I. (2005). Atmospheric visibility trends in an urban area in Taiwan 1961-2003. *Atmospheric Environment, 39*, 5555–5567. doi:10.1016/j.atmosenv.2005.06.012

Winker, D., Pelon, M. J., & McCormick, M. P. (2003). The CALIPSO mission: Spaceborne lidar for observation of aerosols and clouds. *Proceedings SPIE, 4893*, 1–11.

Winker, D. M., Hunt, W. H., & McGill, M. J. (2007). Initial performance assessment of CALIOP. *Geophysical Research Letters, 34*, L19803. doi:10.1029/2007GL030135

Winker, D. M., Pelon, J., Coakley Jr, J. A., Ackerman, S. A., Charlson, R. J., Colarco, P. R., ... Wielicki, B. A. (2010). The CALIPSO mission: A global 3D view of aerosols and clouds. *Bulletin of the American Meteorological Society, 91*, 1211–1229.

Winkler, P. (1988). The growth of atmospheric aerosol particles with relative humidity. *Physical SCR, 37*(2), 223–230. doi:10.1088/0031-8949/37/2/008

X G, L., Zhang, Y. H., Cheng, Y. F., Hu, M., & Han, T. (2012). Aerosol hygroscopicity and its impact on atmospheric visibility and radiative forcing in Guangzhou during the 2006 PRIDE-PRD campaign. *Atmospheric Environment, 60*, 59–67. doi:10.1016/j.atmosenv.2012.06.016

Zhang, R., Wang, M., Sheng, L., Kanai, Y., & Ohta, A. (2004). Seasonal characterization of dust days, mass concentration and dry deposition of atmospheric aerosols over Qingdao, China. *China Particuology, 2*, 196–199. doi:10.1016/S1672-2515(07)60058-X

Zhang, X., & Mcmurry, P. H. (1992). Evaporative losses of fine particulate nitrates during sampling. *Atmos Environment, 26*, 3305–3312. doi:10.1016/0960-1686(92)90347-N

Zhang, Z., Shen, Y., Li, Y., Zhu, B., & Yu, X. (2017). Analysis of extinction properties as a function of relative humidity using a k-EC-Mie model in Nanjing. *Atmos Chemical Physical, 17*, 4147–4157. doi:10.5194/acp-17-4147-2017

Zhou, Y., Cheng, S., Chen, D., Lang, J., Wang, G., Xu, T., ... Yao, S. (2014). Temporal and spatial characteristics of ambient air quality in Beijing, China. *Aerosol Air Qualitative Researcher, 14*, 1–12.

Zujev, V. E., & Krekov, G. M. (1986). Optical models for the atmosphere. *Hidrometeoizdat*, Leningrad (in Russian).

12

Transient effects of nitrogen addition and rainfall suppression on *Vachellia karroo* growth under grass competition in a southern African savanna

Munyaradzi D. Shekede[1]*, Mhosisi Masocha[1] and Amon Murwira[1]

*Corresponding author: Munyaradzi D. Shekede, Faculty of Science, Department of Geography and Environmental Science, University of Zimbabwe, P.O. Box MP167, Mount Pleasant, Harare, Zimbabwe

E-mail: shekede@gmail.com; sheke-de@gis.uz.ac.zw

Reviewing editor: Hazel Chapman, University of Canterbury, NZ, New Zealand

Abstract: It is not clear how woody species, especially nitrogen fixers will respond to the combined effect of increased N fertilisation and reduced rainfall amount in savannas in a changing environment. A field experiment was set up at a southern African savanna site to investigate the interaction effects of increased N deposition and reduced rainfall amount on the growth of *Vachellia karroo* saplings in the presence of grass competition. Rainout shelters were erected around experimental plants to mimic the projected decrease in rainfall in southern Africa while N was added as ammonium nitrate over four growing seasons. The experiment uncovered significant but transient effects of rainfall suppression alone (F_1 = 5.171, P = 0.031) and its interaction with N fertilisation (F_1 = 6.369, P = 0.017) on the height of *V. karroo* saplings in the second growing season but not in the first, third and fourth season (P > 0.05). Rainfall suppression significantly reduced sapling height. The interaction of fertilisation and rainfall suppression increased stem height of the study species. In contrast, the interaction effects of N supply and rainfall suppression significantly (F_1 = 4.213,

ABOUT THE AUTHOR

Munyaradzi D. Shekede was born on the 9th of May 1981 in Marondera, Zimbabwe. He completed his BA Honours degree in Geography at the University of Zimbabwe in 2003 and a Masters degree in Environmental Policy and Planning from the same university in 2005. Currently, Munyaradzi is studying for a PhD at the University of Zimbabwe focusing on the effects of nitrogen deposition and precipitation decline on tree-grass interactions in southern African savannas. From 2007 to date, he has been working as a lecturer at the University of Zimbabwe, Department of Geography and Environmental Science. Some of the courses he teaches include Remote Sensing, Geographic Data and Analysis, Global Environmental Change and Biodiversity Conservation. Munyaradzi is interested in understanding global changes on society and the environment including flood modelling. He has published more than fourteen peer-reviewed articles and three book chapters.

PUBLIC INTEREST STATEMENT

We manipulated rainfall and nitrogen to understand how juvenile Sweet thorn plants will respond to the expected changes in rainfall and nitrogen deposition in savannas of Zimbabwe over four years. The height of plants was affected by rainfall alone as well by the combination of rainfall and nitrogen in the second season but not in the first, third and fourth growing season. Rainfall suppression reduced tree height while the combination of fertilisation and rainfall suppression increased the stem height of target plants. Nitrogen supply and rainfall suppression increased the diameter of juveniles during the first season but not thereafter. In contrast, grass competition did not influence the growth of juveniles either alone or in interaction with rainfall suppression though plants growing in competition with grass had relatively higher growth than the control. Overall, results suggest that N fertilisation may cancel out the predicted negative effect of rainfall decrease on woody species growth thereby enabling their persistence under global environmental changes.

$P = 0.049$) increased diameter of saplings during the first season but not thereafter. Conversely, grass competition did not significantly influence the growth of *V. karroo* either alone or in interaction with the main treatments though saplings growing in competition with grass had relatively higher growth than the control. Overall, results suggest that N fertilisation may cancel out the predicted negative effect of rainfall decrease on woody species growth thereby enabling the persistence of these species under global environmental changes.

Subjects: Plant & Animal Ecology; Conservation – Environment Studies; Ecology – Environment Studies

Keywords: Rainfall reduction; Nitrogen addition; *Vachellia karroo*; southern African savanna

1. Introduction

Rainfall and nitrogen (N) are known to limit woody species recruitment, productivity and diversity in savanna ecosystems (Scholes and Archer, 1997, Vadigi and Ward, 2013, van der Waal et al., 2009). Rainfall, in particular water, is considered the primary determinant of savanna structure (Sankaran, Ratnam, & Hanan, 2008) since water has a direct effect on the physiological processes of photosynthesis, respiration, and plant growth (Huston and Wolverton, 2009). Previous work has shown that maximum woody cover in savannas is regulated by mean annual rainfall (MAR). In semi-arid savannas receiving less than 650 mm of MAR water availability constrains woody cover resulting in a stable coexistence of trees and grasses (Sankaran et al., 2005, Sankaran et al., 2008). In contrast, in humid savannas receiving ≥650 mm of MAR woody cover is not limited by water availability. As a result and in the absence of disturbances such as fire and herbivory, these humid savannas can easily transit into a woodland system, and hence they are considered unstable (Sankaran et al., 2005).

With regard to plant nutrient requirements, N is the most important element required by plants in the largest amounts and savanna vegetation is no exception (Vitousek and Field, 1999, Chapin, 1980). However, in tropical soils where savannas are distributed, the availability of most essential plant nutrients including N is low due to the higher rates of nutrient loss and immobilisation under warm wet conditions, as well as the old age of tropical soils (Huston and Wolverton, 2009). This picture is however changing in the light of the observed global increase in N deposition (WallisDeVries and Bobbink, 2017, Valliere et al., 2017). For example, in semi-arid savannas of southern Africa, atmospheric N deposition has increased three- to four-fold between 1955 and 2003 (Scholes, Scholes, Otter, & Woghiren, 2003) and is likely to accelerate in future (Galloway et al., 2004, Sheffield and Wood, 2008; Miyazaki, Eskes, & K., 2012). The key drivers of this increase include industrialisation (Conradie et al., 2016) and land-use change (Adams et al., 2004, Galy-Lacaux & Delon, 2014). The observed changes in N deposition into savanna systems are occurring against the backdrop of a drying trend in rainfall patterns for southern Africa (Davis & Vincent, 2017, Hoerling, Hurrell, Eischeid, & Phillips, 2006, Hulme, 1996, IPCC, 2007). While there are uncertainties in rainfall projections over southern Africa, decreases of between 10 and 30% are expected in this region (Serdeczny et al., 2017). Currently, not much is known about the effect of the combined decrease in soil moisture availability and increased N deposition on tree–grass interactions in savannas. Yet, this knowledge is critical for savanna management in a changing world.

Although a number of observational and experimental studies investigated the effects of N addition on plant growth and competition interactions of woody and grassy species in savanna ecosystems, conflicting results have been reported. For instance, Sankaran et al. (2008) conducted a meta-analysis of determinants of woody cover in African savannas and found negative effects of increased soil N availability on woody cover owing to increased grass competitiveness. Similarly, in a semi-arid Mediterranean savanna, Trubat et al., (2011) reported a decrease in survival and performance of woody seedlings in an N addition study. In a bush-encroached South African savanna, an experimental study by Kraaij and Ward (2006) detected significant negative effects of N enrichment on woody species growth with amplified negative effects reported in plots with herbaceous competition. Overall, these studies demonstrate that an increase in soil available N enhances the growth of herbaceous plants resulting in the suppression of woody species recruitment and growth (Walker & Langridge, 1997, Moshe et al., 2000). This effect is amplified in N poor environments (Kraaij & Ward, 2006, van der Waal et al., 2009). Contrary to these studies, Ingestad (1980) reported that nitrogen enrichment stimulated woody species growth in a greenhouse experiment. Data from field experimental work in North American savannas by Siemann and Rogers (2003) demonstrated that the growth of woody species increased with N supply. In addition, results from a meta-analysis by Xia and Wan (2008) concluded that N fertilisation stimulates woody plant growth with non-legumes being more sensitive to fertilisation than legumes.

In contrast to the above studies reporting either significant negative or positive effects of increasing N of woody species performance, Holdo (2013) did not detect significant effects of increasing N on the above ground biomass of a dominant woody species. Holdo's experimental work was conducted in a South African savanna targeting a dominant woody species—*Colophospermum mopane*. Similarly, in a short-term greenhouse experiment, van Auken, Gese, and Connors (1985) previously found that N fertilisation did not significantly affect the growth of N fixer woody species—*Acacia smallii*. In addition, Van der Waal et al. (2009) and Mopipi, Trollope, and Scogings (2009) showed that woody seedling growth was not significantly affected by N fertilisation in a nutrient poor semi-arid southern African savanna. Recently, in a study done in a southern African savanna, Barbosa et al. (2014a, 2014b) found out that N addition did not induce significant growth in above-ground biomass production of woody species regardless of whether they were N-fixers or non-fixers. Put together the above studies suggest that N fertilisation could have neutral, positive and negative effect on woody species recruitment. The lack of clear outcome on the effects of increased nitrogen on woody species requires further investigation especially in the context of increased N deposition in savannas across the globe.

So far most experimental studies performed to test the effects of changes in rainfall and N on woody species growth have manipulated either resource in isolation (Beier et al., 2012, Fensham et al., 2005, Ludwig et al., 2001, Vadigi and Ward, 2013, van Auken et al., 1985, February, Higgins, Bond, & Swemmer, 2013, Berry and Kulmatiski, 2017). The few studies that have manipulated both resources have done so over short time scales (Barbosa et al., 2014a, Kraaij & Ward, 2006, Ludwig et al., 2001, Trubat et al., 2011, van der Waal et al., 2009; van Langevelde et al., 2003). In addition, some of the studies have manipulated these two resources in glass houses (van Auken et al., 1985). While glasshouse studies are important in eliminating confounding effects associated with field experiments they often fail to mimic the real world conditions (Limpens et al., 2012). In the light of the shortcomings of these studies, our current understanding of the combined effects of key resources on woody species recruitment and growth remains limited (Higgins et al., 2007). This gap needs to be closed considering that the observed widespread increases in nitrogen deposition and decreasing rainfall amounts are likely to significantly alter tree-grass interactions with far-reaching consequences on savanna structure and functioning.

To bridge this knowledge gap, a field experiment was conducted over four growing seasons to investigate the main and interaction effects of increasing N and reducing water availability on the growth of a common nitrogen fixer common in African savannas (*Vachellia karroo*). Fixed rainout shelters were erected around *V. karroo* saplings to reduce rainfall availability while N was added as ammonium nitrate in the presence and absence of grass competition in a southern African savanna. The study tests the following hypotheses:

(1) Since tree growth is currently limited by rainfall, a 15% decrease will further constrain woody sapling growth.

(2) As grasses are more efficient in utilising resources in the upper soil layer than woody species, N addition will stimulate growth of grass species at the expense of woody species, thereby amplifying their suppressive effects on woody species.

(3) N addition combined with rainfall reduction may dampen the competitive effects of grasses on woody species.

2. Materials and methods

2.1. Study site

The experimental study was performed at Kyle Game Reserve located at 20°11′20.01″ of latitude and 30°58′58.17″ of longitude. The area is located at an average height of 1050 m above sea level with an MAR of 638 mm. The climate of the study area is semi-arid characterised by three seasons: namely (1) hot and wet season from November to April, (2) cool and dry season between May and August and (3) hot and dry season between August and October (Shoko et al., 2015). Average maximum daily temperatures range from 21°C in June to 29°C in October, while the average minimum daily temperatures range between 5°C and 17°C in July and January, respectively (Masocha & Dube, 2017). The study area is characterised by highly variable rainfall and frequent droughts, which occur on annual to decadal scales (Shekede et al., 2016). Soils at the experimental site are mainly clays derived from the basement complex of banded ironstone and gold-belt schist (Wilson, 1964). Woody vegetation at the study site is dominated by *Dichrostachys cinerea, V. karroo* and the invasive *Lantana camara* (Masocha & Skidmore, 2011). The herbaceous layer at the study site mainly consists of *Hyparrhenia filipendula* and *Hyperthelia dissoluta* (Vincent and Thomas, 1960). Key animal species at the study site include *Phacochoerus africanus, Syncerus caffer, Ceratotherium simum, Aepyceros melampus* and *Giraffa camelopardalis*. Our species of interest for this study was *V. karroo* which constitutes a significant proportion of vegetation cover in the Game Reserve and is one of the key encroacher species in southern African savannas (O'connor, Puttick, & Hoffman, 2014).

2.2. Selection of experimental plants

Prior to setting up the field experiment, five candidate sites dominated by *V. karroo* adult trees interspersed with a continuous herbaceous layer were identified. At each candidate site measuring approximately 100 m by 100 m, 1-m belt transects were laid out in October 2012. Transects were placed 10 m apart. In each transect, three observers identified and counted the number of *V. karroo* seedling and saplings considered to belong to the same age cohort on the basis of similarity in height. An individual was classified as a seedling if it was less than 5 cm in height, whereas those above 5 cm and with a diameter at the base greater than 1 cm were classified as saplings. The field experiment targeted seedlings or saplings because these growth stages experience intense competition with grasses whose outcome can be mediated by resource availability and environmental changes leading to either woody species recruitment success or failure. Due to the failure to identify sufficient number of seedlings (>50 individuals at each candidate site), a decision was made to use *V. karroo* saplings as the experimental units. One site with more than 50 *A. karroo* saplings was randomly chosen to guarantee sufficient number of replicates for the experiment. The site was fenced off to exclude herbivory by small and large mammals.

Prior to starting the experiment, each *V. karroo* sapling identified was assigned a unique label and measured for stem height, basal stem diameter and number of secondary branches. Then, a total of 48 saplings that were similar in size located at least 2 m from the edge of fence were selected at random to minimize the influence of treatments among plots. To reduce the confounding effect of size, one-way analysis of variance (ANOVA) was used to test and confirm that there were no significant differences in sapling basal diameter and height. ANOVA results indicated that there were no significant differences in the stem basal diameter ($F_{40} = 0.402$, $P = 0.895$) as well as sapling height ($F_{40} = 0.768$, $P = 0.617$) at the start of the experiment. An 1 m × 1 m plot with the focal sapling centrally located was established around each selected sapling before random assignment of treatments.

2.3. Experimental design

To investigate how nitrogen (N) fertilisation and changes in rainfall reduction affect the growth of *V. karroo*, a field experiment in which N and rainfall amount were manipulated in the presence and absence of grass competition was set up. The field experiment consisted of three main treatments each with two levels: (1) rainfall suppression versus natural rain, (2) nitrogen addition versus no nitrogen and (3) grass competition versus no grass competition. The experiment followed a fully crossed design to yield a total of eight different treatment combinations:

(1) Adding N in the presence of competition from herbaceous vegetation;
(2) Adding N in the absence of competition from herbaceous vegetation;
(3) Reducing rainfall amount by 15% in the presence of competition from herbaceous vegetation;
(4) Reducing rainfall amount by 15% in the absence of competition from herbaceous vegetation;
(5) Herbaceous vegetation left intact to allow competition;
(6) Simultaneously adding N and reducing rainfall in the absence of competition from herbaceous vegetation;
(7) Simultaneously adding N and reducing rainfall In the presence of competition from herbaceous vegetation; and the
(8) Control in which no N was added, rainfall was not reduced and herbaceous vegetation was removed.

2.4. Treatments

Each treatment combination was replicated six times to yield a total of 48 experimental units at the start of the experiment. N was applied as a top dressing at a rate of 4 g of N m^{-2} at the start of the rainfall season in October. Nitrogen fertiliser was added in the form of ammonium nitrate (NH_4NO_3) granules evenly sprinkled within the plot by hand. This application rate is three times the N deposition rates projected for southern Africa for the 2050s (Phoenix et al., 2006).

To mimic IPCC (2007) rainfall projections for southern Africa, rainfall was reduced by 15%. This was achieved by constructing a fixed-location rainout shelters around each 1 m × 1 m plot experimental plot following the design used by Yahdjian and Sala (2002). The rainout shelter with had arcyllic bands on the roof designed to direct excess rainfall to a storage container that was emptied after rainfall event. The sides of the rainout shelter were left open to allow free air circulation and to minimise the confounding effects of temperature and humidity (Fiala, Tůma, & Holub, 2009). The rainout shelter was maintained throughout the duration of the experiment. Control plants had no rainout shelters and thus received normal rainfall. During the experimental period, the amount of rainfall suppressed was variable with as little as 50 mm (for a rainfall total = 337 mm) in 2012 to as much as 125 mm in 2014 (for a rainfall total 832 mm)

Competition between *A. karroo saplings* and herbaceous vegetation was manipulated by clipping to ground level all above-ground herbaceous biomass present in control plots at the start of the experiment and monthly thereafter. The experiment ran from October 2012 to October 2016. However, some 11 saplings died during the course of the experiment resulting in an unequal number of experimental units across treatments.

2.5. *Monitoring* Vachellia karroo *sapling growth*

Vachellia karroo saplings were monitored from October 2012 to September 2016. To quantify the treatment effects on *V. karroo* growth, basal diameter and stem height were measured on each individual at the end of every month. Here, only the results for the month of May of each of the four growing season are presented as these represent the growth accumulated during the growing season. At the study site, the growing season stretches from November to April. Basal diameter was measured using a veneer caliper at a height of 2 cm from the ground level. Sapling height was measured using a tape measure and height was considered as the highest point of the plant from the ground surface.

2.6. *Statistical analysis*

Factorial ANOVA was used to test whether the means differed significantly among the eight treatments. Post-hoc multiple comparisons based on the Fischer's LSD test were done when the null hypothesis that the means were the same was rejected by ANOVA. Statistical analysis was done separately for stem height and basal diameter. Prior to statistical analyses data on stem height and diameter data were log transformed and square rooted, respectively, to reduce skewness. The data were then tested for normality using Shapiro–Wilk's test. All statistical analyses were performed in STATISTICA version 10 (StatSoft, Inc., Tusla, Oklahoma, USA) and the results are reported for four growing seasons separately. The seasonal approach was adopted in this study to reduce the confounding effects of negative growth that may occur during dry season.

3. Results

3.1. *Effects of N addition, rainfall suppression and grass competition on the variation in height of* Vachellia karroo *saplings*

Figure 1 illustrates the effects of N fertilisation, rainfall suppression and grass competition as well as their interactions on height increment of *V. karroo* saplings across four growing seasons. We observe that the height of *V. karroo* saplings exposed to different treatments generally increased across the four growing seasons. N fertilisation in the absence of grass competition had negative but non-significant effect on the growth of *V. karroo* saplings across the four seasons (Table 1). In contrast, N fertilisation in the presence of grass competition resulted in larger height increment

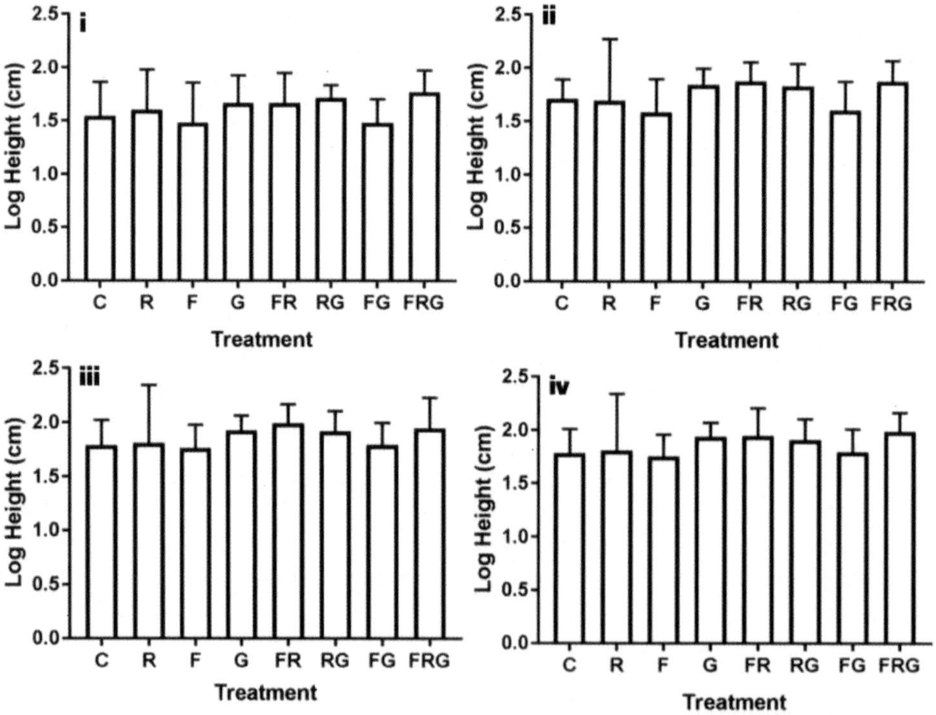

Figure 1. Effects of N fertilisation, rainfall suppression and grass competition on the mean log height of *Vachellia karroo* saplings after (i) first growing season, (ii) second growing season, (iii) third growing season and (iv) fourth growing season C = control, F = N addition, G = grass competition and R = rainfall suppression. Error plots represent log mean ±95%confidence interval.

Table 1. Effects of N addition, rainfall suppression, grass competition and their interaction on height increment of *Vachellia karroo* saplings.

	Start of experiment		1st season		2nd season		3rd season		4th season	
Effect	**F**	**P**	**F**	**P**	**F**	**P**	**F**	**P**	**F**	**P**
Intercept	9923.54	0.000	1993.85	0.000	3586.89	0.000	4811.70	0.000	4816.85	0.000
Fertiliser (F)	0.110	0.742	0.214	0.647	0.333	0.568	0.052	0.821	0.045	0.833
Rainfall (R)	3.638	0.066	4.012	0.055	5.171	**0.031**	3.306	0.079	3.142	0.087
Grass (G)	0.010	0.922	0.210	0.650	1.576	0.219	2.192	0.149	2.404	0.132
F*R	0.273	0.605	1.609	0.215	6.369	**0.017**	3.006	0.094	3.216	0.083
F*G	2.873	0.101	1.342	0.256	1.081	0.307	0.614	0.439	0.601	0.444
R*G	0.010	0.921	0.155	0.696	0.005	0.946	0.002	0.964	0.043	0.837
F*R*G	0.002	0.963	0.104	0.749	0.008	0.931	0.060	0.808	0.071	0.792

than fertilisation alone. However, the height differences in plots with and without grass completion were not statistically significant across the duration of the experiment (Table 1).

Results for ANOVA illustrate that rainfall suppression alone did not have significant effects on the height of *V. karroo* saplings across all growing seasons except in the second growing season (Tables 5-1). Specifically, in the second season, rainfall suppression significantly and positively influenced the stem height of *V. karroo* saplings with plants grown under rainfall suppression having a greater stem height (log mean 1.81 ± 0.12 cm (standard error) than those in the control (log mean 1.79 ± 0.08 cm). Nonetheless, the effects of these treatments are not detectable at the end of the third and fourth growing season (Tables 5-1). A comparison of growth patterns of *V. karroo* saplings illustrates that rainfall suppression induced more height increment in the presence of grass competition than without. In particular, plants exposed to rainfall suppression in the presence of grass competition increased in height 6% more than those that were exposed to rainfall suppression alone.

The interactive effects N fertilisation and rainfall suppression had significant effects on the height of *V. karroo* saplings only in the second season. However, these effects were more enhanced in plots with grass competition than without (Figure 1 and Table 1). Table 2 shows the results of the post-hoc analysis performed to test which means significantly differ from each other.

3.2. Effects of N addition, rainfall suppression and grass completion on diameter increment of **Vachellia karroo** *saplings*

The variations in diameter growth of *V. karroo* saplings exposed to N addition, rainfall suppression, grass competition and their interactions are shown in Figure 2. Results illustrate that, similar to height growth patterns, N addition alone and in combination with grass competition did not significantly affect the diameter of *V. karroo* saplings though plants subjected to N addition had larger diameter than those under a combination of fertilisation and grass competition (Tables 5-2)

Furthermore, we observe that rainfall suppression alone and in combination with grass competition did not appear to significantly affect the diameter of *V. karroo* saplings. A comparison of the growth patterns of these plants illustrates that saplings growing under rainfall suppression had smaller diameter than those exposed to a combination of rainfall suppression and grass competition. We also observe that by the end of the first growing season, the interaction of rainfall suppression and N addition is the only treatment that negatively and significantly affected diameter of *V. karroo* saplings. Specifically, the diameter of *V. karroo* saplings exposed to the interaction of rainfall suppression and N treatment had significantly smaller diameter

Table 2. Results of the post-hoc analysis based on the least significant difference test on tree height growth in the second season.

	C	G	Rl	F	R &G	F R	F G	F R G
C								
G								
R								
F		•						
RG				•				
F R				•				
F & G		•			•	•		
F, R G				•		•		

Notes: The black dots indicate treatments that significantly differ from each other.

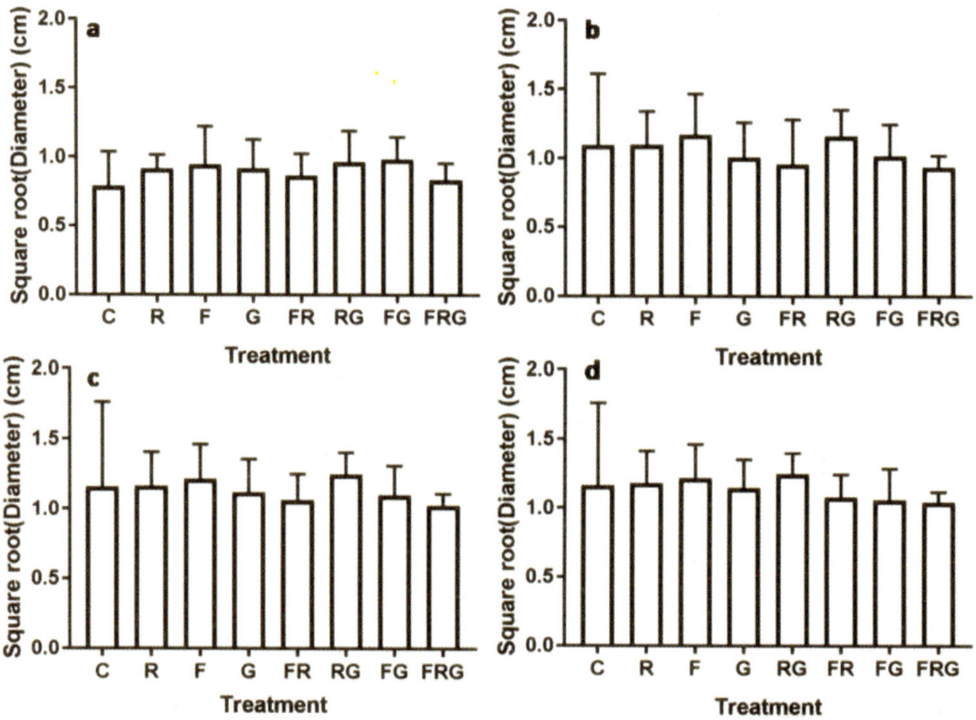

Figure 2. Effects of N fertilisation, rainfall suppression and grass competition on diameter of *Vachellia karroo* saplings after (a) first growing season,(b) second growing season, (c) third growing season and (d) fourth growing season. C = control, F = N addition, G = grass competition and R = rainfall suppression. Error plots represent mean ± CI.

(1.14 ± 0.12 cm) than the control (1.24 ± 0.19 cm). Moreover, we did not observe any significant effects of main treatments or their interactions thereafter (Table 3).

4. Discussion

The aim of this study was to test whether N fertilisation and rainfall suppression mediated by grass competition have significant effects on the growth of *V. karroo* saplings growing in a semi-arid savanna ecosystem. Our results indicate that interactions between N fertilisation and rainfall suppression have significant but transient effects on the growth of woody saplings. In addition, rainfall suppression significantly reduced height but not diameter of our focal species. On the other hand, grass competition did not significantly influence the growth of our focal species either alone or in interaction with the main treatments. However, saplings growing in competition with grass had relatively higher growth than the control.

Table 3. Effects of N addition, rainfall suppression, grass competition and their interaction on stem diameter growth of *Vachellia karroo* saplings.

	Start of experiment		1st growing season		2nd growing season		3rd growing season		4th growing season	
Effect	**F**	**P**	**F**	**P**	**F**	**P**	**F**	**P**	**F**	**P**
Intercept	374.27	0.000	331.60	0.000	290.20	0.000	383.40	0.000	404.8	0.000
Fertiliser (F)	0.202	0.656	0.022	0.885	1.298	0.264	1.694	0.203	2.331	0.138
Rainfall (R)	0.965	0.334	0.094	0.762	0.305	0.585	0.179	0.675	0.051	0.822
Grass (G)	0.017	0.898	1.264	0.270	0.571	0.456	0.194	0.662	0.378	0.543
F*R	0.137	0.714	4.213	0.049	3.202	0.084	2.414	0.131	1.510	0.229
F*G	0.340	0.565	0.812	0.375	0.386	0.539	0.589	0.449	0.886	0.355
R*G	0.721	0.403	0.463	0.501	1.075	0.308	0.699	0.410	0.737	0.398
F*R*G	0.157	0.695	0.007	0.934	0.028	0.867	0.054	0.819	0.007	0.936

4.1. *Effects of nitrogen on* Vachellia karroo *sapling growth*

Results of our short-term experiment indicate that N fertilisation alone had negative but non-significant effects on the growth of *V. karroo*, a nitrogen fixer. This is contrary to the expectation that nitrogen addition would significantly increase the growth of our focal species based on the premise that the ready availability of N would result in the reallocation of energy from N fixation to other processes such as growth. Results of this study are consistent with previous studies that observed a 20% reduction in woody species growth as a result of N addition (Schuster and Dukes, 2017). Similar studies (Mopipi et al., 2009, Van Der Waal et al., 2009, Xia and Wan, 2008) found no significant effects of N addition on the growth of woody seedlings in semi-arid savannas. In recent studies, Barbosa et al. (2014a, 2014b) found out that N addition did not induce significant growth in above-ground biomass production of woody species regardless of whether they were N-fixers or non-fixers in a southern African savanna. Indeed, the growth of leguminous woody species especially *Vachellia* can be insensitive to N addition but may respond positively to other mineral nutrients (van Auken et al., 1985).

Although the mechanism explaining the growth pattern found in this study is not clear, we speculate that under increased N conditions, other factors such as water and phosphorous may be limiting resulting in investment of energy towards the root system of the plants to increase water and nutrient foraging capability (Mokany, Raison, & Prokushkin, 2006, Vitousek et al., 2002, Xia and Wan, 2008). These results are confirmed by previous research that have documented negative relationship between soil available N and woody biomass (Sankaran et al., 2008, Walker & Langridge, 1997). The negative effects of increased N availability on woody biomass growth have been explained through competition-based mechanism. Through this mechanism, herbaceous species are hypothesised to directly pre-empt N or indirectly reduce critical resources such as light or water available to woody species as a result of enhanced herbaceous species growth (Kraaij & Ward, 2006, Sankaran et al., 2008). However, in plots with N addition alone, we removed grass competition thereby ruling out the competitive effects of grasses on *V. karroo* saplings.

Nitrogen fertilisation led to the increase in the basal diameter of *V. karroo* species, though the effects were not significant. Our results are confirmed by previous studies that observed investments in basal diameter under conditions of increased nutrient availability (Vadigi and Ward, 2013). Conversely, nitrogen addition had negative effects on the stem height of our focal species. These two contrasting results of the response of *V. karroo* saplings to increased nitrogen availability could be reflecting unique and time-variant resource allocation strategy aligned to addressing limitations to plant growth imposed by the environment. Previous studies have demonstrated a change in biomass allocation to shoots or storage organs in response to increased nitrogen availability (Ingestad & Agren, 1991, Miranda, Kozovits, & Bustamante, 2014, Poorter & Nagel, 2000, Poorter et al., 2012). Our results fit well with the optimal partitioning models that suggest that plants act in response to environmental variation through partitioning biomass among organs to optimise resource acquisition and maximise growth (McCarthy and Enquist, 2007).

4.2. *Effects of rainfall suppression on the growth of* Vachellia karroo *saplings*

Contrary to Hypothesis 1 that expected rainfall suppression to reduce the growth of *V. karroo* saplings, the result from this study indicate that rainfall suppression significantly increased height but not diameter of *V. karroo* saplings in the second season after the start of the experiment. Specifically, saplings growing under 15% rainfall suppression had a larger mean height but smaller mean diameter than the control suggesting some compensatory behaviour in the growth patterns of our focal species. A possible mechanism explaining the larger height increment in the second season of saplings under rainfall suppression compared with the control could be increased water use efficiency. Recent studies have shown that increased water use efficiency of woody species associated with increasing aridity helps in overcoming precipitation-driven constraints (Devine, Mcdonald, Quaife, & Maclean, 2017). Consequently, plants growing in environments with moderate reduction in precipitation can exhibit similar growth patterns as the control as long as the reduction in precipitation does not result in severe moisture stress (Padilla, Miranda, Jorquera, &

Pugnaire, 2009, Poorter et al., 2012). In fact, aridity is cited as a key factor in explaining variance in nitrogen-fixer abundance across the globe (Pellegrini, Staver, Hedin, Charles-Dominique, & Tourgee, 2016). Our results are also consistent with findings by February et al. (2013) who observed that woody species establishment was more pronounced during periods of short-term droughts when grass competition is reduced. Overall, our finding implies the possibility of persistence of nitrogen-fixing woody plants in a drying environment that is anticipated in southern African savannas in future.

4.3. *Interactive effects of rainfall suppression and nitrogen addition on* Vachellia karroo *saplings' growth*

Results of this study have shown that N fertilisation coupled with rainfall suppression significantly and positively increased sapling height and diameter of *V. karroo* saplings. This suggests that the potential negative effects of relative moisture stress imposed on *V. karroo* saplings' growth could be counteracted by nitrogen fertilisation. In fact, N fertilisation under reduced rainfall could increase plant tolerance to soil moisture stress (Kinugasa, Tsunekawa, & Shinoda, 2012) through, for example, increasing below ground biomass (Van den Driessche, 1992). In addition, studies elsewhere have shown that N-fixing plants are capable of sustaining high foliar N concentrations leading to efficient water use (Adams, Turnbull, Sprent, & Buchmann, 2016). This may result in larger growth rates for sapling growing in water-suppressed environment compared to plants growing under natural rainfall. Our results are confirmed by previous research that observed significant interactive effects of nutrient addition and water manipulation on woody seedling growth suggesting mediating effect of rainfall reduction on the growth of *V. karroo* saplings (Adams et al., 2016).

4.4. *Effect of grass competition on* Vachellia karroo *sapling growth*

Contrary to findings from previous studies, our results showed that grass competition alone and in interaction with either N fertilisation or rainfall suppression did not have significant effects on the growth (diameter and height increment) of *V. karroo* saplings. Similar studies have observed the suppressive effects of grass competition on woody species especially at the early stages of growth when grasses and woody species have intense competition for resources in the upper layers as predicted by Walter's competition model (Cramer, Chimphango, VAN CAUTER, Waldram, & Bond, 2007, Knoop & Walker, 1985, Vadigi and Ward, 2013, Van Der Waal et al., 2009). However, it is not unusual to observe lack of suppressive effects of grasses on woody species especially N-fixers as previously documented by O'connor (1995). In fact, several ecological studies have emphasised the increasing importance of facilitation over competition under conditions of increased abiotic stress (Callaway, 2007, Dohn et al., 2013, He, Bertness, & Altieri, 2013, Maestre et al., 2009); with the largest facilitative effects being realised at intermediate stressful conditions (Holmgren & Scheffer, 2010). A recent study on understanding tree–grass interactions along a rainfall gradient by Mazía et al. (2016) has shown that facilitation could even extend up to 1800 mm. In our case, a 15% reduction in rainfall might have been marginally stressful and thus enhanced facilitation between grasses and woody saplings for nitrogen since N-fixers have been shown to transfer fixed N to non-legume species (Sierra and Nygren, 2006, Zhang et al., 2016). Our study could be providing new insights into the role of facilitation in shaping plant communities even under unstressful conditions. In fact, our study could be confirming that facilitation does not necessarily need to occur under severe environments as previously thought.

The lack of significant interactive effects of N addition and grass competition on *V. karroo* saplings could also suggest the scale dependence nature of the growth of our experimental plants. Previous studies have demonstrated that plants respond differently to coarse versus fine-scale nutrient application (Hutchings et al. 2003, Birch and Hutchings 1994; Fransen et al. 1998, Kume et al. 2006, van der Waal, 2010). For the N addition and grass competition treatment, fertiliser was sprinkled by hand, thereby making it available as scattered grains instead of large patches. Consequently, we speculate that grasses might have been more efficient in exploiting fertiliser applied at fine scale application than woody species thereby lowering growth rate of saplings exposed to N addition in the presence of grass competition than N addition alone.

An important finding of this study is that while rainfall suppression and its interaction with N addition significantly affected the growth of *V. karroo* saplings, these effects may be transient. In fact, results indicate that 15% rainfall suppression alone and its interaction with fertiliser addition had significant effects on the diameter of our focal species only in the second growing season. Similarly, the interaction of rainfall suppression and N addition only had significant effect on the diameter of *V. karroo* saplings in the first growing season but could not be detected thereafter. To the best of our knowledge, our study is among the few that have reported transient effects of rainfall suppression and N addition woody nitrogen fixers.

The novelty of this study lies in manipulating rainfall, nitrogen and grass competition under field conditions to test whether these factors significantly influence the growth of *V. karroo* saplings in a semi-arid savanna. Previous studies on effects on nitrogen deposition have either been carried out on relatively short time scales (Barbosa et al., 2014a, Holdo, 2013, Kraaij & Ward, 2006, Mopipi et al., 2009, Van Der Waal et al., 2009) or were performed in green houses, and thus may not provide useful insights into the response of savanna woody species to changes in key resources. This therefore underscores the importance of long-term studies in understanding the effects of global nutrient enrichment and changes in rainfall patterns on savanna ecosystem structure and function over longer time scales (Bhadouria, Singh, Srivastava, & Raghubanshi, 2016). In addition, our study is among a few that has manipulated key resources limiting growth of key woody species in the presence and absence of grass competition. We assert that such interactive studies are critical to the understanding of the combined effects of global environmental change and herbaceous competition on tree–grass ratios in tropical ecosystems.

While useful insights have been gained from assessing the effects of key limiting resources and grass competition in savannas based on a leguminous woody species, it is important that future work focuses on the response of multiple functional species typical of savannas. This is especially important considering that non-leguminous plant species are incapable of fixing nitrogen and may thus respond differently to un/availability of key resources. For instance, it has often been hypothesised that nitrogen-fixer plants may lose their comparative advantage over non-nitrogen-fixer plants in the event of increased nitrogen deposition. In addition, nitrogen fixers are also predicted to increase their water use efficiency under a drying climate as a result of increase in carbon fertilisation. It may therefore be important to manipulate multiple resources on several plant species with different functional traits in order to enhance our understanding of tree–grass interactions in savannas that are undergoing transformation due to global environmental changes.

Funding
This work was supported by the University of Zimbabwe through the Vice Chancellor's Research Grant.

Competing Interest
No conflicts of interests were declared.

Author details
Munyaradzi D. Shekede[1]
E-mail: shekede@gmail.com
Mhosisi Masocha[1]
E-mail: masocha@gis.uz.ac.zw
Amon Murwira[1]
E-mail: murwira@gis.uz.ac.zw
[1] Department of Geography and Environmental Science, University of Zimbabwe, P.O. Box MP167, Mount Pleasant, Harare, Zimbabwe.

References

Adams, M., Ineson, P., Binkley, D., Cadlsch, G., Tokuchi, N., Scholes, M., & Hicks, K. (2004). Soil function responses to excess nitrogen inputs at global scale. *Ambio, 33*, 530–536. doi:10.1579/0044-7447-33.8.530

Adams, M. A., Turnbull, T. L., Sprent, J. I., & Buchmann, N. (2016). Legumes are different: Leaf nitrogen, photosynthesis, and water use efficiency. *Proceedings of the National Academy of Sciences of the United States of America, 113*, 4098–4103. doi:10.1073/pnas.1523936113

Barbosa, E. R., Tomlinson, K. W., Carvalheiro, L. G., Kirkman, K., De Bie, S., Prins, H. H., & Van Langevelde, F. (2014a). Short-term effect of nutrient availability and rainfall distribution on biomass production and leaf nutrient content of savanna

tree species. *PLoS One*, *9*, e92619. doi:10.1371/journal.pone.0092619
Barbosa, E. R., van Langevelde, F., Tomlinson, K. W., Carvalheiro, L. G., Kirkman, K., de Bie, S., & Prins, H. H. (2014b). Tree species from different functional groups respond differently to environmental changes during establishment. *Oecologia*, *174*(4), 1345–1357.
Beier, C., Beierkuhnlein, C., Wohlgemuth, T., Penuelas, J., Emmett, B., Körner, C., & Hansen, K. (2012). Precipitation manipulation experiments-challenges and recommendations for the future. *Ecology Letters*, *15*(8), 899-911.
Berry, R. S., & Kulmatiski, A. (2017). A savanna response to precipitation intensity. *PloS one*, *12*(4), e0175402.
Bhadouria, R., Singh, R., Srivastava, P., & Raghubanshi, A. S. (2016). Understanding the ecology of tree-seedling growth in dry tropical environment: A management perspective. *Energy, Ecology and Environment*, *1*, 296–309. doi:10.1007/s40974-016-0038-3
Birch, C. P. D., & Hutchings, M. (1994). Exploitation of patchily distributed soil resources by the clonal herb Glechoma hederacea. *Journal of Ecology*, 653–664.
Callaway, R. M. (2007). *Positive interactions and interdependence in plant communities*. Dordrecht: Springer.
Chapin III, F. S. (1980). The mineral nutrition of wild plants. *Annual Review of Ecology and Systematics*, *11* (1), 233-260.
Conradie, E., Van Zyl, P., Pienaar, J., Beukes, J., Galy-Lacaux, C., Venter, A., & Mkhatshwa, G. (2016). The chemical composition and fluxes of atmospheric wet deposition at four sites in South Africa. *Atmospheric Environment*, *146*, 113–131. doi:10.1016/j.atmosenv.2016.07.033
Cramer, M. D., Chimphango, S. B., Van Cauter, A., Waldram, M., & Bond, W. (2007). Grass competition induces N2 fixation in some species of African Acacia. *Journal of Ecology*, *95*, 1123–1133. doi:10.1111/jec.2007.95.issue-5
Davis, C. L., & Vincent, K. (2017). *Climate risk and vulnerability: A handbook for southern Africa* (2nd ed.) Pretoria, South Africa: CSIR.
Devine, A. P., Mcdonald, R. A., Quaife, T., & Maclean, I. M. (2017). Determinants of woody encroachment and cover in African savannas. *Oecologia*, *183*, 939–951. doi:10.1007/s00442-017-3807-6
Dohn, J., Dembélé, F., Karembé, M., Moustakas, A., Amévor, K. A., & Hanan, N. P. (2013). Tree effects on grass growth in savannas: Competition, facilitation and the stress-gradient hypothesis. *Journal of Ecology*, *101*, 202–209. doi:10.1111/1365-2745.12010
February, E. C., Higgins, S. I., Bond, W. J., & Swemmer, L. (2013). Influence of competition and rainfall manipulation on the growth responses of savanna trees and grasses. *Ecology*, *94*, 1155–1164.
Fensham, R. J., Fairfax, R. J., & Archer, S. R. (2005). Rainfall, land use and woody vegetation cover change in semi-arid Australian savanna. *Journal of Ecology*, *93*(3), 596-606.
Fiala, K., Tůma, I., & Holub, P. (2009). Effect of manipulated rainfall on root production and plant belowground dry mass of different grassland ecosystems. *Ecosystems (New York, N.Y.)*, *12*, 906–914. doi:10.1007/s10021-009-9264-2
Fransen, B., de Kroon, H., & Berendse, F. (1998). Root morphological plasticity and nutrient acquisition of perennial grass species from habitats of different nutrient availability. *Oecologia*, *115*(3), 351-358.
Galloway, J. N., Dentener, F. J., Capone, D. G., Boyer, E. W., Howarth, R. W., Seitzinger, S. P., ... Vörösmarty, C. (2004). Nitrogen cycles: Past, present, and future. *Biogeochemistry*, *70*, 153–226. doi:10.1007/s10533-004-0370-0
Galy-Lacaux, C., & Delon, C. (2014). Nitrogen emission and deposition budget in West and Central Africa. *Environmental Research Letters*, *9*, 125002. doi:10.1088/1748-9326/9/12/125002
He, Q., Bertness, M. D., & Altieri, A. H. (2013). Global shifts towards positive species interactions with increasing environmental stress. *Ecology Letters*, *16*, 695–706. doi:10.1111/ele.12080
Higgins, S. I., Bond, W. J., February, E. C., Bronn, A., Euston-Brown, D. I., Enslin, B.,... & Scheiter, S. (2007). Effects of four decades of fire manipulation on woody vegetation structure in savanna. *Ecology*, *88* (5), 1119-1125.
Hoerling, M., Hurrell, J., Eischeid, J., & Phillips, A. (2006). Detection and attribution of twentieth-century northern and southern African rainfall change. *Journal of Climate*, *19*, 3989–4008. doi:10.1175/JCLI3842.1
Holdo, R. M. (2013). Effects of fire history and N and P fertilization on seedling biomass, Specific Leaf Area, and root: Shoot ratios in a South African savannah. *South African Journal of Botany*, *86*, 5–8. doi:10.1016/j.sajb.2013.01.005
Holmgren, M., & Scheffer, M. (2010). Strong facilitation in mild environments: The stress gradient hypothesis revisited. *Journal of Ecology*, *98*, 1269–1275. doi:10.1111/jec.2010.98.issue-6
Hulme, M. (1996). Recent climatic change in the world's drylands. *Geophysical Research Letters*, *23*, 61–64. doi:10.1029/95GL03586
Huston, M. A., & Wolverton, S. (2009). The global distribution of net primary production: Resolving the paradox. *Ecological Monographs*, *79*(3), 343-377.
Hutchings, M. J., John, E. A., & Wijesinghe, D. K. (2003). Toward understanding the consequences of soil heterogeneity for plant populations and communities. *Ecology*, *84*(9), 2322-2334.
Ingestad, T. (1980). Growth, nutrition, and nitrogen fixation in grey alder at varied rate of nitrogen addition. *Physiologia Plantarum*, *50*(4), 353-364.
Ingestad, T., & Agren, G. I. (1991). The influence of plant nutrition on biomass allocation. *Ecological Applications*, *1*, 168–174. doi:10.2307/1941809
IPCC. 2007. Climate change 2007: The physical science basis. In S. Solomon, D. Qin, M. Manning, Z. Chen, M. Marquis, K. B. Averyt, M. Tignor & H. L. Miller (Eds.), *Contribution of Working group I to the Fourth Assessment Report of the Intergovernmental Panel on Climate Change* (pp. 847-940). Cambridge and New York, NY.
Kinugasa, T., Tsunekawa, A., & Shinoda, M. (2012). Increasing nitrogen deposition enhances post-drought recovery of grassland productivity in the Mongolian steppe. *Oecologia*, *170*, 857–865. doi:10.1007/s00442-012-2354-4
Knoop, W., & Walker, B. (1985). Interactions of woody and herbaceous vegetation in a southern African savanna. *The Journal of Ecology*, 235–253. doi:10.2307/2259780
Kraaij, T., & Ward, D. (2006). Effects of rain, nitrogen, fire and grazing on tree recruitment and early survival in bush-encroached savanna, South Africa. *Plant Ecology*, *186*, 235–246. doi:10.1007/s11258-006-9125-4
Kume, T., Sekiya, N., & Yano, K. (2006). Heterogeneity in spatial P-distribution and foraging capability by Zea mays: Effects of patch size and barriers to restrict root proliferation within a patch. *Annals of Botany*, *98* (6), 1271-1277.

Limpens, J., Granath, G., Aerts, R., Heijmans, M. M., Sheppard, L. J., Bragazza, L.,... & Rochefort, L. (2012). Glasshouse vs field experiments: Do they yield ecologically similar results for assessing N impacts on peat mosses? *New Phytologist*, *195*(2), 408-418.

Ludwig, F., de Kroon, H., Prins, H. H., & Berendse, F. (2001). Effects of nutrients and shade on tree-grass interactions in an East African savanna. *Journal of Vegetation Science*, *12*(4), 579-588.

Maestre, F. T., Bowker, M. A., Puche, M. D., Hinojosa, M. B., Martinez, I., Garcia-Palacios, P., ... Escudero, A. (2009). Shrub encroachment can reverse desertification in semi-arid Mediterranean grasslands. *Ecology Letters*, *12*, 930–941. doi:10.1111/j.1461-0248.2009.01352.x

Masocha, M., & Dube, T. (2017). Relationship between native and exotic plant species at multiple savannah sites. *African Journal of Ecology*, *56*(1), 81–90.

Masocha, M., & Skidmore, A. K. (2011). Integrating conventional classifiers with a GIS expert system to increase the accuracy of invasive species mapping. *International Journal of Applied Earth Observation and Geoinformation*, *13*, 487–494. doi:10.1016/j.jag.2010.10.004

Mazía, N., Moyano, J., Perez, L., Aguiar, S., Garibaldi, L. A., & Schlichter, T. (2016). The sign and magnitude of tree–Grass interaction along a global environmental gradient. *Global Ecology and Biogeography*, *25*, 1510–1519. doi:10.1111/geb.2016.25.issue-12

McCarthy, M. C., & Enquist, B. J. (2007). Consistency between an allometric approach and optimal partitioning theory in global patterns of plant biomass allocation. *Functional Ecology*, *21*(4), 713-720.

Miranda, V. T., Kozovits, A. R., & Bustamante, M. M. (2014). Competition alters responses of juvenile woody plants and grasses to nitrogen addition in Brazilian savanna (Cerrado). *Nitrogen Deposition, Critical Loads and Biodiversity*, Springer.

Miyazaki, K., Eskes, H. J., & K., S. (2012). Global NOx emission estimates derived from an assimilation of OMI tropospheric NO2 columns. *Atmospheric Chemistry and Physics*, *12*, 2263–2288. doi:10.5194/acp-12-2263-2012

Mokany, K., Raison, R., & Prokushkin, A. S. (2006). Critical analysis of root: Shoot ratios in terrestrial biomes. *Global Change Biology*, *12*, 84–96. doi:10.1111/gcb.2006.12.issue-1

Mopipi, K., Trollope, W. S. W., & Scogings, P. F. (2009). Effects of moisture, nitrogen, grass competition and simulated browsing on the survival and growth of Acacia karroo seedlings. *African Journal of Ecology*, *47*, 680–687. doi:10.1111/j.1365-2028.2008.01022.x

Moshe, D., Bailey, C. L., & Scholes, R. J. (2000). The effect of elevated atmospheric carbon dioxide on selected savanna plants. In *Proceedings: Natural forests and savanna woodlands symposium II: towards sustainable management based on scientific understanding of natural forests and woodlands. Department of Water Affairs and Forestry, Knysna, South Africa* (pp. 142–144).

O'connor, T. (1995). Transformation of a savanna grassland by drought and grazing. *African Journal of Range & Forage Science*, *12*, 53–60. doi:10.1080/10220119.1995.9647864

O'connor, T. G., Puttick, J. R., & Hoffman, M. T. (2014). Bush encroachment in southern Africa: Changes and causes. *African Journal of Range & Forage Science*, *31*, 67–88. doi:10.2989/10220119.2014.939996

Padilla, F. M., Miranda, J. D., Jorquera, M. J., & Pugnaire, F. (2009). Variability in amount and frequency of water supply affects roots but not growth of arid shrubs. *Plant Ecology*, *204*, 261–270. doi:10.1007/s11258-009-9589-0

Pellegrini, A. F., Staver, A. C., Hedin, L. O., Charles-Dominique, T., & Tourgee, A. (2016). Aridity, not fire, favors nitrogen-fixing plants across tropical savanna and forest biomes. *Ecology*, *97*, 2177–2183. doi:10.1002/ecy.1504

Phoenix, G. K., Hicks, W. K., Cinderby, S., Kuylenstierna, J. C. I., Stock, W. D., Dentener, F. J., ... Ineson, P. (2006). Atmospheric nitrogen deposition in world biodiversity hotspots: The need for a greater global perspective in assessing N deposition impacts. *Global Change Biology*, *12*, 470–476. doi:10.1111/j.1365-2486.2006.01104.x

Poorter, H., & Nagel, O. (2000). The role of biomass allocation in the growth response of plants to different levels of light, CO2, nutrients and water: A quantitative review. *Functional Plant Biology*, *27*, 1191. doi:10.1071/PP99173

Poorter, H., Niklas, K. J., Reich, P. B., Oleksyn, J., Poot, P., & Mommer, L. (2012). Biomass allocation to leaves, stems and roots: Meta-analysesof interspecific variation and environmental control. *New Phytologist*, *193*, 30–50. doi:10.1111/j.1469-8137.2011.03952.x

Sankaran, M., Hanan, N. P., Scholes, R. J., Ratnam, J., Augustine, D. J., Cade, B. S.,... & Ardo, J. (2005). Determinants of woody cover in African savannas. *Nature*, *438*(7069), 846.

Sankaran, M., Ratnam, J., & Hanan, N. P. (2008). Woody cover in African savannas: The role of resourcs, fire and herbivory. *Global Ecology and Biogeography*, *17*, 236–245. doi:10.1111/j.1466-8238.2007.00360.x

Scholes, M. C., Scholes, R. J., Otter, L. B., & Woghiren, A. J. (eds.). (2003). *Biochemistry: The cycling of elements*. Washington: The Kruger experience Island Press.

Scholes, R. J., & Archer, S. R. (1997). Tree-grass interactions in savannas. *Annual Review of Ecology and Systematics*, *28*(1), 517–544.

Schuster, M. J., & Dukes, J. S. (2017). Rainfall variability counteracts N addition by promoting invasive Lonicera maackii and extending phenology in prairie. *Ecological Applications*, *27*, 1555–1563. doi:10.1002/eap.2017.27.issue-5

Serdeczny, O., Adams, S., Baarsch, F., Coumou, D., Robinson, A., Hare, W., ... Reinhardt, J. (2017). Climate change impacts in Sub-Saharan Africa: From physical changes to their social repercussions. *Regional Environmental Change*, *17*, 1585–1600. doi:10.1007/s10113-015-0910-2

Sheffield, J., & Wood, E. F. (2008). Projected changes in drought occurrence under future global warming from multi-model, multi-scenario, IPCC AR4 simulations. *Climate Dynamics*, *31*, 79–105. doi:10.1007/s00382-007-0340-z

Shekede, M. D., Murwira, A., Masocha, M., & Zengeya, F. M. (2016). Decadal changes in mean annual rainfall drive long-term changes in bush-encroached southern African savannas. *Austral Ecology*, *41*, 690–700. doi:10.1111/aec.12358

Shoko, C., Masocha, M., & Dube, T. (2015). A new potential method to estimate abundance of small game species. *African Journal of Ecology*, *53*, 406–412. doi:10.1111/aje.12211

Siemann, E., & Rogers, W. E. (2003). Changes in light and nitrogen availability under pioneer trees may indirectly facilitate tree invasions of grasslands. *Journal of Ecology*, *91*, 923–931. doi:10.1046/j.1365-2745.2003.00822.x

Sierra, J., & Nygren, P. (2006). Transfer of N fixed by a legume tree to the associated grass in a tropical

silvopastoral system. *Soil Biology and Biochemistry, 38*, 1893–1903. doi:10.1016/j.soilbio.2005.12.012
Trubat, R., Cortina, J., & Vilagrosa, A. (2011). Nutrient deprivation improves field performance of woody seedlings in a degraded semi-arid shrubland. *Ecological Engineering, 37*(8), 1164-1173.
Vadigi, S., & Ward, D. (2013). Shade, nutrients, and grass competition are important for tree sapling establishment in a humid savanna. *Ecosphere, 4*, 1–27. doi:10.1890/ES13-00239.1
Valliere, J. M., Irvine, I. C., Santiago, L., & Allen, E. B. (2017). High N, dry: Experimental nitrogen deposition exacerbates native shrub loss and nonnative plant invasion during extreme drought. *Global Change Biology, 23*(10), 4333-4345.
Van Auken, O. W., Gese, E. M., & Connors, K. (1985). Fertilization response of early and late successional species: Acacia smallii and Celtis laevigata. *Botanical. Gazzette, 1*, 564–569. doi:10.1086/337561
Van den Driessche, R. (1992). Changes in drought resistance and root growth capacity of container seedlings in response to nursery drought, nitrogen, and potassium treatments. *Canadian Journal of Forest Research, 22*, 740–749. doi:10.1139/x92-100
Van der Waal, C., De Kroon, H., De Boer, W. F., Heitkänig, I., Skidmore, A. K., De Knegt, H. J., ... Page, B. R. (2009). Water and nutrients alter herbaceous competitive effects on tree seedlings in a semi-arid savanna. *Journal of Ecology, 97*, 430–439. doi:10.1111/j.1365-2745.2009.01498.x
van der Waal, C. (2010). *Nutrients in an African Savanna: the consequences of supply heterogeneity for plants and animals*. (Doctoral dissertation). Retrieved from https://library.wur.nl/WebQuery/wurpubs/fulltext/146149.
Van Langevelde, F., Van De Vijver, C. A., Kumar, L., Van De Koppel, J., De Ridder, N., Van Andel, J.,... & Prins, H. H. (2003). Effects of fire and herbivory on the stability of savanna ecosystems. *Ecology, 84*(2), 337-350.
Vincent, V., & Thomas, R. G. (1960). *An agricultural survey of Southern Rhodesia: Part I: Agro-ecological survey*. Salisbury: Government Printers.
Vitousek, P. M., & Field, C. B. (1999). Ecosystem constraints to symbiotic nitrogen fixers: a simple model and its implications. In *New Perspectives on Nitrogen Cycling in the Temperate and Tropical Americas* (pp. 179-202). Springer, Dordrecht.
Vitousek, P. M., Cassman, K., Cleveland, C., Crews, T., Field, C. B., Grimm, N. B., & Rastetter, E. B. (2002a). Towards an ecological understanding of biological nitrogen fixation. *Biogeochemistry, 57*, 1–45. doi:10.1023/A:1015798428743
Walker, B. H., & Langridge, J. L. (1997). Predicting savanna vegetation structure on the basis of plant available moisture (PAM) and plant available nutrients (PAN): A case study from Australia. *Journal of Biogeography, 24*, 813–825. doi:10.1046/j.1365-2699.1997.00123.x
WallisDeVries, M. F., & Bobbink, R. (2017). Nitrogen deposition impacts on biodiversity in terrestrial ecosystems: Mechanisms and perspectives for restoration. *Biological Conservation, 212*, 387-389.
Wilson, J. F. (1964). The geology of the country around Fort Victoria. *Bulletin – Rhodesia Geological Survey*, 58.
Xia, J., & Wan, S. (2008). Global response patterns of terrestrial plant species to nitrogen addition. *New Phytologist, 179*, 428–439. doi:10.1111/j.1469-8137.2008.02488.x
Yahdjian, L., & Sala, O. E. (2002). A rainout shelter design for intercepting different amounts of rainfall. *Oecologia, 133*, 95–101. doi:10.1007/s00442-002-1024-3
Zhang, H.-Y., Yu, Q., Lü, X.-T., Trumbore, S. E., Yang, J.-J., & Han, X.-G. (2016). Impacts of leguminous shrub encroachment on neighboring grasses include transfer of fixed nitrogen. *Oecologia, 180*, 1213–1222. doi:10.1007/s00442-015-3538-5

13

Land cover data as environmentally sensitive decision-making mediator in territorial and administrative reform

Artan Hysa[1]* and Fatma Ayçim Türer Başkaya[2]

*Corresponding author: Artan Hysa, Graduate School of Science, Engineering, and Technology, Istanbul Technical University, Istanbul, Turkey

E-mail: artanh@gmail.com

Reviewing editor: Ehsan Sharifi, School of Architecture and Built Environment, University of Adelaide, Australia

Abstract: This paper presents a multi-criteria conceptual framework for decision-making processes during Territorial and Administrative Reform (TAR) relying on sustainable development principles. In general, TAR processes highly consider socio-cultural and economic factors, but they lack responsiveness to environmental dynamics of the context. While this practice achieves a fair allocation of the new administrative units' centers, the border defining criteria are indistinct. Thus, we turn a spotlight on the environmental factors as fundamental criteria in TAR decision-making processes, especially the boundary definition stage. Topography, watershed, land cover, and natural conservation areas are among the proposed environmental measures. The research makes a real case of land cover data utilization as environmentally sensitive decision-making mediator. First, the CORINE Land Cover (CLC) data of 2012 serves as an evaluation criteria of the recent TAR (2014) in Albanian territory, based on the landscape fragmentation caused by the new spatial division. Second, the CLC data generates an alternative municipal boundary of the Albanian capital. The results show that a TAR process not taking into account the environmental criteria leads to functionally disconnected territories which in the long run may lead to physically fragmented natural landscapes.

ABOUT THE AUTHORS

Artan Hysa is an architect by education. He is a PhD candidate at Istanbul Technical University, Department of Landscape Architecture, currently being at the very final stage of his doctoral studies. This article is prepared within the scope of his PhD dissertation entitled "Utilizing CORINE Land Cover in Diverse Spatial Decision Making and Management Processes." His research activity is driven by an interest on the utility of cross-disciplinary knowledge in problem-oriented research.

Fatma Aycim Turer Baskaya is an associate professor at Istanbul Technical University, Department of Landscape Architecture. She is the coordinator of the landscape architecture master and PhD programs at the Graduate School of Science, Engineering and Technology, ITU. Her research areas are Coastal Landscape Planning and Design, Disaster Sensitive Landscape Architecture, Landscape Reclamation and Restoration, and GIS Based Landscape Studies.

PUBLIC INTEREST STATEMENT

TAR is a crucial decision-making process, having direct impact on social, economic, and environmental aspects in both national and local levels. A tangible outcome of TAR is the territorial subdivision of the national lands into local administrative units. In general, during decision-making process of TAR, the social and economic factors are being highly considered, but the environmental ones are neglected. This practice leads to conflictual relationship between the administrative border line and natural landscapes. The split administrative competences, over the same natural surface separated by the border line, may cause landscape fragmentation, which leads to natural lands degradation. This study, first, analyzes the recent TAR in Albania quantifying landscape fragmentation caused by the current territorial subdivision. Furthermore, it presents a method to use land cover data as a guide to define new local administrative borders to minimize natural landscapes' fragmentation.

Furthermore, the new alignment for municipal borders of Tirana shows a successful result in minimizing natural landscape fragmentation caused by local administrative boundaries. The proposed multi-criteria conceptual framework and the application via CLC utilization presents a methodical approach which may assist decision-making processes of TAR in other developing countries, conform sustainable territorial management principles.

Subjects: Resource Management—Environmental Studies; Environmental Policy; GIS, Remote Sensing & Cartography

Keywords: cross-border landscapes; landscape fragmentation; environmental conservation; sustainable landscape management; environmental boundary; CORINE land cover; Albania

1. Introduction

The conflict between the political borders and cross-boundary ecoregions is a well-known issue in conservation studies (Kark et al., 2015). The main contradiction relies on the transboundary character of ecological services (Hoffman, Varady, Flessa, & Balvanera, 2009), while specific governmental bodies perform the conservation planning practice within the boundary jurisdiction. This practice significantly brings uncoordinated interventions (Jantke & Schneider, 2010) and sometimes even interfering/overlapping/duplicating each other (Gordon, Bastin, Langford, Lechner, & Bekessy, 2013). Certainly, there is an emphasis on the collaboration among a variety of actors as well as across spatial borders (political, regional, municipal, and others) (Mace et al., 2000). Moreover, the natural conservation issue is of a multi-range character including a variety of shareholders acting at different spatial scales (Guerrero, Mcallister, & Wilson, 2014) such as, global, regional, or local. In the global scale, the international collaboration in natural conservation agendas might be the only solution toward cross-border conservation goals. This is due to the fact that the national borders apparently are a product of the most extreme conflicts (wars) among the humankind (Harvey & Most, 1976), and their readjustment is the last considerable alternative on the table.

Differently, in the local scale, the administrative boundaries within the national borders and the subdivision of the territory into local administrative units are a vibrant topic continuously consuming a serious discussion. One of the most tangible platforms of the debate is the Territorial and Administrative Reform (TAR), which is still scholarly under-estimated (Kuhlmann & Wollmann, 2011). The reforms of the local administration and the territorial reorganization were first applied during the '60s and the '70s (Norton, 1994) for the western European countries, and after the '90s for the post-socialist countries (Wollmann, 2010). The TAR process follows certain criteria to perform the territorial subdivision. Generally, the socio-cultural and economic factors are seriously considered during the decision-making processes of TAR. However, the case is not the same with the environmental criteria which are substantially ignored, despite being advocated to be the third pillar of sustainable development (Giovannoni & Fabietti, 2013; Hens, 1996).

This practice results in justified decisions on the amount and centers' location (urbanized areas) of the new administrative units. But, it remains insufficient in reasonably defining the administrative boundary line, especially while crossing through natural lands. As a consequence, the borders defined, based only on the socio-cultural and economic factors, result in functionally divided natural landscapes, which later might lead to permanent physical landscape fragmentation. In this study, the landscape fragmentation stands for the phenomenon not caused by physical barriers; it is a breakup of the interest and the consumption methods toward a specific common pool resource (CPR) (Ostrom, 2010) split among different, local administrative units (Swianiewicz, 2010). In the long-run, the split management and administration of a specific natural landscape surface may lead to physical landscape fragmentation.

In this context, the study focuses on the local scale, first, by conceptually discussing the multi-criteria character of the decisions within TAR processes. Furthermore, we propose a conceptual framework based on sustainable landscapes principles (Buttimer, 2001). Even though, it still relies on three fundamental pillars (social, economic, and environmental) of sustainable development, sustainable landscapes notion brings a decoupling between "sustainability" and "development" by implying that the former already bears the latter within itself (Blaschke, 2006). In consequence, we structured the criteria of the proposed framework into three main groups: the socio-cultural, the economic, and the environmental criterion. Each category consists of several sub-criteria being organized by their role and scope within the TAR process.

Beyond the conceptual multi-criteria approach, the paper presents a demonstrative application of the framework by utilizing land cover data as an environmental sub-criteria in the decision-making process of TAR, especially during local administrative boundary definition phase. The experiment relies on CORINE Land Cover (CLC) data of 2012 applied upon the recent TAR of Albanian territory (2014). CLC data are currently labeled as a well-established and reliable information source in support of environmental policy-making (Steenmans, 2016). The spatial relationship between CLC data and the current, local administrative borders provides pieces of evidence of landscape fragmentation caused by the reform. Furthermore, CLC data are empirically put forward as local administrative boundary definition mean as shown here for the case of Tirana. The experiential stages of this study are performed utilizing ArcGIS 10.2.2 software.

2. Case study: territorial and administrative reform in Albania

In 2014, Albania ratified the administrative and territorial reform as a demand by the European Union within the decentralization processes of local administration (Reçi & Ymeri, 2016). The new divisions, referred to as development zones, are responsible for the funds' management on economic and social development of candidate countries (Ghinea & Moraru, 2006). The territory is re-organized from 374 (65 municipalities and 309 communes within 36 localities joining into 12 prefectures) (Figure 1(b)) administrative units to 61 municipal districts (Ndreu, 2016) (Figure 1(c)). The local administration reform of decentralization reduced distinctly the institutional ties and dependency between central and local governments. Moreover, the new reform annihilated a core administrative unit of the pre-reform version such as "rreth," reported as administratively

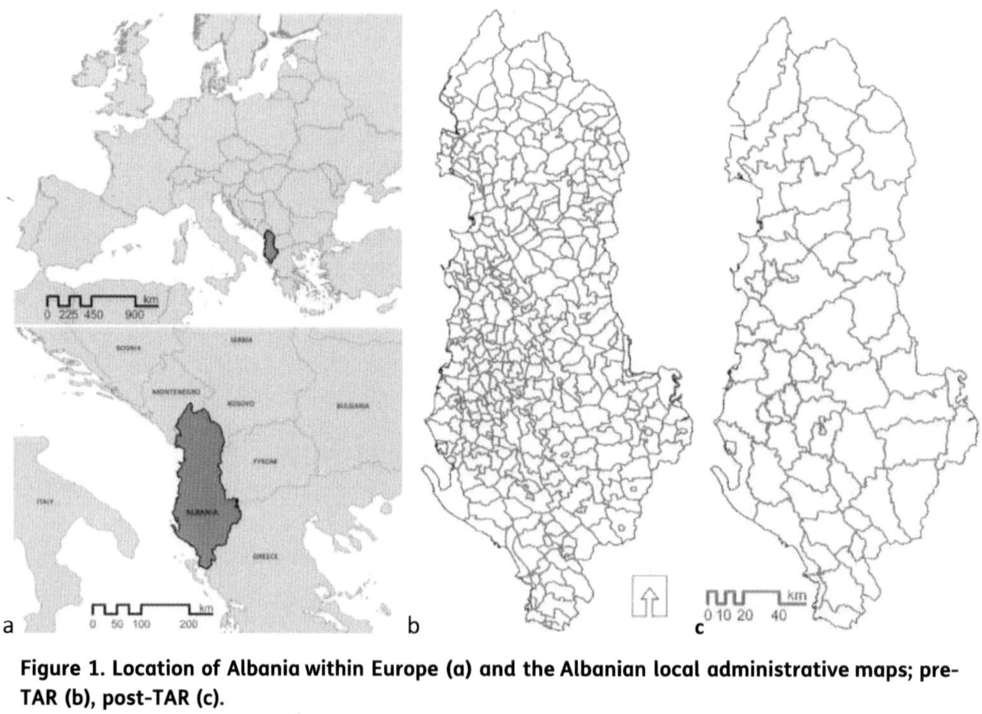

Figure 1. Location of Albania within Europe (a) and the Albanian local administrative maps; pre-TAR (b), post-TAR (c).

ineffective (Saltmarshe, 2001). According to the new local administrative map of Albania, provided via the geoportal by the Albanian State Authority of Geospatial Information, the new layout decreased the local administrative borders cumulative length by more than 50%, specifically from 7766 km to 3583 km (ASIG, 2015).

The recent Albanian TAR is criticized for being highly driven by electoral interest (Krasniqi, 2014), which is a well-known concern having earlier been addressed by different scholars even in western European regions such as Flanders (De Peuter, Pattyn, & Wayenberg, 2011). This is due to the fact that the reform process may lead to alterations of political posts, job security, and prestige (Swianiewicz, 2010). Beyond the political discourse, the official technical report of TAR in Albania highlights to have considered a variety of social and economic criteria during the decision-making process. For example, the concept of Functional Zones defines a territorial area characterized by dense and substantial interaction between citizens and institutions with economic, social development, and cultural interest (MLA, 2014). Any area that has a relatively distinctive and considerable level of economic and unique cultural identity, deserves to become a local administrative unit under the new division. Furthermore, the ethnic minority rights have been an important drive leading to independent local administration units for certain minority groups (example of Greek minorities) located in the southern Albanian territory. Ethnic minority rights is a fragile issue and should be carefully considered during TAR decision-making processes, especially in South Eastern European context, as experiences from multi-ethnic geographies such as FYROM has shown before (Kreci & Ymeri, 2010).

However, referring to the same report, the environmental criteria are absent from the decision-making process during the recent TAR in Albania. The consequences of this fact are reported to have led to a less fair distribution of the core natural resources comparing spatial division schemes of pre (36 rreth) and post reform (61 municipality) (Hysa, 2018). The study shows that broad-leaved forest surfaces (clc-311), considered as a crucial part of the natural capital (Costanza et al., 2016), head the least fair distributed natural surfaces among local administrative units. The anthropocentric approach, predominantly focusing on the economic and socio-cultural assets, might delineate the number of the new administrative units and their corresponding center locations. However, this approach fails to address the environmentally sensitive definition of the administrative boundaries at the rural and at the natural landscape level.

While the recent administrative reform in Albania has apparently neglected the environmental concerns, the new local administration law 115/2014, dated 31.07.2014 (QBZ, 2014), has raised the natural assets' value. The new law exclusively assigns the management of Forested Lands to the municipality (as the new local administrative unit) lodging them. Moreover, referring to the Directive by the Albanian Ministry of Environment, the local government can give for rent areas from the Forest and Pastures Fund contributing directly to their institutional budgetary income (MEA, 2016).

Commercializing the ecological services (Costanza et al., 2011), brings new financial opportunities for local government based on the natural attributes of the lands under their jurisdiction. Yet, it boosts a serious conflict between development interest and conservation goals. Furthermore, the HELVETAS (2016) organization monitoring the TAR implementation, reports on an official transfer of the Fire Safety and Evacuation (FSE) services, from the central government's responsibility to the local administration.

The responsibility dissolution of the Forest and Pasture Fund management and the FSE service among local government units implies uncertainty for natural landscapes. For instance, a natural land surface patch crossed by administrative borders functionally splits into parts of distinct local administrations. Thus, uncoordinated decisions on the usage of Forest and Pasture fund may lead to over consumption of resources. Similarly, in case of a wildfire in a disrupted forest surface, as the firefighting service responsibility ends within the administrative boundary, it might cause an uncontrolled wildfire situation beyond the border. To minimize the split of natural landscape patches, the proposal highlights the land cover pattern of the territory as decisive criteria in local administrative boundary definition phase of TAR process.

3. Methods and materials

This study follows a multi-criteria methodological approach in accordance with the sustainable development principles cited previously. We highlight the environmental factors as a significant, yet neglected, aspect of the TAR processes. Our objective is to test the land cover data set as an environmental factor which might serve as a measure to criticize and improve the current, local administrative reform in Albania (Figure 2(a)). The study utilizes Geographic Information Systems (GIS) technologies as a tool for the data management and analysis.

In the first part of this empirical study, we assess the landscape fragmentation resulting from the new local administrative map of Albania. Furthermore, the numerical data on the count and surface area of intersected CLC patches split via local administrative boundaries present information about spatial coverage distribution of fragmented land cover surfaces (Figure 3). This step aims to validate our hypothesis: a TAR process neglecting the environmental concerns during the decision-making stages, leads to function-wise fragmented natural landscapes.

In the second part, we propose a new alignment for the administrative boundary of the Albanian capital, Tirana. Assuming that the TAR process has significantly considered the socio-cultural and economic criteria, we put forward the land cover pattern as a boundary defining reference to

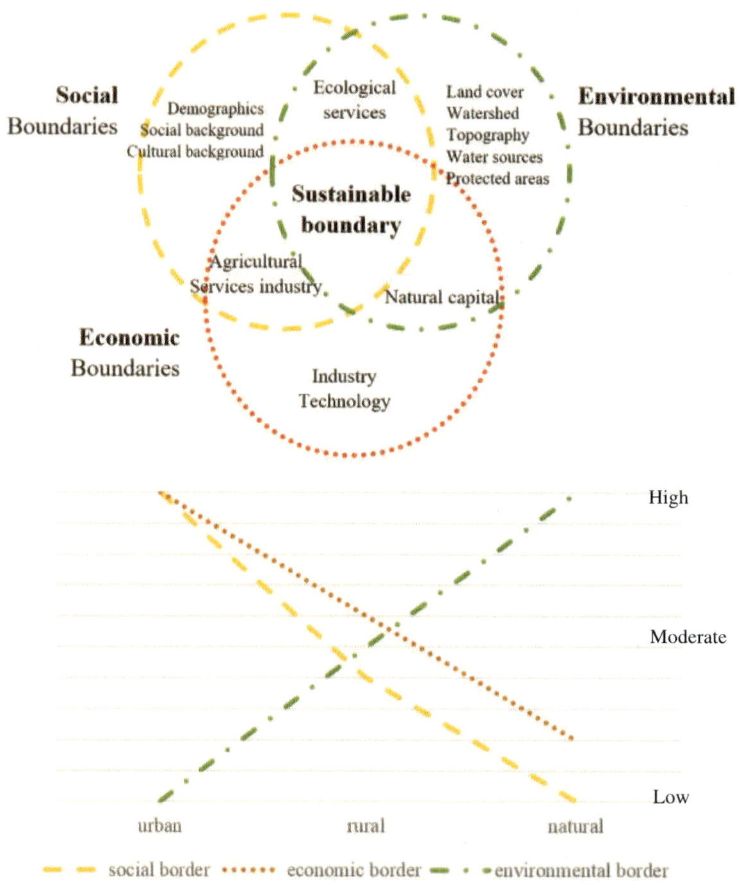

a

b

Figure 2. Conceptual diagram of sustainable boundary derived from multi-criteria boundary dynamics (a); and Hypothetical impact rates of social, economic, and environmental factors at urban, rural and natural context (b).

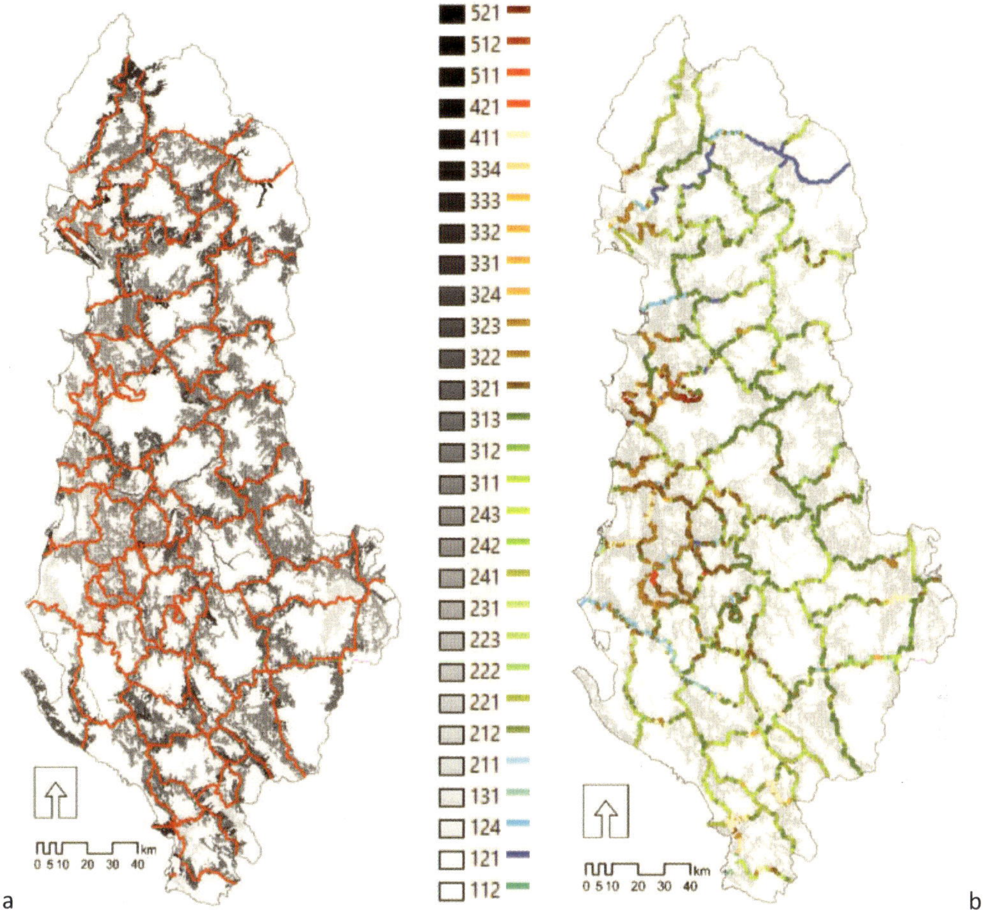

Figure 3. CLC patches inter-secting with local administrative borders (a); and local administrative borders by their land cover type matching (b).

reduce the landscape fragmentation. The landscape patches intersecting with the actual borders are the core focus of the study, leading to a modified administrative boundary line. The alteration follows two rules: (i) the new border should track the land cover patch boundary it overlaps with and (ii) the smaller (in terms of area) parts of split patches, should spatially transfer to the neighboring administrative unit already containing the largest portion (Figure 4).

3.1. Multi-criteria framework for sustainable territorial and administrative reform

The environmental factors complement the socio-cultural and economic criteria, all of which comprise the sustainable management of the territory. However, the TAR decision-making processes either refer minimally or totally neglect the environmental ones. We posit that the environmental criteria when taken in consideration in a TAR process leads to a sustainable management

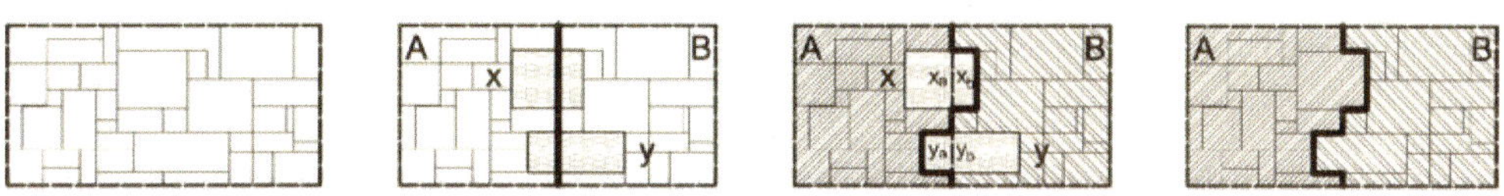

Figure 4. Conceptual model for boundary definition via CLC; (a) landscape mosaic in a territory facing TAR, (b) border defined via socio-economic criteria between administrative units A and B and fragmentation of patch x and y, (c) transference of smaller split patches, and (d) the final boundary.

of cross-border natural landscapes. Our approach (Figure 2) follows the "sustainable landscapes" notion, standing as a holistic as well as anthropocentric model, considering the social, economic, and environmental dimensions of the landscapes all at once (Blaschke, 2006). Landscape sustainability is strongly related with both the ecological and community scales. Yet, an obstacle in achieving sustainable landscapes is the discordance between responsibility boundaries among agencies or stakeholders (Saunders & Briggs, 2002).

Figure 2(a) presents a conceptual scheme of sub-criteria distribution per each main category (social, economic, environmental) according to their shared fields of relevancy. For example, we categorize the demographical data, social, and cultural background within the social criteria boundaries. Similarly, industry, commerce, and technology fall under the economic factors. Furthermore, we consider the land cover, watershed, topography, water features, and protected areas as environmental features of the territory. Besides, certain shared criteria fall within the common ground of two main sets. For instance, a common criterion across the economic and environmental factors is the natural capital. Similarly, we propose the ecological services as a shared factor of social and environmental, whereas the agricultural activity falls within the common ground of social and economic criteria.

Further analyzing the criteria by their relative significance weight at different spatial scales and context, like urban, rural, and natural, we state that the socio-cultural and economic factors are primarily more decisive in the urban context (Figure 2(b)). The aforementioned factors, predominantly associate with anthropogenic dynamics, which are spatially concentrated around the urbanized areas of the local administrative units, whereas the environmental criteria are more influential in the rural and natural environments extending in the municipal territories' peripheral areas. Although our hypothetical assumption might motivate future studies on the issue, it stands on the same line with the distribution scheme of each criterion to TAR process stages as proposed in Table 1.

The proposal outlines three main stages of the TAR process: (i) defining the number and the local administrative units' centers locations, as a macro-scale judgment, (ii) defining each unit's hinterland, as a mezzo-scale finding, and (iii) defining the local, administrative boundary lines, as a micro-scale decision. Referring to Table 1, the socio-cultural and economic criteria are concentrated in the Stage 1 and the Stage 2, supporting the hypothesis in Figure 2(b), while, the environmental factors are predominantly present in the Stage 2 and the Stage 3 of TAR process. Finally, by taking into account all the criteria, we promote a condition which defines new administrative borders that conform the sustainability principles.

Table 1. Multi-criteria framework presented by their relevance to decision-making process stages of TAR

	OBJECTIVE	**SOCIAL**	**ECONOMIC**	**ENVIRONMENTAL**
STAGE 1	*1-a. Define the amount of new administrative units 1-b. Define the center of each administrative unit*	S1-Settlements S2-Cultural (collective memory S4—Minority rights	E2-Economic zones E3-Infrastructure	
STAGE 2	*2-a. Define the Hinterland of each administrative unit*	S3- Accessibility/ Transport	E1-Agriculture activity	EN1-Natural capital distribution
STAGE 3	*3-a. Define the border line 3-b. Define the boundary buffer*	S5- Property		EN2-Watershed/ TopographyB EN3-Land Cover EN4-Watercourses/Water Bodies

3.2. Measuring landscape fragmentation caused by territorial division of administrative reform

The recent Albanian TAR process is performed within the scope of the decentralization policies in post-socialist countries. It generated a new map of local administrative organization by reducing the number of units from 374 to 61. Principally, the shrinkage in number of the administrative units, thus their amalgamation or territorial upscaling (Askim, Klausen, Vabo, & Bjurstrøm, 2017) in the scope of TAR, stands effective in reducing the landscape fragmentation within the territory (Ojima & Chuluun, 2008). However, if TAR does not consider the environmental features of the territory, it cannot guarantee the minimum fragmentation among natural lands. In this study, we experiment with CLC data as a mean to assess the landscape fragmentation caused by the TAR. We analyze the spatial relation among the land cover patches and the local administrative borders as illustrated in Figure 3.

Land cover properties of a territory may give reliable facts on the state of its natural assets. Previous studies (Zhang, Zhao, Liu, Liu, & Li, 2015) use the land cover statistics to analyze the changes of the natural capital through certain temporal sequences due to land use alterations. In our study, we employ the land cover statistics to assess and discuss the landscape fragmentation caused by the new administrative reform in Albania. The fragmentation analysis relies on the intersection condition between the CLC data and post-reform local administrative map as illustrated in Figure 3(a). We ground our findings on the statistical results from the cross-split operation between both data. First, the statistics about the length of local administrative boundaries split through the overlapping CLC patches (Figure 3(b)) highlight the borders in conflict with natural land surfaces. The assessment expands by measuring the surface areas of split land cover patches overlapping with the local administrative boundaries.

3.3. Land Cover Data as Environmental Criterion in Deciding the Borders of the Local Administrative Units

The geography literature defines landscape as the external surface of the earth beneath the atmosphere, and merely being an outward manifestation of the factors at work in the area (Hartshorne, 1939). The study introduces land cover data as an evidence of the landscape surface properties, which might be utilized as an environmental parameter during decision-making processes of TAR.

Our experiment relies on CLC data, which provide structured typological spatial information on the land cover classified under specific categories. CLC nomenclature is organized in three hierarchical levels of surface cover types. The first divides the land surfaces into five main categories; artificial surfaces, agricultural areas, forests and semi-natural areas, wetlands, and waterbodies (Büttner & Kosztra, 2017). Finally, CLC provides information about land coverage properties under 44 classes of the third level. Open source spatial data derived from EIONET database provide useful spatial information about the land cover types of 6 years intervals. In this study, we decided to proceed with the data of 2012, as they coincide with the timeline of the recent TAR decision-making process performed in Albania, between 2013 and 2014. As presented in the multi-criteria framework, we applied the land cover data especially during the local administrative boundary definition phase.

In this study, we discuss the boundary concept referring to Finch's articulation. The first one is based on the natural features of the context, which he calls "eternal," caused by the "physical geography of the soil." It is equivalent to the "antecedent boundary" concept theorized by Hartshorne in his evolution of boundaries theory (Dikshit, 1975). The second is the "variable" as caused by the "power of man" (Finch, 1833). In our research, we first borrow the dichotomy between natural and manmade, during the division of CLC classes into two groups of land cover surfaces. The artificial (CLC-100) and agricultural lands (CLC-200) are classified as surfaces transformed via anthropogenic factors, whereas forests and semi-natural areas (CLC-300), wetlands (CLC-400), and water bodies (CLC-500) are categorized as natural surfaces demarcated by eternal boundaries.

The conceptual dichotomy leads the study to highlight two aspects of the multi-criteria frame, as the most definitive input while deciding the boundary line between two local administrative units. First, the boundary condition between two artificial land cover patches belonging to their respective adjacent administrative units can be defined as a border line based on the socio-cultural and economic criteria, such as private property. We accept the property lines as definitive indicators of the local administrative boundary dividing the artificial surfaces. As we assume that TAR considers all the socio-cultural and economic factors, including the property rights, the actual municipal borders passing through artificial land cover patches are considered successful, thus, outside the scope of this work.

However, we propose that the border line passing among natural lands should track according to the land cover pattern, following certain rules. The procedure starts with the identification of the existing overlapping condition between the municipal borders of the reference map (post-reform local administrative map of Albania), and the natural landscape patches as derived from CLC data of 2012 (Figure 4). Via a split operation, we subdivide the cross-border patches into at least two components (x and y) falling inside different adjacent administrative units (A and B) (Figure 4(b)). Figure 4 illustrates diagrammatically the fundamental rule of the proposal: the smaller (in surface area) split patches are combined with their larger sibling patch, thus being spatially transferred to the neighboring administrative unit (Figure 4(c)). Finally, the new local administrative border should follow the boundary line of the patch it previously split through (Figure 4(d)).

3.4. *Utilizing GIS technology in boundary definition of local administrative units*

The study utilizes ArcGIS 10.2.2 software during the empirical steps of the research. Initially, we assess the landscape fragmentation generated by the new layout of administrative map, via spatial statistics in ArcMap. Apart from the default toolbox of ArcGIS package, the research applied two extension tools. As a first step, the ModelBuilder utility automates the process of generating the map of split cross-boundary landscape patches. ModelBuilder utilization varies; however, in this research, we customize it as a unique tool according to the goals of our study (Allen, 2011). The workflow of the model relies on two input variables: (i) the local administrative map and (ii) CLC data (shp). The model produces a new polyline shapefile presenting an improved version of the reference administrative borders' map following the rules explained before. Furthermore, we applied the Matrix Green toolbox (Bodin & Zetterberg, 2010) while preparing the island-free patches of transferred landscape surfaces, as the patches to be reclassified may consist of other patches within them. In such cases, all island patches face the same transfer with the patch they fall within.

4. Results & discussions

We organize the results of this research into two main tiers. First, it delivers information about the landscape fragmentation caused by the current, local administrative borders with an emphasis on the natural landscapes. Second, it presents a revised version of the municipal borders of Tirana, based on the unique pattern of the natural land cover mosaics. Finally, we discuss significant drawbacks as further development of the work presented in this article.

4.1. *Measuring landscape fragmentation caused by TAR in Albania*

The overlapping condition between administrative boundaries and land cover patches provides ground for the assessment of functional landscape fragmentation caused by the new TAR in Albania. First, we use the "selection by location" tool in ArcMap, to filter CLC patches that intersect with the local borders' lines. As a result, we derive numerical information about the number of patches (count) and their surface area as illustrated in the Table 2. Next, local boundary polyline feature split by the geometry of the filtered CLC patches provides statistical evidence about the total border length per land cover type. Further on, the filtered CLC polygons split by the polyline feature of the local borders, prepare the spatial basis for the local boundary definition method as explained in Figure 4.

Table 2. Statistics of landscape patches being fragmented by administrative borders

CORINE Land Cover Class	CLC	Length [km]	Length %	Count	Count %	Average [km]	Area [ha]	Area %
Discontinuous urban fabric	**112**	51	1.42	56	2.64	0.9	18,844	1.56
Industrial or commercial units	**121**	8	0.21	7	0.33	1.1	1173	0.10
Airports	**124**	2	0.07	2	0.09	1.2	602	0.05
Mineral extraction sites	**131**	12	0.35	4	0.19	3.1	1769	0.15
Non-irrigated arable land	**211**	157	4.38	67	3.15	2.3	73,953	6.11
Permanently irrigated land	**212**	5	0.14	3	0.14	1.7	1100	0.09
Vineyards	**221**	5	0.15	7	0.33	0.8	763	0.06
Fruit trees and berry plantations	**222**	22	0.60	22	1.04	1.0	2884	0.24
Olive groves	**223**	51	1.41	46	2.16	1.1	10,816	0.89
Pastures	**231**	53	1.49	63	2.96	0.8	13,811	1.14
Annual crops associated with permanent crops	**241**	2	0.06	1	0.05	2.0	117	0.01
Complex cultivation patterns	**242**	308	8.60	113	5.32	2.7	145,189	12.00
Land principally occupied by agriculture	**243**	254	7.08	195	9.18	1.3	79,076	6.54
Broad-leaved forest	**311**	691	19.27	249	11.72	2.8	372,227	30.77
Coniferous forest	**312**	96	2.67	66	3.11	1.4	25,625	2.12
Mixed forest	**313**	58	1.61	47	2.21	1.2	13,285	1.10
Natural grasslands	**321**	448	12.51	304	14.31	1.5	135,648	11.21
Moors and heathland	**322**	22	0.60	11	0.52	2.0	6268	0.52
Sclerophyllous vegetation	**323**	367	10.23	261	12.28	1.4	98,347	8.13
Transitional woodland-shrub	**324**	399	11.13	312	14.68	1.3	108,923	9.00
Beaches, dunes, sands	**331**	69	1.92	45	2.12	1.5	11,327	0.94
Bare rocks	**332**	9	0.26	8	0.38	1.2	6538	0.54
Sparsely vegetated areas	**333**	243	6.78	181	8.52	1.3	53,612	4.43
Burnt areas	**334**	13	0.36	15	0.71	0.9	3281	0.27
Inland marshes	**411**	2	0.04	4	0.19	0.4	540	0.04
Salt marshes	**421**	3	0.10	6	0.28	0.6	1527	0.13
Water courses	**511**	113	3.14	13	0.61	8.7	8529	0.71
Water bodies	**512**	119	3.31	14	0.66	8.5	11,851	0.98
Coastal lagoons	**521**	3	0.08	2	0.09	1.5	2174	0.18
Sea and ocean	**523**	1	0.02	1	0.05	0.7	0	0.00
Total		3584		2125			1,209,798	

Table 2 illustrates that 74% of new administrative borders overlap with natural surfaces. Referring to the same evidence, the list of border length per land cover type is headed by boundaries crossing, broad-leaved forests (CLC-311), natural grasslands (CLC-321) and transitional woodland-shrub (CLC-324), comprising about 43% of the total boundary length. Comparing percentage distribution of borders length and land cover patch count values, the highest difference belongs to broad-leaved surfaces (19.27%—11.72%), water courses (3.14%—0.61%), and water bodies (3.31%—0.66%). These numbers are on the same line with the average length values per surface type, respectively, being: 2.8, 8.7, 8.5 km. The high rates of broad-leaved surfaces indicate that the forested landscape patches are larger in surface area in comparison with other land cover types, resulting in longer border lines they overlap with.

The numerical data of Table 2 in accordance with the visual information in Figure 3(b), point out a possible consideration of the water resources as decision-making criteria during the last TAR process in Albania. According to the statistical data in Table 2 and Figure 3(b), 113 km of administrative borders overlap with watercourses (CLC-511, southern Albania) and 119 km are tracking through waterbodies (CLC-512, northern Albania). This fact implies a sharing of the freshwater resources among neighboring administrative units. Although water sources are among the only environmental criteria relatively considered during the reform, unfortunately it is not explicitly included in the technical report of TAR.

While, three quarters of the total administrative borders are overlapping with the natural land cover surfaces, the remaining 26% of local administrative boundaries cross over artificial (CLC-100) and agricultural (CLC-200) surfaces. More precisely, agricultural surfaces dominated by complex cultivation patterns (CLC-242) and non-irrigated arable land (CLC-211), overlap with 24% of total administrative borders. The remaining 2% of the total administrative borders overlap with the artificial surfaces, which are dominated by discontinuous urban fabric (CLC-112), industrial and commercial units (CLC-121), and mineral extraction sites (CLC-131).

The dominance of natural landscapes and the minimal presence of artificial surfaces intersecting administrative borders might serve as a validation to the hypothesis illustrated in Figure 2(b) and the Table 1. Accordingly, we might assert that the environmental criterion is the most significant drive during the Stage 3 or boundary definition phase, specifically at the rural context. As a consequence, the results show that the TAR processes of Albania which apparently neglected the environmental properties of the territory have led to extensive functional fragmentation, particularly among natural landscapes.

4.2. CLC data as reference for defining the boundaries of local administrative units

Beyond the conceptual approach and the assessment of landscape fragmentation caused by the new administrative division under the recent TAR in Albania, the study makes a demonstrative case of readjusting the municipal borders of the capital city, Tirana (Figure 5). The experiment stands on the assumption that if the existing version of local administrative map was an output of a process that has significantly considered the socio-cultural and economic features of the context during the decision-making process of TAR, then the land cover data as an environmental criteria are proposed as the final step of local boundary definition/refinement stage as conceptually targeted in the Table 1. Based on this postulation, the boundaries dividing two artificial surfaces are considered successful, consequently, out of the scope of the experiment.

The focus of the alteration are the border segments of the recent administrative map of Tirana municipality (Figure 5) that overlap with the natural lands of CLC data. The borders crossing over artificial land surfaces are excluded, since, as proposed in the multi-criteria framework, the boundaries of artificial and agricultural land cover surfaces are dictated by the cadastral property line rather than the land cover pattern of their mosaic. Furthermore, if urbanized areas (CLC-112), as shown in Figure 6, are considered within the readjustment process, it results in transferring of the adjacent smaller urbanized zones causing a total cancellation of the neighboring administrative unit. Since the consequence is in conflict with the socio-cultural criteria of the Stage 1 and the Stage 2 of the multi-criteria framework (Table 1), the urbanized and artificial land cover surfaces are excluded from the readjustment procedure.

According to the overlapping operation between CLC data of 2012 and Tirana municipality border line, there are 143 cross-border land cover patches shared among Tirana and other neighboring municipalities (Figure 6). Among patches split by local administrative border, 49 are artificial surfaces covering an area of 38,265 ha (31%). The remaining 94 patches are natural lands of a total area of

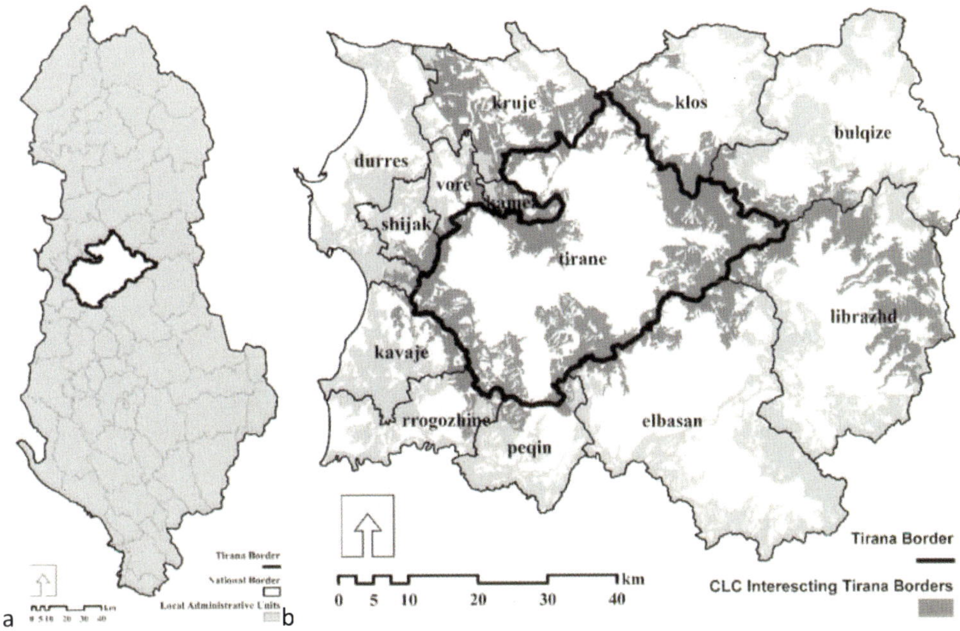

Figure 5. Location of Tirana municipality (a) and the shared municipal borders overlapping with CLC patches (b).

84,707 ha (69%). According to the above-mentioned rationale, in the further steps of the experiment, the artificial surfaces are excluded. First, the natural land cover surfaces (CLC-311; CLC-312; CLC-321; CLC-323; CLC-324; CLC-331; CLC-333; CLC-512) are filtered from the set of patches intersecting the administrative borders of Tirana (Figure 6). The selected natural patches are split via the current administrative border line. A comparison among split parts of each cross-border natural land cover patch is performed, to identify the smaller portions to be deported to the neighboring administrative unit consisting the larger part. This condition is based on the rules of the proposed method, where the smaller fragments of split patches have to be joined with the largest sibling portion (Figure 7).

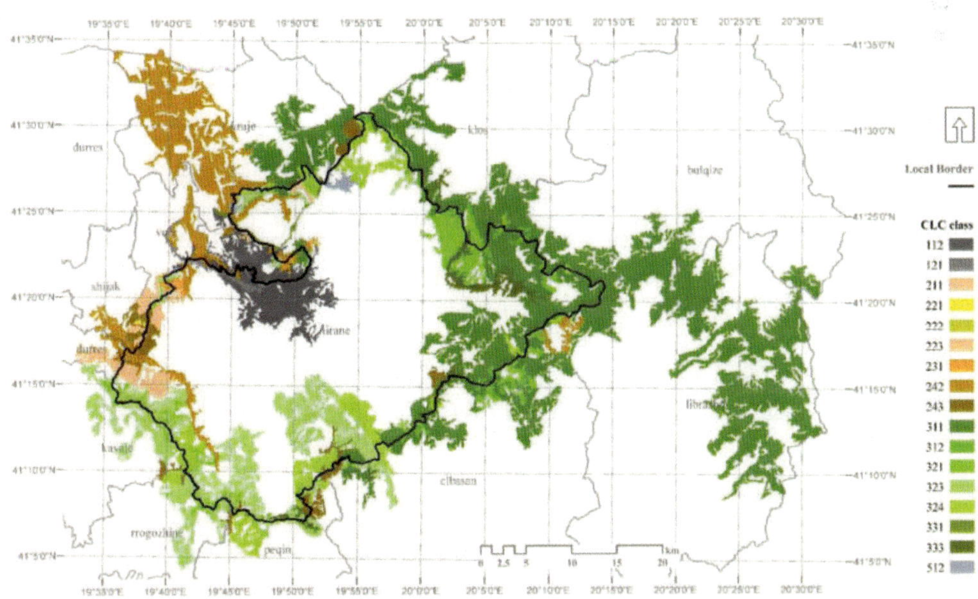

Figure 6. CLC patches inter-secting with the current muni-cipality borders of Tirana.

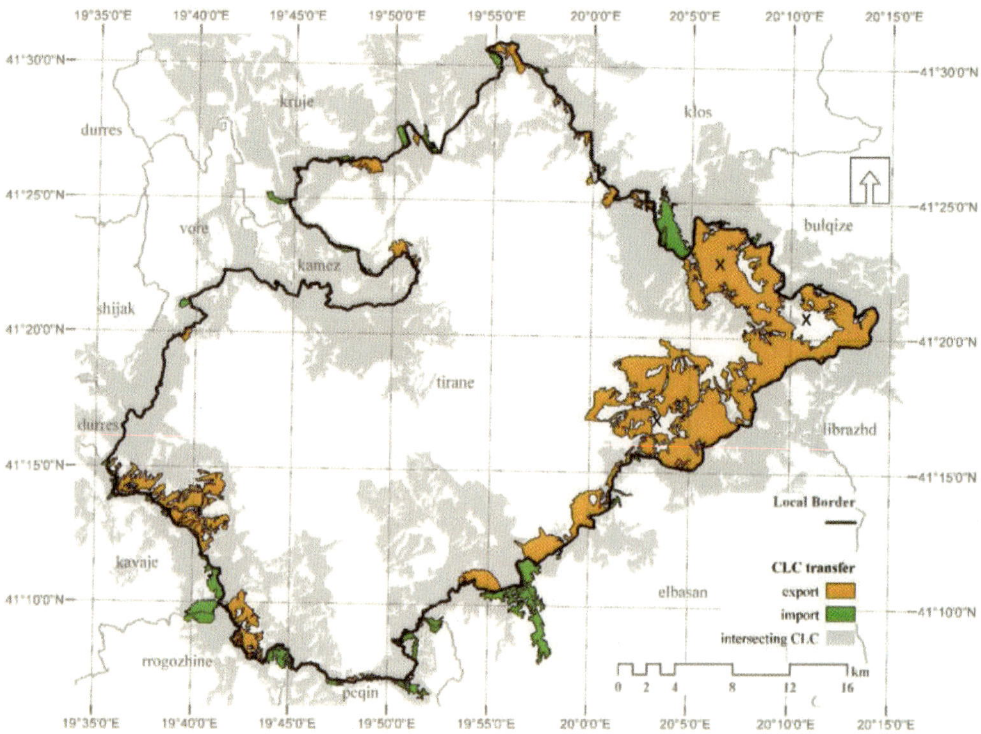

Figure 7. Import and export of cross-border natural landscape patches in case of Tirana.

According to the map shown in Figure 7 and numerical data included in Table 3, Tirana is exporting 38 patches (orange) of natural land cover with a total area of 13,825 ha. On the other hand, Tirana is importing 56 patches (green) covering 3003 ha of surface area. As it is shown in Figure 7, the largest patch to be exported located in the eastern corner of the municipal border consists of certain island patches within it (indicated by "x" in Figure 7). These areas are accepted to be transferred following the destination of the same patch they spatially fall within. Thus, if the island areas are included, the total exported area is updated to 15,103 ha, implying for an island lands of 1300 ha. Consequently, according to the proposed methodology, Tirana municipal surface area is decreased by 12,100 ha, which makes about 11% of the total municipal surface area.

5. Discussion and further improvements

First, considering that the process results in patches to be imported and exported simultaneously, one may expect a certain balance in the cumulative change of the local administrative territories' surface areas. In this context, referring to the results of the study, the reduction of the surface area of Tirana municipal territory by 11%, in principle, is beyond the expectation. This dominant

Table 3. Statistics of imported and exported landscape patches according to new proposal of Tirana Municipality borders

	CLC	311	312	321	323	324	331	333	total
import	area (ha)	0	28	784	787	1080	140	184	3003
	count	0	3	26	9	10	4	4	56
	%	0	0.9	26	26.2	36	4.7	6.1	100
export	area (ha)	11,54	0	233	1953	89	0	45	13,825
	count	9	0	13	9	6	0	1	38
	%	83.21	0	1.7	14.1	0.64	0	0.3	100

difference between imported and exported patches is dictated by a special case of the largest exported surface in the eastern part consisting of broad-leaved forested areas (CLC-311). The patch alone has an area of 9677 ha (about 9% of the total municipal area). At the same time, it is a smaller portion of a large cross-border patch (ID = “AL-10,469”), which covers a total surface area of 29,115 ha and is concurrently split among four neighbor administrative units. Moreover, a major part of this patch is part of national protected areas being subject of central government management. Considering these reasons, the patch “AL-10,469” and similar other patches might be excluded from the transferring procedure, consequently improving the results of the study.

Furthermore, the modified municipal border might result in conflict with the other criteria, the socio-cultural or economic properties of the territory. For instance, such a conflict emerges in case when the smallest (in terms of surface area) portion of a cross-border natural landscape patch has a linear layout perpendicular to the splitting border by continuing deep into the central territory of the administrative unit (Figure 8). This is due to the fact that the transferred patch may overlap with crucial elements such as local transportation network, cultural heritage sites, rural settlements that traditionally may be connected with the other urban center. Consequently, the paper calls for further research on analyzing these conflicting cases. To avoid similar divergences, the methodology might improve by bringing other restricting parameters such as surface area, shape of patch, the national scale interest, and status of large patches.

Finally, we acknowledge a possible contradiction between the rigid form of the administrative boundaries and the evolving character of land cover properties of the territory. As previous studies have shown, the transformation of land cover pattern of the territory within a relatively short period of time does not only result out of anthropogenic reasons, but also by changes in the systemic cycles in the natural environment such as global climate change (Davidson et al., 2012) or ecological succession (Alonso & Sole, 2000; Parker, Mans, Janssen, Hoffmann, & Deadman, 2003). The concept of dynamic local administrative boundary might be elaborated as a further step of the study.

6. Conclusions

This paper demonstrates a case of the local administrative boundaries re-organization, based on land cover data introduced as a territory's environmental feature. Principally, the methodology

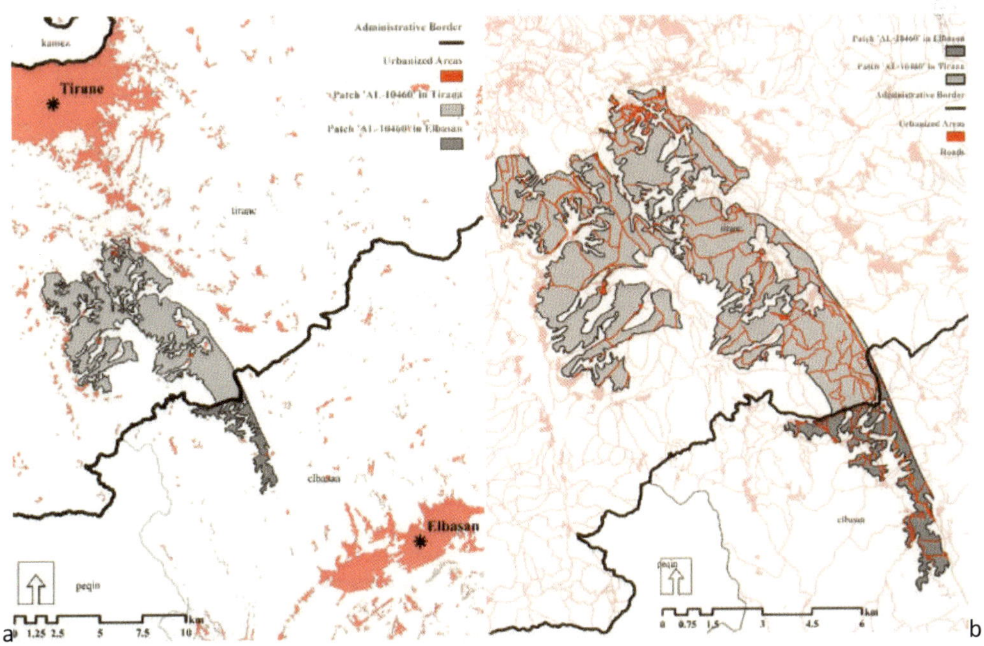

Figure 8. Patch ID = AL-10,460 in relation with urban centers (a) and with rural settlements and transportation network (b).

relies on a proposed multi-criteria framework, which consists of socio-cultural, economic, and environmental features of the context. The multi-criteria framework aims at providing a comprehensive structure for considering the complete properties of a territory undergoing the process of TAR.

The results of the study show that a local administrative division process which disregards the environmental factors of the territory leads to functionally fragmented landscapes. The fragmentation consists of a division of responsibilities and interest on a certain landscape patch as a natural capital asset, among different administrative units. In the long run, this practice is forecasted to lead to uncoordinated changes in the landscape structure, thus potentially resulting in physically fragmented landscapes.

The study has shown that the fragmentation among the landscape patches minimizes when TAR processes consider the land cover properties of the territory, thus the landscape mosaic structure is a decisive criteria during the local administrative boundary definition phase. The proposed method generates a different administrative division which avoids any condition of cross-border natural landscape patches. Hence, the landscape fragmentation significantly reduces at the CLC data scale. While the method results successful in most of the small- and moderate-sized land cover patches, it is debatable in conflicting cases between large natural landscape patches and the socio-economic priorities of region.

The study proposes a multi-criteria framework highlighting the environmental features of the territory facing TAR processes, along with the socio-cultural and economic, converging with the principles and goals of sustainable development. The proposed method for defining the local administrative borders based on multi-criteria basis by utilizing land cover data can be applied in future TAR processes, especially in post-socialist countries. Finally, the CLC data provided an appropriate spatial information to define the local boundary based on sustainable land management principles, within the scope and objectives of TAR.

Acknowledgements
The authors are grateful to the European Environment Agency, for providing the open source Corine Land Cover data utilized as the raw material of the presented model. Further on, the authors are grateful to Hayriye Eşbah Tunçay and B. Niyami Nayim for their continuous guidance. This research study has been made in support of the PhD thesis of the contact author in the Graduate School of Science, Engineering, and Technology at Istanbul Technical University.

Funding
The authors received no direct funding for this research.

Author details
Artan Hysa[1]
E-mail: artanh@gmail.com
ORCID ID: http://orcid.org/0000-0002-0952-6303
Fatma Ayçim Türer Başkaya[2]
E-mail: turerfat@itu.edu.tr
[1] Graduate School of Science, Engineering and Technology, Istanbul Technical University, Istanbul, Turkey.
[2] Department of Landscape Architecture, Istanbul Technical University, Istanbul, Turkey.

References

Allen, D. W. (2011). *Getting to know ArcGIS modelBuilder*. Redlands: Esri Press.

Alonso, D., & Sole, R. V. (2000). The DivGame simulator: A stochastic cellular automata model of rainforest dynamics. *Ecological Modeling, 133*(1/2), 131–141. doi:10.1016/S0304-3800(00)00287-8

ASIG. (2015, June 1). *Administrative Units Layer*. Retrieved September 14, 2017, from: https://geoportal.asig.gov.al/map/

Askim, J., Klausen, J. E., Vabo, S. I., & Bjurstrøm, K. (2017). Territorial upscaling of local governments: A variable-oriented approach to explaining variance among Western European countries. *Local Government Studies, 43*(4), 555–576. doi:10.1080/03003930.2017.1310102

Blaschke, T. (2006). The role of the spatial dimension within the framework of sustainable landscapes and natural capital. *Landscape and Urban Planning, 75*(2), 198–226. doi:10.1016/j.landurbplan.2005.02.013

Bodin, Ö., & Zetterberg, A. (2010). *MatrixGreen user's manual: Landscape ecological network analysis tool*. Retrieved September 2016, from http://www.stockholmresilience.org/.

Buttimer, A. (Ed.). (2001). *Sustainable landscapes and lifeways: Scale and appropriateness*. Cork: Cork University Press.

Büttner, G., & Kosztra, B. (2017). *CLC2018 technical guidelines*. Wien: European Environment Agency.

Costanza, R., d'Arge, R., de Groot, R., Farber, S., Grasso, M., Hannon, B., & van den Belt, M. (2016). The value of the world's ecosystem services and natural capital. In P. Newell (Ed.), *The globalization and environment reader* (pp. 117–134). West Sussex: John Wiley & Sons.

Costanza, R., Kubiszewski, I., Ervin, D., Bluffstone, R., Boyd, J., Brown, D., ... Yeakley, A. (2011). Valuing ecological

systems and services. *F 1000 Biology Reports*. doi:10.3410/B3-14
Davidson, E. A., de Araújo, A., Artaxo, P., Balch, J. K., Brown, F., Bustamante, M. M., ... Wofsy, S. C. (2012). The amazon basin in transition. *Nature, 481*(7381), 321–328. doi:10.1038/nature10717
De Peuter, B., Pattyn, V., & Wayenberg, E. (2011). Territorial reform of local government: Evaluation criteria underpinning decisions and debate in Flanders. *Local Government Studies, 37*(5), 533–552. doi:10.1080/03003930.2011.604546
Dikshit, R. D. (1975). *Political geography: The spatiality of politics*. New York: McGraw-Hill Publishing Company.
Finch, I. (1833). Power of man over boundaries. In I. Finch (Ed.), *Travels in the United States of America and Canada: Containing some account of their scientific institutions and a few notices of the geology and mineralogy of those countries* (pp. 419-421). London: Longman.
Ghinea, A., & Moraru, A. (2006). *Aspects regarding decentralization process in Romania*, The *Administrative Territorial Reform*. Bucharest: Institute for Public Policy. Retrieved September 7, 2017, from http://www.ase.md/files/catedre/mgs/Decentralization_Process.pdf
Giovannoni, E., & Fabietti, G. (2013). What is sustainability? A review of the concept and its applications. In C. Busco, M. L. Frigo, A. Riccaboni, & P. Quattrone (Eds.), *Integrated reporting; Concepts and cases that redefine corporate accountability* (pp. 21–40). Switzerland: Springer International Publishing.
Gordon, A., Bastin, L., Langford, W. T., Lechner, A. M., & Bekessy, S. A. (2013). Simulating the value of collaboration in multi-actor conservation planning. *Ecological Modelling, 249*, 19–25. doi:10.1016/j.ecolmodel.2012.07.009
Guerrero, A. M., Mcallister, R. R., & Wilson, K. A. (2014). Achieving cross-scale collaboration for large scale conservation initiatives. *Conservation Letters, 8*(2), 107-117.
Hartshorne, R. (1939). *The nature of geography: A critical survey of current thought in the light of the past*. Lancaster, Pennsylvania: Association of American Geographers.
Harvey, S., & Most, B. A. (1976). The substance and study of borders in international relations research. *International Studies Quarterly, 20*(4), 581–620. doi:10.2307/2600341
HELVETAS. (2016). *The Inter-sectorial Strategy for Decentralisation of Local Government: Monitoring Report for July 2015- August 2016*. Tirana: The Minister of State for Local Affairs. Retrieved September 26, 2017, from http://www.ceshtjetvendore.gov.al/files/pages_files/Raporti_i_Monitorimit
Hens, L. (1996). The Rio Conference and thereaft. In L. Hens & D. Devuyst (Eds.), *Sustainable development European Centre for Pollution Research* (pp. 81–109). Brussels: Free Univer- sity of Brussels and UN Educational Scientific and Cultural Organisation VUB Pres. doi: 10.1016/S0940-2993(96)80099-6
Hoffman, L. L., Varady, R. G., Flessa, K. W., & Balvanera, P. (2009). Ecosystem services across borders: A framework for transboundary conservation policy. *Frontiers in Ecology and the Environment, 8*(2), 84–91. doi:10.1890/070216
Hysa, A. (2018). Land cover statistics as a measure of natural capital distribution fairness among altered administrative territorial divisions. *International Journal of Applied Statistics and Econometrics, 1*(1), 48–61.
Jantke, K., & Schneider, U. A. (2010). Multiple-species conservation planning for European wetlands with different degrees of coordination. *Biological Conservation, 143*(7), 1812–1821. doi:10.1016/j.biocon.2010.04.036
Kark, S., Tulloch, A., Gordon, A., Mazor, T., Bunnefeld, N., & Levin, N. (2015). Cross-boundary collaboration: Key to the conservation puzzle. *Current Opinion in Environmental Sustainability*, 12–24. doi:10.1016/j.cosust.2014.08.005
Krasniqi, A. (2014, July 19). *ORANEWS*. RetrievedSeptember 07, 2017, from http://www.oranews.tv/analiza/reforma-territoriale-produkt-i-pazareve-politike/
Kreci, V., & Ymeri, B. (2010). The impact of territorial re-organisational policy interventions in the republic of Macedonia. *Local Government Studies, 36*(2), 271–290. doi:10.1080/03003930903560646
Kuhlmann, S., & Wollmann, H. (2011). The evaluation of institutional reforms at sub-national government levels: A still neglected research Agenda. *Local Government Studies, 37*(5), 479–494. doi:10.1080/03003930.2011.604542
Mace, G., Balmford, A., Boitani, L., Cowlishaw, G., Dobson, A., Faith, D. P., & Gaston, K. J. (2000). It's time to work together and stop duplicating conservation efforts. *Nature, 405*, 393–493. doi:10.1038/35013247
MEA. (2016, June 09). *Forest fund legislation*. Retrieved February 25, 2017, from Ministry of Environment of Albania: http://www.mjedisi.gov.al/files/userfiles/Pyjet/Udhezimi_Nr._1_Date_09.06.2016_merged.pdf
MLA. (2014, March 06). *albanian administrative- territorial reform; Technical criteria*. Ministry of Local Affairs. Tirana: Author. Retrieved March 06, 2017, from http://www.reformaterritoriale.al/images/presentations/KRITERETTEKNIKEPROPOZIMIPERKOMISIONIN_28Prill2014.pdf
Ndreu, A. (2016). Implementation of the administrative-territorial reform for a sustainable development of the economy in Albania. *Social and Natural Sciences, 10*(1), 1–4.
Norton, A. (1994). *International handbook of local and regional government: A comparative analysis of advanced democracies*. Cheltenham, UK: Edward Elgar Pub.
Ojima, D., & Chuluun, T. (2008). Policy changes in Mongolia: Implications for land-use and landscapes. In K. A. Galvin, R. S. Reid, R. H. Behnke Jr, & T. N. Hobbs (Eds.), *Fragmentation in Semi-Arid and Arid landscapes: Consequences for human and natural systems* (pp. 179–193). Dordrecht, The Netherlands: Springer.
Ostrom, E. (2010). The challenge of common-pool resources. *Science and Policy for Sustainable Development, 50*(4), 8–21.
Parker, D. C., Mans, S. M., Janssen, M. A., Hoffmann, M. J., & Deadman, P. (2003). Multi-agent systems for the simulation of land-use and land-cover change: A review. *Annals of the Association of American Geographers, 93*(2), 314–337. doi:10.1111/1467-8306.9302004
QBZ. (2014, September 1). Retrieved from Center of Official Publications: http://www.qbz.gov.al/botime/fletore_zyrtare/2014/PDF-2014/137-2014.pdf
Reçi, S., & Ymeri, I. (2016). Administrative and territorial reform in Albania and the issue of decentralization: Comparative overview of relevant international legisla-

tion. *Academic Journal of Business, Administration, Law and Social Sciences, 2*(1), 175–183.

Saltmarshe, D. (2001). Developments in Albanian Local Government. *Local Government Studies, 27*(2), 111–129. doi:10.1080/714004100

Saunders, D. A., & Briggs, S. V. (2002). Nature grows in straight lines—Or does she? What are the consequences of the mismatch between human-imposed linear boundaries and ecosystem boundaries? An Australian example. *Landscape and Urban Planning, 61*, 71–82. doi:10.1016/S0169-2046(02)00103-2

Steenmans, C. (2016). Foreword. In J. Feranec, T. Soukup, G. Hazeu, & G. Jaffrain (Eds.), *European landscape dynamics: CORINE land cover data* (pp. xi–xii).

Swianiewicz, P. (2010). If territorial fragmentation is a problem is Amalgamation a solution an east european perspective. *Local Government Studies*, 183–203. doi:10.1080/03003930903560547

Wollmann. (2010). Territorial local level reforms in the East German Regional States (Länder): Phases, patterns, and dynamics. *Local Government Studies, 36* (2), 251–270. doi:10.1080/03003930903560612

Zhang, Y., Zhao, L., Liu, J., Liu, Y., & Li, C. (2015). The impact of land cover change on ecosystem service values in Urban Agglomerations along the coast of the Bohai Rim, China. *Sustainability, 7*(8), 10365–10387. Retrieved 2017 doi: 10.3390/su70810365

Permissions

All chapters in this book were first published in CES, by Cogent OA; hereby published with permission under the Creative Commons Attribution License or equivalent. Every chapter published in this book has been scrutinized by our experts. Their significance has been extensively debated. The topics covered herein carry significant findings which will fuel the growth of the discipline. They may even be implemented as practical applications or may be referred to as a beginning point for another development.

The contributors of this book come from diverse backgrounds, making this book a truly international effort. This book will bring forth new frontiers with its revolutionizing research information and detailed analysis of the nascent developments around the world.

We would like to thank all the contributing authors for lending their expertise to make the book truly unique. They have played a crucial role in the development of this book. Without their invaluable contributions this book wouldn't have been possible. They have made vital efforts to compile up to date information on the varied aspects of this subject to make this book a valuable addition to the collection of many professionals and students.

This book was conceptualized with the vision of imparting up-to-date information and advanced data in this field. To ensure the same, a matchless editorial board was set up. Every individual on the board went through rigorous rounds of assessment to prove their worth. After which they invested a large part of their time researching and compiling the most relevant data for our readers.

The editorial board has been involved in producing this book since its inception. They have spent rigorous hours researching and exploring the diverse topics which have resulted in the successful publishing of this book. They have passed on their knowledge of decades through this book. To expedite this challenging task, the publisher supported the team at every step. A small team of assistant editors was also appointed to further simplify the editing procedure and attain best results for the readers.

Apart from the editorial board, the designing team has also invested a significant amount of their time in understanding the subject and creating the most relevant covers. They scrutinized every image to scout for the most suitable representation of the subject and create an appropriate cover for the book.

The publishing team has been an ardent support to the editorial, designing and production team. Their endless efforts to recruit the best for this project, has resulted in the accomplishment of this book. They are a veteran in the field of academics and their pool of knowledge is as vast as their experience in printing. Their expertise and guidance has proved useful at every step. Their uncompromising quality standards have made this book an exceptional effort. Their encouragement from time to time has been an inspiration for everyone.

The publisher and the editorial board hope that this book will prove to be a valuable piece of knowledge for researchers, students, practitioners and scholars across the globe.

List of Contributors

Kassahun Tassie Wegedie
Department of Agricultural Economics, Bahir Dar University, College of Agriculture and Environmental Science, Bahir Dar, Ethiopia

Michel Soto Chalhoub
Department of Civil and Environmental Engineering, Notre Dame University, Louaize, Lebanon

Tehseen Zafar
State Key Laboratory of Ore Deposit Geochemistry, Institute of Geochemistry, Chinese Academy of Sciences, Guiyang, China
University of Chinese Academy of Sciences, Beijing, China

Abiola Oyebamiji
State Key Laboratory of Ore Deposit Geochemistry, Institute of Geochemistry, Chinese Academy of Sciences, Guiyang, China
University of Chinese Academy of Sciences, Beijing, China
Department of Science Laboratory Technology, Ekiti State University, Ado-Ekiti, Nigeria

Amobichukwu Amanambu
University of Chinese Academy of Sciences, Beijing, China
State Key Laboratory of Desert and Oasis Ecology, Xinjiang Institute of Ecology and Geography, Chinese Academy of Sciences, Urumqi, China

Damilare Stephen Akinyemi
State Key Laboratory of Vegetation and Environmental Change, Institute of Botany, Chinese Academy of Sciences, Beijing, China
Department of Botany, Obafemi Awolowo University, Ile Ife, Nigeria

Kenneth Balikoowa and David M. Tumusiime
School of Forestry, Environment and Geographical Sciences, College of Agricultural and Environmental Sciences, Makerere University, Kampala, Uganda

Gorettie Nabanoga
School of Agricultural Sciences, College of Agricultural and Environmental Sciences, Makerere University, Uganda

Pogisego Dinake, Onneetse Maphane, Karabo Sebogisi and Obed Kamwi
Department of Chemical and Forensic Sciences, Botswana International University of Science and Technology, Biust, Palapye, Botswana
Department of Crop Science and Production, Botswana University of Agriculture and Natural Resources, Private Bag 0027, Gaborone, Botswana
Safety Health and Environment, Botswana Defence Force, Bdf, Sir Seretse Khama Barracks, Private Bag x06, Mogoditshane, Botswana

Manickam Nithyanandan, Lewis Le Vay, Dinesh Kumar Raja, Ratheesh Kesavan and Don Pereira
Environmental Department, Sabah Al Ahmad Sea City, La Ala Al Kuwait Real Estate Co. K.S.C., Souk Al Kuwait 6th Floor, Office No. 613, Safat, Kuwait 13090, Kuwait
Ecosystem Based Management of Marine Living Resources Program, Environment and Life Sciences Research Center, Kuwait Institute for Scientific Research, Salmiya 22017, Kuwait
Centre for Applied Marine Sciences, College of Natural Sciences, Bangor University, Menai Bridge, Anglesey LL59 5AB, Wales, UK
Environment World Company, Suite 26, 3rd Floor, Al-Mubarak Commercial Complex, Street 50, Block 1, Jleeb Al-Shuyoukh, Kuwait

Sojung Claire Kim
Department of Communication, George Mason University, 4400 University Drive, Fairfax, VA 22030, USA

Sandra L. Cooke
Department of Biology, High Point University, One University Parkway, High Point, NC 27268, USA

Marcel Achkar Borras, Gustavo Seoane, Mariana Gomez-Camponovo, Eleuterio Umpierrez Vazquez and Nicolas Perez
Department of Sciences, Territorial Organization and Sustainable Development Laboratory, University of the Republic, Montevideo, UY, Uruguay
Department of Organic Chemistry, University of the Republic, Montevideo, UY, Uruguay
Social Medicine Unit, University of the Republic, Paysandu, UY, Uruguay
Department of Chemistry's Technological Institute, Environmental, Drug & Doping Unit, University of the Republic, Montevideo, UY, Uruguay
Department of Engineering, Chemical Engineering Institute-Electric Engineering Institute, University of the Republic, Montevideo, UY, Uruguay

Aboubakar Sako
Université de Dédougou, BP. 139, Dédougou, Burkina Faso

Mamadou Nimi
BUMIGEB, Bureau des Mineset de la Géologie du Burkina Faso, Direction Régionale de Bobo, Bobo-Dioulasso, Burkina Faso

A. Waris Abu-Safianu, S. J. Cobbina and E. Yeleliere
Department of Ecotourism and Environmental Management, Faculty of Natural Resources and Environment, University for Development Studies, Tamale, Ghana

Nnaemeka Dom Onyeuwaoma, Theo Chidiezie Chineke, Okeychukwu Kelechi Nwofor, Ian Crandell, Olushina Olawale Awe, Ajileye Olasumbo, Alex I Opara, Nwachukwu Pius, Mba Tochukwu and Nwachukwu Joy
Cooperative Information Network, NASRDA, PMB, 022, Obafemi Awolowo University, Ile-Ife, Nigeria
Atmospheric Physics Group, Department of Physics, Imo State University, Owerri, Nigeria
Department of Mathematics, Virginia Tech Institute, Blacksburg, VA, USA
Department of Mathematics, Obafemi Awolowo University, Ile-Ife, Nigeria
Department of Geology, Federal University of Technology, Owerri, Nigeria
Advanced Space Applications Laboratory, University of Uyo, Uyo, Nigeria
Regional Center for Space Science Technology Education in English, Obafemi Awolowo University, Ile-Ife, Nigeria

Munyaradzi D. Shekede, Mhosisi Masocha and Amon Murwira
Department of Geography and Environmental Science, University of Zimbabwe, Mount Pleasant, Harare, Zimbabwe

Adeniyi JohnPaul Adewumi
Department of Geological Sciences, Achievers University, Owo, Nigeria

Artan Hysa
Graduate School of Science, Engineering and Technology, Istanbul Technical University, Istanbul, Turkey

Fatma Ayçim Türer Başkaya
Department of Landscape Architecture, Istanbul Technical University, Istanbul, Turkey

Index